Proceedings of the First European Astronomical Meeting
Athens, September 4-9, 1972

Volume 2

Stars and the Milky Way System

Edited by

L. N. Mavridis

Springer-Verlag
Berlin Heidelberg New York 1974

With 169 Figures

ISBN 3-540-06383-8 Springer-Verlag Berlin · Heidelberg · New York
ISBN 0-387-06383-8 Springer-Verlag New York · Heidelberg · Berlin

 Library of Congress Catalog Card Number 73-9108.

Typesetting, printing and binding: Universitätsdruckerei H. Stürtz AG Würzburg

Editors' Note

The present second volume of the Proceedings of the First European Astronomical Meeting held under the auspices of the International Astronomical Union in Athens, Greece from 4th to 9th September 1972 contains the papers presented at this meeting which are related to Stellar Astronomy and the Milky Way System.

The Editor would like to thank all those who have contributed to the publication of this volume and especially:

a) The Executive Committee of the International Astronomical Union for sponsoring the meeting.

b) The Hellenic Ministry of Culture and Science for its financial support for the organization of the meeting.

c) The members of the Scientific Organizing Committee of the meeting, i.e.
A. Blaauw, European Southern Observatory (Chairman),
H. A. Brück, Royal Observatory, Edinburgh, United Kingdom,
G. Contopoulos, Astronomical Department, University of Thessaloniki, Greece,
Ch. Fehrenbach, Observatoire de Marseille, France,
L. Gratton, Laboratorio di Astrofisica, Frascati, Italy,
P. O. Lindblad, Stockholm Observatory, Sweden,
E. R. Mustel, Astronomical Council, USSR Academy of Sciences, Moscow, U.S.S.R.,
L. Perek, Astronomical Institute, Czechoslovak Academy of Sciences, Prague, Czechoslovakia,
H. H. Voigt, Universitäts-Sternwarte, Göttingen, GFR,
who took care of the scientific organization and selected the papers to be presented at the meeting.

d) The members of the Local Organizing Committee of the meeting, i.e.
J. Xanthakis, Research Center for Astronomy and Applied Mathematics, Academy of Athens (Chairman),
B. Barbanis, Department of Astronomy, University of Patras,
G. Contopoulos, Astronomical Department, University of Thessaloniki,
D. Kotsakis, Department of Astronomy, University of Athens,
C. Macris, Research Center for Astronomy and Applied Mathematics, Academy of Athens,

L. N. Mavridis, Department of Geodetic Astronomy, University of Thessaloniki,
S. Svolopoulos, Department of Astronomy, University of Ioannina,
who took care of the local organization of the meeting.

e) The authors for their contributions and

f) The officials and the staff of the Springer-Verlag for the excellent care they have taken in printing this volume.

L. N. MAVRIDIS
Editor

Introductory Remarks

A need for closer contacts between the astronomers in Europe has been increasingly felt in the course of the past years. This need is well understood by anyone who is aware of the necessity of internationalization of research on our continent. Progress in astronomical research requires pooling of manpower and financial means in order that efforts in Europe may lead to achievements comparable to those elsewhere, particularly to those in the United States. Typical for this development are such international projects as ESRO, ESO, and possibly in the near future, JOSO and CESRA. At the same time, we observe a trend for internationalization of large national projects in both radio astronomy and optical astronomy.

Whereas in these large projects specially oriented groups have already arrived at an encouraging degree of integration across the national borders in Europe, it is felt by many an astronomer that the relations with colleagues in other countries outside these large projects still leave much to be desired. In most of the astronomical institutes in the European countries there is relatively little acquaintance with the research programmes on a small and intermediate scale in progress in the institutes of other countries. The language barrier undoubtedly is an important cause of this regrettable situation, but it would not seem to be the only one. Equally important seems to be the circumstance that there is little opportunity, especially for the younger, not yet internationally recognized, astronomers, to have contacts with their colleagues of other countries outside the occasional symposia and colloquia. With the tendency for restricted participation, which is desirable in order to render such meetings fruitful for a follow-up by participants, they certainly do not suffice for the establishment of good contacts between the coming generation of astronomers.

Whereas in the United States the American Astronomical Society plays an essential role in promoting such contacts, such a common organization is lacking on the European continent. Therefore, a most important step towards remedying this situation was the initiative of the International Astronomical Union to create the institute of Regional Meetings. This initiative was supplemented admirably by the Greek National Committee for Astronomy, when early in 1972 it proposed to the IAU to hold a first European regional meeting in Athens. The present three volumes containing the proceedings of these meetings represent the fruit of this excellent effort.

For those who were present at the meetings in Athens, the memory will remain of an inspiring series of lectures and discussions, many of these latter very lively, and with that participation from astronomers from all over Europe (and also from across the ocean) which the International Astronomical Union's Executive

Committee must have had in mind when it initiated these meetings. Hopefully, it may induce colleagues from other states in Europe to organize similar meetings in the future. For the Scientific Organizing Committee, it was a privilege to be able to share in this first "regional" effort. It is a pleasure, on behalf of this committee, to acknowledge the excellent help it has had from the part of the Greek National Committee.

A. Blaauw
Chairman,
Scientific Organizing Committee

Contents

Variable Stars

KIPPENHAHN, R.: Variable Stars and Stellar Evolution (Invited Lecture) 1
ASTERIADIS, G., MAVRIDIS, L. N., and TSIOUMIS, A.: On the Stability of the Light Curves of Galactic Cepheids 17
NIKOLOV, N., and TSVETKOV, Ts.: On the Pulsation Amplitude of Cepheid Variables . . 31
DENIS, J., and SMEYERS, P.: The Influence of an Axial Rotation and a Tidal Action on the Non-Radial Oscillations of a Star, with Reference to the β Canis Majoris Stars . . . 36
SEITTER, W.: An Atlas of Nova Delphini 1967 39

Binary Stars

TODORAN, I.: Apsidal Motion in Close Binaries with Variable Orbital Period 42
POPOVICI, C., and DUMITRESCU, A.: The Radius-Luminosity Relation in Eclipsing Binaries . 53
BAKOS, G. A.: Kinematic Properties of Selected Visual Binaries 60

Space Distribution and Motions of the Stars

LINDBLAD, P. O.: Gould's Belt (Invited Lecture) 65
SUNDMAN, A.: On the Distribution of Stars in the Carina-Centaurus Region 76
FRICKE, W.: Conclusions on Galactic Kinematics from Proper Motions with Respect to Galaxies 83

Interstellar Matter

MEZGER, P. G.: Molecules as Probes of the Interstellar Matter (Invited Lecture) 88
DAVIES, R. D., COHEN, R. J., and WILSON, A. J.: Molecules in the Widely Distributed Interstellar Clouds 115
HACHENBERG, O., and MEBOLD, U.: On the Homogeneity and the Kinetic Temperature of the Intercloud HI-Gas 120
DAVIES, R. D.: The Outer Spiral Structure of the Galaxy and the High-Velocity Clouds . 124
PAPADOPOULOS, G. D.: Very Long Baseline Interferometer Measurements of H_2O Masers in Orion A, W49N, W30H and VY CMA (Abstract) 129
UTIGER, H. E.: High Resolution Search for Interstellar Lithium 131

Galactic Center

MEZGER, P. G., CHURCHWELL, E. B., and PAULS, T. A.: Ionized Gas in the Direction of the Galactic Center I: Kinematics and Physical Conditions in the Nuclear Disk 140
SANDQVIST, AA.: Results of Lunar Occultations of the Galactic Center Region in HI, OH and H_2CO Lines and in the Nearby Continua 157
SABBATA, V. DE, FORTINI, P., and GUALDI, C.: A Model for the Galactic Center (Abstract) . 164

Chemical Evolution of the Galaxy

BASINSKA, E., and IWANOWSKA, W.: Diffusion of Elements in the Galaxy 165
WILLIAMS, P. M.: Narrow-Band Observations of Super-Metal-Rich Stars 174
WOLF, B.: Fine Analytic Abundance Determination of Magellanic Cloud A-Supergiants and its Importance for the Discrimination of Theories for the Chemical Evolution of the Galaxy 178
HUCHT, K. A. VAN DER, and LAMERS, H. J.: High Resolution Ultraviolet Stellar Spectra the Obtained with the Orbiting Stellar Spectrophotometer S 59 182

Infrared Astronomy

BORGMAN, J.: Infrared Astronomy (General Lecture) 188
POTTASCH, S. R.: Interpretation of Far Infrared Observations (Invited Lecture) 209
SMYTH, M. J.: Infrared Techniques (Invited Lecture) 232
OLTHOF, H., WIJNBERGEN, J. J., HELMERHORST, TH. J., and DUINEN, R. J. VAN: Infrared Photometry of Galactic HII Regions 243
AITKEN, D., and JONES, B.: 8—13 μm Spectrum of the Galactic Centre (Abstract) . . . 248

Instruments

REDDISH, V. C.: High Speed Evaluation of Photographic Plates (Invited Lecture) . . . 249
HAWKINS, M. R. S.: Analysis of Electronographic Images 255
WYLLER, A. A.: URSIES and the Bartol Coudé Observatory 265

Three-Body Problem

SZEBEHELY, V.: The Problem of Three Bodies (Invited Lecture). 273
CONTOPOULOS, G., and ZIKIDES, M.: Periodic Orbits of the Restricted Problem for Various Values of the Mass-Ratio 279
DANIELSSON, L., and MEHRA, R.: On the Possibility of a Resonance Capture of the Asteroid Toro by the Earth 283
ROY, A. E.: Commensurable Mean Motions as a Tool in Solar System Dynamical Studies (Abstract) 290

Galactic Dynamics

EINASTO, J.: Galactic Models and Stellar Orbits (Invited Lecture) 291
WIELEN, R.: The Gravitational N-Body Problem for Star Clusters (Invited Lecture) . . 326
MARTINET, L., and MAYER, F.: A Comparative Study of Periodic Orbits ($m<4$) in Various Three-Dimensional Models of Our Galaxy 355

Author Index 358

Subject Index 366

List of Contributors to Volume 2

Aitken, D., Department of Physics and Astronomy, University College London, Great Britain

Asteriadis, G., Department of Geodetic Astronomy, University of Thessaloniki, Thessaloniki, Greece

Bakos, G. A., Department of Physics, University of Waterloo, Waterloo, Ontario, Canada

Basinska, E., Polish Academy of Sciences, Institute of Astronomy, Laboratory of Astrophysics, Ul. Sienkiewicza 30, P-87100 Toruń, Poland

Borgman, J., Kapteyn Observatory, Mensingeweg 20, Roden, The Netherlands

Churchwell, E. B., M.P.I. für Radioastronomie, D-5300 Bonn, Argelanderstraße 03, Germany

Cohen, R. J., University of Manchester, Nuffield Radio Astronomy Laboratories, Jodrell Bank, Macclesfield, Cheshire SK 11 9DL, Great Britain

Contopoulos, G., University of Thessaloniki, Department of Astronomy, Thessaloniki, Greece

Danielsson, L., Royal Institute of Technology, Stockholm, Sweden

Davies, R. D., University of Manchester, Nuffield Radio Astronomy Laboratories, Jodrell Bank, Macclesfield, Cheshire SK 11 9DL, Great Britain

Denis, J., Astronomisch Instituut, Katholieke Universiteit Leuven, Naamsenstraat 61, B-3000 Leuven, Belgium

Duinen, R. J. van, Department of Space Research, Kapteyn Astronomical Institute, University of Groningen, P.O. Box 800, Groningen, The Netherlands

Dumitrescu, A., Academia R.S. România, Observatorul Astronomic, Str. Cuţitul de Argint Nr. 5, Bucureşti 28, R.S. Romania Nr. 1293, Rumania

Einasto, J., W. Struve Astrophysical Observatory, Tartu-Töravere, Estonia, U.S.S.R.

Fortini, P., Osservatorio Astronomico, I-00136 Roma, Italy

Fricke, W., Astronomisches Rechen-Institut, 6900 Heidelberg, Mönchhofstraße 12 bis 14, Germany

Gualdi, C., Instituto di Fisica, Universitá di Bologna, I-Bologna, Italy

Hachenberg, O., M.P.I. für Radioastronomie, D-5300 Bonn, Argelanderstraße 03, Germany

Hawkins, M. R. S., The Observatories, Cambridge University, Madingley Road, Cambridge, Great Britain

Helmerhorst, Th. J., Department of Space Research, Kapteyn Astronomical Institute, University of Groningen, P.O. Box 800, Groningen, The Netherlands

HUCHT, K. A. van der, Space Research Laboratory, Astronomical Institute Utrecht, The Netherlands
IWANOWSKA, W., Polish Academy of Sciences, Institute of Astronomy, Laboratory of Astrophysics, Ul. Sienkiewicza 30, P-87100 Toruń, Poland
JONES, B., Department of Physics and Astronomy, University College London, Great Britain
KIPPENHAHN, R., Sternwarte der Universität Göttingen, D-3400 Göttingen, Geismarlandstraße 11, Germany
LAMERS, H. J., Space Research Laboratory, Astronomical Institute Utrecht, The Netherlands
LINDBLAD, P. O., Stockholm Observatory, S-13300, Saltsjöbaden, Sweden
MARTINET, L., Observatoire de Genève, Sauverny, Switzerland
MAVRIDIS, L. N., Department of Geodetic Astronomy, University of Thessaloniki, Thessaloniki, Greece
MAYER, F., Observatoire de Genève, Sauverny, Switzerland
MEBOLD, U., M.P.I. für Radioastronomie, D-5300 Bonn, Argelanderstraße 03, Germany
MEHRA, R., Royal Institute of Technology, Stockholm, Sweden
MEZGER, P. G., M.P.I. für Radioastronomie, D-5300 Bonn, Argelanderstraße 03, Germany
NIKOLOV, N., Department of Astronomy, Faculty of Physics, University of Sofia, Sofia, Bulgaria
OLTHOF, H., Department of Space Research, Kapteyn Astronomical Institute, University of Groningen, P.O. Box 800, Groningen, The Netherlands
PAPADOPOULOS, G. D., Research Laboratory of Electronics, Massachusetts Institute of Technology, Cambridge, Mass. 02139, U.S.A
PAULS, T. A., M.P.I. für Radioastronomie, D-5300 Bonn, Argelanderstraße 03, Germany
POPOVICI, C., Academia R.S. România, Observatorul Astronomic, Str. Cuţitul de Argint Nr. 5, Bucureşti 28, R.S. Romania Nr. 1293, Rumania
POTTASCH, S. R., Kapteyn Astronomical Institute, Hoogbouw W.S.N., Postbus 800, Groningen 8002, The Netherlands
REDDISH, V. C., Royal Observatory Edinburgh, Great Britain
ROY, A. E., Department of Astronomy, University of Glasgow, Great Britain
SABBATA, V. DE, Instituto di Fisica, Universitá di Bologna, I-Bologna, Italy
SANDQVIST, AA., Stockholm Observatory, S-13300 Saltsjöbaden, Sweden
SEITTER, W., Observatorium Hoher List, D-5560 Daun/Eifel, Germany
SMEYERS, P., Astronomisch Instituut, Katholieke Universiteit Leuven, Naamsenstraat 61, B-3000 Leuven, Belgium
SMYTH, M. J., University of Edinburgh, Department of Astronomy, Royal Observatory, Edinburgh EH9 3HJ, Scotland, Great Britain
SUNDMAN, A., Stockholm Observatory, S-13300 Saltsjöbaden, Sweden
SZEBEHELY, V., Department of Aerospace Engineering, The University of Texas, Austin, Texas 78712, U.S.A
TODORAN, I., Astronomical Observatory, Cluj, Rumania
TSIOUMIS, A., Department of Geodetic Astronomy, University of Thessaloniki, Thessaloniki, Greece

TSVETKOV, Ts., Department of Astronomy, Faculty of Physics, University of Sofia, Sofia, Bulgaria

UTIGER, H. E., Department of Physics, Middle East Technical University, Ankara, Turkey

WIELEN, R., Astronomisches Rechen-Institut, 6900 Heidelberg, Mönchhofstraße 12 bis 14, Germany

WIJNBERGEN, J. J., Department of Space Research, Kapteyn Astronomical Institute, University of Groningen, P.O. Box 800, Groningen, The Netherlands

WILLIAMS, P. M., The Observatories, Madingley Road, Cambridge CB 3 OHA, Great Britain

WILSON, A. J. (now A. J. DANIEL), University of Manchester, Nuffield Radio Astronomy Laboratories, Jodrell Bank, Macclesfield, Cheshire SK 11 9DL, Great Britain

WOLF, B., ESO, Casilla 16317 Correo 9, Santiago de Chile, Chile

WYLLER, A. A., Stockholm Observatory, S-13300 Saltsjöbaden, Sweden

ZIKIDES, M., Laboratory of Astronomy, University of Athens, Panepistimioupolis, Athens (621), Greece

Variable Stars and Stellar Evolution

(Invited Lecture)

By R. Kippenhahn, Göttingen University Observatory, F.R.G.

With 14 Figures

Abstract

Observations of T Tauri stars are compared with theoretical models for stars surrounded by absorbing clouds and especially with recent computations of the formation of protostars. The problem of vibrational stability at the point where the instability strip crosses the main sequence is discussed as well as the possible influence of the direction of evolution on the properties of RR Lyr stars. Finally, models for R CrB stars are discussed and compared with observations.

I. Introduction

The computation of sequences of stellar models has explained a number of observable features of stars like the main sequence, the formation of red giants and supergiants after the main sequence stage, and the crossings of the Cepheid strip. But if one follows a star on the computer into further and further evolutionary stages the computations not only become more and more difficult because the chemical structure gets more complicated, but the underlying physics also becomes uncertain. Thermal pulses of shell sources appear and one is unable to follow these rapid changes which might repeat more than hundred times in the star. Therefore, we do not know what the final result will be when thermal pulses in shell sources occur. Another problem is that of chemical mixing. Will the nuclei which are formed in different layers by nuclear reactions remain where they are being formed or are there mixing or diffusion processes which redistribute the chemical composition? We do not know whether stars evolve with constant mass throughout most of their lifetime or whether continuous mass loss plays an important role in the evolution. We have no idea where to put the helium stars into the evolutionary scheme, and we are rather helpless explaining stars with other unusual chemical compositions.

While stellar evolution after hydrogen ignition is fortunately independent of the details of the formation of the stars, the pre-main-sequence evolution is not. And even if we were able to give precisely the initial state at the moment of star formation we would still be at a loss since star formation is certainly not a spherically symmetric phenomenon, the results of which are difficult to predict. Therefore, our knowledge of the pre-main-sequence stages of stellar evolution is rather poor. In this situation we need more information from observation.

Can the variable stars teach us more about stellar evolution? I think they can give us information about processes we normally neglect or even forget when we are collecting stellar model after stellar model from our printouts. Furthermore, I think they can give us hints how evolution goes during phases which are still not accessible to evolutionary calculations.

I will give a few examples here: T Tauri stars, pulsating variables on and near the main sequence, RR Lyrae stars and R Coronae Borealis stars.

II. T Tauri Stars

Today, there is no doubt that stars which have T Tauri characteristics in their spectra or in their light curves are in the stage of pre-main-sequence evolution. The classical example is WALKER's cluster NGC 2264 (WALKER, 1956). We think we understand its HR-diagram. Isochrones derived from stellar evolution theory agree roughly with the position of the pre-main-sequence branch of the cluster and give an age of the order of one to three million years. Stellar evolution theory has even been extended to explain the scatter in the diagram by differences in the ages of the stars of this cluster (IBEN and TALBOT, 1966). The result was that the stars of high mass are being formed at the end. This agreed with what has been suggested

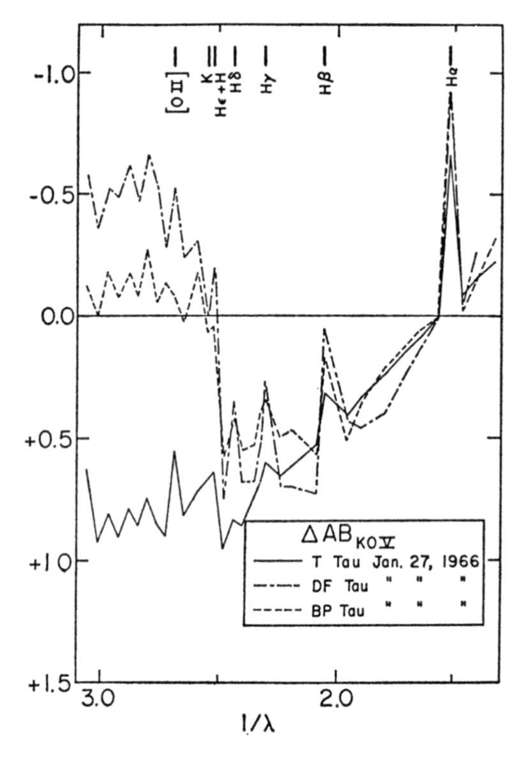

Fig. 1. Energy distribution of some T Tauri stars relative to that of a K0V star (horizontal line at 0.0). The UV excess is seen for

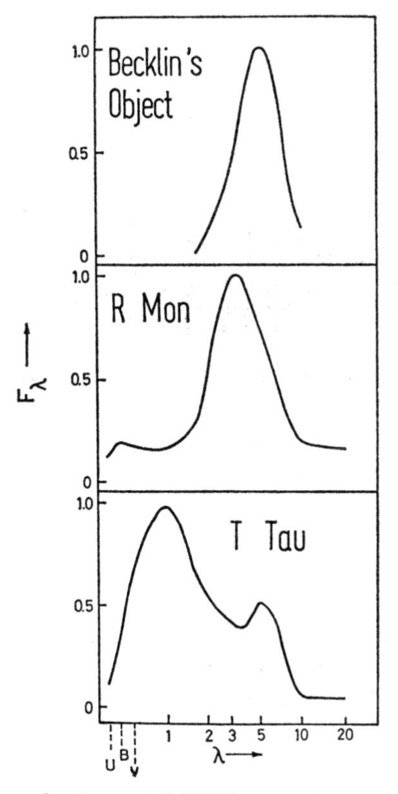

Fig. 2. Objects with IR excess after Low *et al.* (1970)

previously from observational grounds and one, therefore, believed this explanation of this scatter until recent doubts came up (STROM *et al.*, 1971, 1972; BREGER, 1972a). That the whole story of the scatter in the HR-diagram cannot lie in age differences is obvious if one takes into account stars like W 90 which in the HR-diagram are below the main sequence (Fig. 7).

But the T Tauri stars have other strange properties. It is known that they have an UV excess. This can be seen from Fig. 1 which is due to KUHI (1970). It seems that the UV excess is larger for later spectral types and that it is correlated with the strength of the Hα emission.

It is also known that these stars have an IR excess (LOW *et al.*, 1970). In Fig. 2 energy distribution functions are given for several objects. In case of T Tau and even more in the case of R Mon the infrared luminosity is comparable to or even higher than the visual luminosity. It seems that there is a sequence from BECKLIN's object where all the energy is in the IR via R Mon where the spectrum of a star becomes visible through T Tau where the star's radiation becomes more and more pronounced to the main sequence where the infrared radiation is gone and the radiation which is observed comes from the surface of the star only. The shells which radiate in the IR are in the order of 2–4 AU for T Tauri stars and in the order of 10–30 AU for BECKLIN's objects.

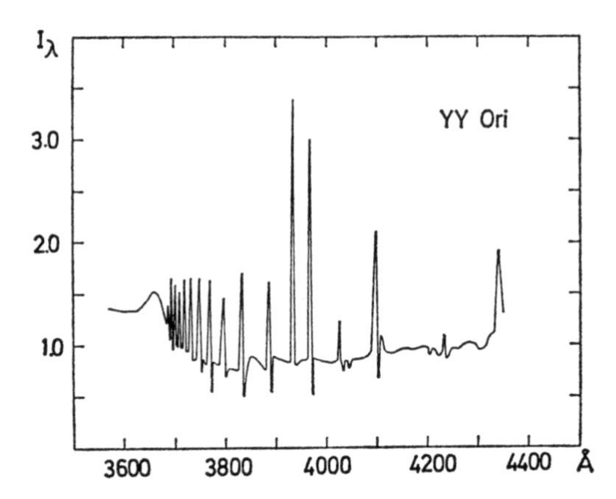

Fig. 3. The energy curve of YY Ori according to WALKER (1966) showing (blue-shifted) emission lines and (red-shifted) absorption lines

It is well known that the T Tauri stars have emission lines which normally are blue shifted indicating an outflow of matter of the order between $-90 \cdots -160$ $km \cdot s^{-1}$. The line shift was used to estimate the mass loss of T Tau stars (KUHI, 1964; GAHM, 1970). But in some cases there are also absorption lines which are red shifted. They indicate an inflow of material of the order of $150 \cdots 300\,km \cdot s^{-1}$. This can be seen in Fig. 3 in the case of YY Ori measured by WALKER (1966). The infalling material is observed only in stars which have UV excess and it seems that there is a correlation between the redshifted lines and the UV excess from which WALKER (1969) concludes that UV excess is caused by the infalling material.

The energy radiated in the UV and in the IR is in some cases much higher than the energy radiated in the visual range. For the case Lk Hα 190, a T Tauri star, the total integrated luminosity, including the radiation in the UV and in the IR makes this star an A supergiant (KUHI 1972, private communication). I think one, therefore, can conclude that the energy which comes in UV and IR is not just energy taken away from the radiation of the star itself and transformed into other wavelength ranges by some non-thermal mechanisms. One must instead conclude that there is an additional source of energy. One might think that the energy comes from the kinetic energy of the infalling material. In this picture the star would be surrounded by a cloud of infalling gas and dust. The energy of the infalling material is radiated away in the UV by the gas and in the IR by the dust. The optical depth of the dust clouds seems to be of the order of one. We, therefore, might think that the optical variability of the T Tauri stars is caused by the variation of the transparency of the dust cloud. This would make the T Tauri stars to a special type of eclipsing variables (BREGER, 1972a).

The problem certainly is more complicated. We probably have deviations from spherical symmetry. Already the simultaneous infall and outflow of material at the same time suggests that we are watching a non-spherical phenomenon. Polarisation (BREGER and DYCK, 1972; SERKOWSKI, 1969) suggests that there is a circumstellar disk like the disks in Be stars. The polarisation is variable in strength

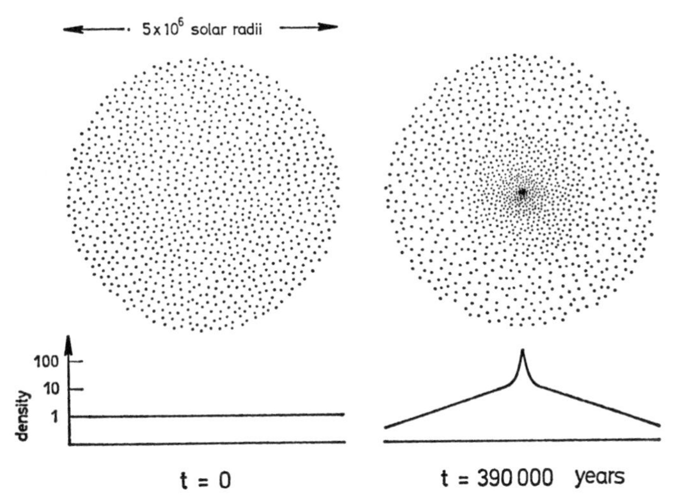

Fig. 4. The initial state of LARSON's collaps calculations (left). The density is constant in the initial cloud. Later (right) a density concentration has been formed in the center

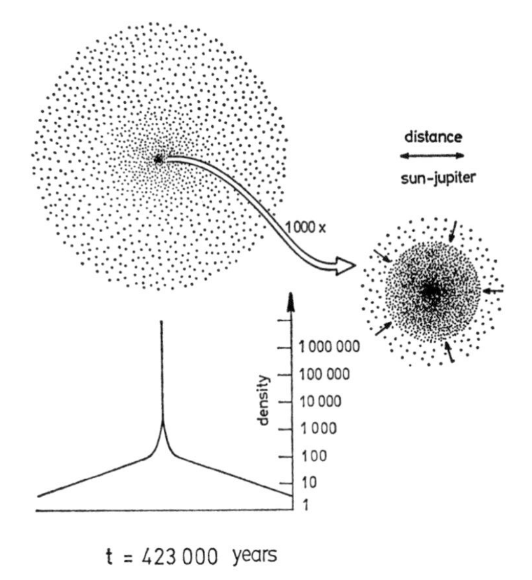

Fig. 5. LARSON's model some times later. The innermost part became optically thick and the pressure which raised with temperature stops the collaps in the central region. A shock wave is being formed around the inner core which is sketched separately on a larger scale

as well as in position angle. But let us assume the picture of the light variability caused by varying transparency of the surrounding material to be correct. Then one would also expect the circumstellar disk to be partially obscured by moving dust clouds. This could cause the variation of the polarisation. The circumstellar shells, infalling material, and additional energy by thermal and non-thermal processes seem to be related to present models of star formation (LARSON, 1969a, b).

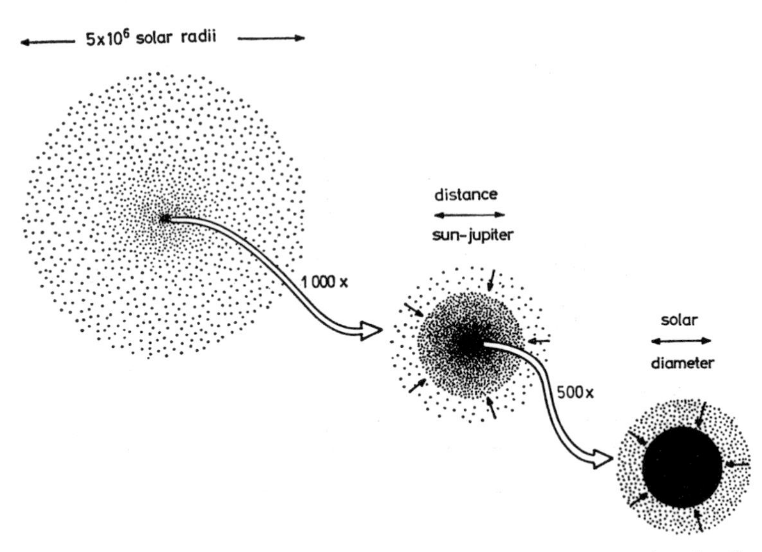

Fig. 6. The core which was plotted in Fig. 5 becomes dynamically unstable when the dissoziation of hydrogen molecules sets in. It collapses again and forms a new core. On the right hand side the cores are plotted on enlarged scales

A main result of LARSON's computations was: while the innermost part of the cloud is fairly rapidly forming a dense core which later becomes a normal star the outermost layers fall slowly. After the star in the center is already formed matter from the outside is still raining on top of it. Some stages of that process are given in Figs. 4–6. When a star is being formed the slowly falling envelope obscures completely the condensed object in the center. Therefore, the collapsing object might have properties like BECKLIN's object. In later stages the outer regions become transparent and the newly born star shines through. This suggests an explanation of the sequence of IR objects of Fig. 2 as an evolutionary sequence.

Unfortunately, LARSON's computations are valid only for non-rotating objects. More recent calculations by LARSON (1972) indicate that a rotating interstellar cloud undergoes fragmentation long before a star has been formed out of the collapsing medium. One, therefore, must be careful to use the calculations of the non-rotating case for quantitative comparison with the observation. But what certainly remains is that dust shells surrounding the very young stars with the fullness of their observed properties might give new ideas to the theory of the evolution of protostars.

III. Pulsating Stars on or Near the Main Sequence

In 1959 E. HOFMEISTER and myself (KIPPENHAHN, 1965) investigated the region where the Cepheid instability strip extended to lower luminosities crosses the main sequence. We checked the stars there for their vibrational stability and found the models to be slightly unstable. Pulsations of a period of about one hour were

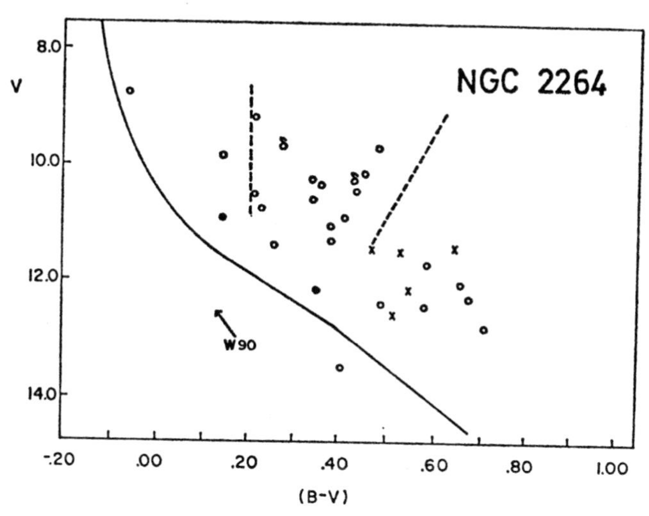

Fig. 7. Variable stars in the HR-diagram of NGC 2264 by BREGER (1972a). Open circles indicate non-variable stars, filled circles stars which are suspected to be variable, crosses irregular variables, the letters δ give the position of the two pulsating variables found by BREGER. The dashed lines give the border of the instability strip. The arrow denotes the variation of W 90 between 1954 and 1972

maintained by the κ-mechanism. Since then the problem has become more interesting. Variables have been found there among the main sequence stars; the number of δ Scuti stars has been increased and recently BREGER (1972a) found in two cases that stars which come through the instability strip during their pre-main-sequence contraction become pulsating variables (Fig. 7). From the stability theory this is not surprising. The fact to wonder about is that there are pulsating stars as well as non-pulsating stars in the same area of the HR-diagram (Fig. 8).

We think we understand pulsating stars there. δ Scuti stars seem to agree with stellar evolution theory (CHEVALIER, 1971). But why are there variables and non-variables in the HR-diagram so close together? We are faced with the following problem. In this area of the HR-diagram the mass is a smooth function of the position in the diagram; therefore, the surface gravity g and the effective temperature T_{eff} are smooth functions in the HR-diagram. But as one knows from pulsation theory for a given chemical composition the two parameters g, T_{eff} alone determine the stability! However, we can observe stars here which have almost the same values of g and T_{eff}, but the one is variable while the other is not. There must be another parameter which decides whether there is pulsation or not. And, indeed, there is an indication from the observations. BREGER (1972b) has found the following correlation:

slow rotation – metallicity – stability

Slow rotators in general seem to have more metals and do not pulsate while fast rotators have less metals and have a preference for pulsation. The correlation between slow rotation and metal content suggests that diffusion is important since

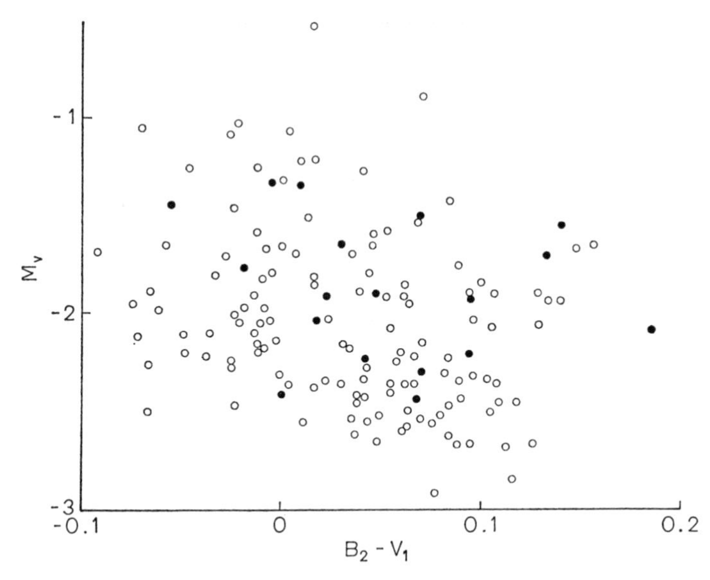

Fig. 8. Non-pulsating Am stars (filled circles) and pulsating stars (open circles) in the same region of the HR-diagram. (BAGLIN, 1972)

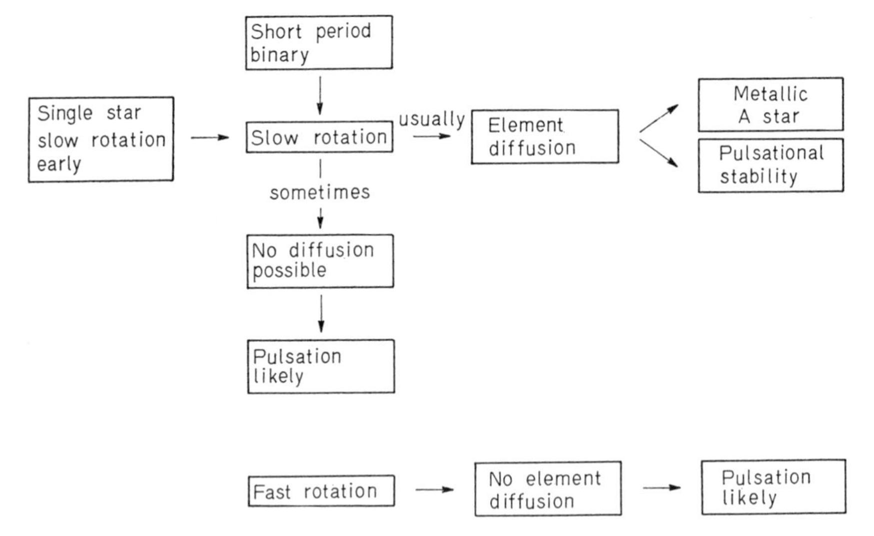

Fig. 9. BREGER's logical scheme for his correlation between rotation, metal content, and stability (BREGER, 1972b)

for slow rotators the meridional circulation currents driven by rotation which can mix the material are slow and, therefore, gravitational settling or separation of different elements by radiation pressure might become important. BREGER, therefore, suggests a logical scheme which is reproduced in Fig. 9. Unfortunately, it is always rather unclear how diffusion can overcome the mixing caused by rotationally driven circulation. I must say I always have difficulties in seeing how the circulation currents can be surpressed to make the diffusion process work. Recently BAGLIN (1972) tried to give an explanation for BREGER's correlation. She suggested that the circulation pattern near the surface of the stars might be laminar for slow rotators while for rapid rotators it might be turbulent. In the latter case circulation would be even more active and would mix the material completely. These stars then should have normal metal and normal helium content in the outer layers and since helium considerably contributes to the κ-mechanism she suggests that these stars are unstable. For slow rotators the circulation is laminar. Therefore, mixing would be much less important and chemical abnormalities can be built up by diffusion processes: Metals are enriched in the outer layers while helium goes down. In this case the lack of helium—according to BAGLIN—reduces the effect of the κ-mechanism and the objects are stable. I have difficulties with this picture despite the fact that already the laminar circulation velocities are several powers of ten higher than the diffusion velocity. The decision whether the circulation flow is laminar or turbulent depends on the RICHARDSON'S number. But BAGLIN has used the REYNOLD's number which does not take the stabilizing effect of buoyancy forces into account.

But I think BREGER's observational results should be taken seriously by theorists and one should try to rediscuss the effect of circulation in surface layers. In almost all cases where meridional circulation caused by rotation has been discussed one started with a given angular velocity distribution (in most cases with solid-body rotation) and then derived the velocity field of the meridional circulation. But this circulation field immediately changes the originally given angular velocity distribution. There might exist special angular velocity distributions in certain regions of a star for which the circulation is practically surpressed. Circulation-free angular velocity distributions have been discussed (SCHWARZSCHILD, 1942; KIPPENHAHN, 1963; ROXBURGH, 1964) but their importance had become doubtful after the GOLDREICH-SCHUBERT-FRICKE criterium was found (GOLDREICH and SCHUBERT, 1967; FRICKE, 1968) and they turned out to be secularly unstable. But the question is still open whether in certain subregions of a star an angular velocity distribution can exist which is circulation free and at the same time is secularly stable.

IV. RR Lyrae Stars

It is not yet possible to follow a star of about one solar mass from the main sequence through helium flash into the horizontal branch of globular clusters. Thermal pulses mentioned at the beginning are partially responsible for it. But we are sure now that the horizontal branch stars have a non-degenerate helium core where helium is burning. The core is surrounded by a hydrogen rich envelope. But since the models are rather sensitive to the chemical composition and to other uncertain parameters one can only guess how the evolution in the horizontal branch goes. We, therefore, do not know very much of the evolutionary history of the RR Lyrae stars.

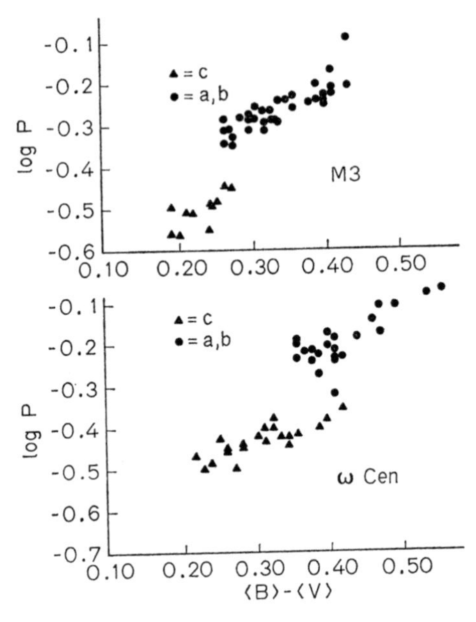

Fig. 10. The transition from *c* type variables to *a*, *b* type variables in M3 and in *ω* Cen after VAN ALBADA and BAKER (1972). In the two clusters the transition takes place at different colors (temperatures)

In the RR Lyrae gap on the blue side are the *c*-type stars which pulsate in the first overtone and on the right-hand side of the gap are the *a*, *b* types which oscillate in the fundamental mode. The transition from variables of one type of oscillation to those of the other type is characterised by certain effective temperatures. This transition temperature is different for different clusters as can be seen in the case of M3 and *ω* Cen in Fig. 10. The problem of oscillations in the fundamental and in the first overtone has been discussed by CHRISTY (1966). He used an initial value method. I.e. he started with a model at a given effective temperature and with a given initial oscillation and computed through many many periods in order to find out whether in the final stage the fundamental oscillation or the first overtone oscillation will take over. The difficulty is that in the RR Lyrae gap both the fundamental and the first overtone are excited in the linear theory and only non-linear theory can decide which type of oscillation will really be preferred. CHRISTY found that, indeed, on the left-hand side of the gap the first overtone seems to be preferred, while on the right-hand side the fundamental is more probable. He also gave an expression for the transition temperature as a function of luminosity. Although M3 contains more metals than *ω* Cen the difference in transition temperature cannot be explained by difference in chemical composition unless one assumes extreme differences in helium content or in mass of the stars in the two clusters (STOBIE, 1971). In all cases these computations are rather cumbersome since one has to compute through many periods in order to find out whether a transition from one type of oscillation to the other takes place or not (it takes

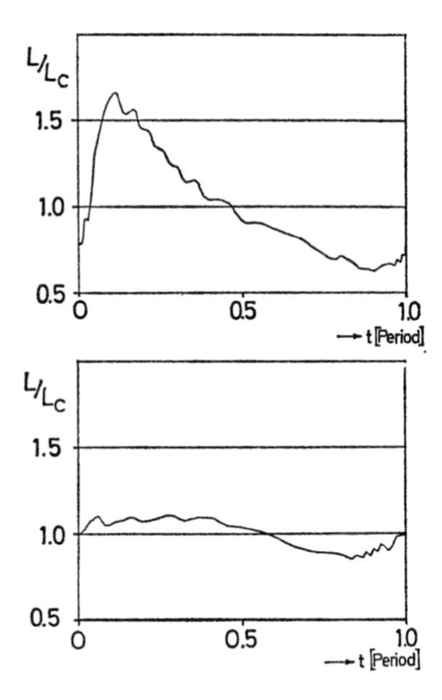

Fig. 11. Non-linear periodic pulsation of RR Lyrae models, obtained with the code of BAKER and VON SENGBUSCH. Periodic pulsations are possible for the fundamental (above) as well as for the first overtone (below). The question in which mode the star really will pulsate can be given by investigating the stability of one mode with respect to perturbations of the period of the other mode. The result of this perturbation procedure is given in Fig. 12

especially long time if one wants to compute until the transition does not take place).

During the last years BAKER and VON SENGBUSCH (1972) have developed a new method of solving the equations of non-linear pulsation theory. They do not have to determine the final stage of oscillation by starting with an initial problem. They immediately obtain the periodic solutions. In the case of the RR Lyrae gap they can produce the fully developed periodic solutions either in the fundamental or in the first overtone. The kinetic energy E_{kin} which is contained in the non-linear pulsation is a measure for the strength of the pulsation. It is a measure of the amplitude. As an example in Fig. 11 the pulsation in the fundamental and in the first overtone is given for an RR Lyrae model but this does not yet tell whether in the real case the final state of an oscillation will be in the first overtone or in the fundamental. In order to find this out BAKER and VON SENGBUSCH (1972) have investigated the stability of the modes. One has to keep in mind that both modes, the fundamental as well as the first overtone are excited. But if the star is oscillating in the fundamental one can ask whether a small perturbation of the type of the first overtone will grow and one might get a transition to the first overtone. Therefore, for each model in the gap BAKER and VON SENGBUSCH can determine whether a given mode is unstable or not with respect to perturbations of the type of the other mode. The result is summerized in Fig. 12. As one can see there is a region on the left where the fundamental is unstable and the first overtone is stable and

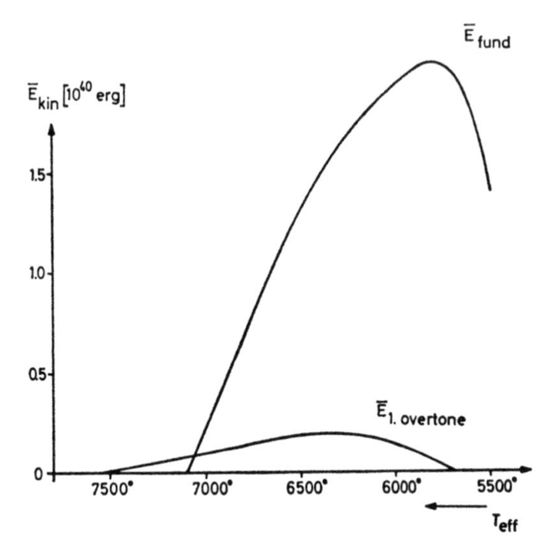

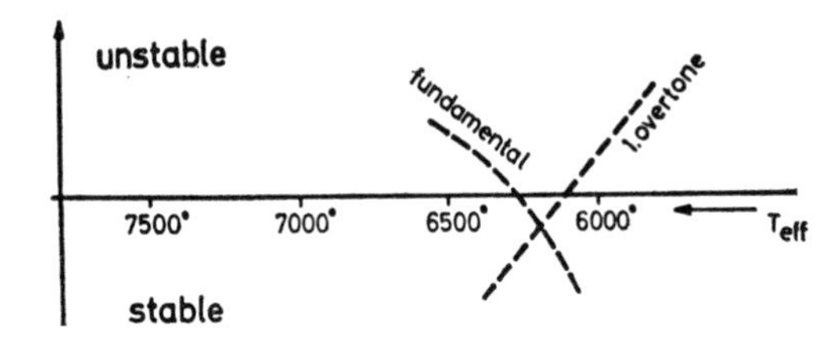

Fig. 12. For RR Lyrae models of different effective temperature periodic solutions as those in Fig. 11 were obtained and the mean kinetic energy (as a measure of the amplitude of the pulsation) is plotted for the fundamental as well as for the first overtone. In the lower part the stability of the two modes is indicated (the ordinate is a measure for the stability of the mode). (BAKER and VON SENGBUSCH, 1972)

one must, therefore, expect the stars to pulsate in the first overtone. On the right-hand side we have the opposite case and we will expect oscillations in the fundamental. But there is an overlapping region in between in which both types of oscillations are stable. In which mode will a star oscillate if it is in the middle of the gap in the overlapping region? The answer is given by VAN ALBADA and BAKER (1972). If the evolution carries the star through the RR Lyrae gap from the left to the right the star first goes through the left-hand region in which it is pulsating in the first overtone; and while the evolution is carrying the object into the overlapping region the star continues to oscillate in that mode because this mode is stable. Only when the star reaches the right-hand border of the overlapping region does the first overtone become unstable and the star switches to oscillations in the fundamental. If in another cluster evolution brings the star from the right to the left through the RR Lyrae gap then for the same reasons the transition will be at the left-hand border of the overlapping region. Although there is not yet quantitative agreement VAN ALBADA, BAKER, and VON SENGBUSCH believe that the difference in transition temperature for different clusters is caused by different directions of the evolutionary tracks: in M 3 the evolution goes from the right to the left and in ω Cen into the opposite direction. During the transition from one mode to the other the star will be an irregular variable for a certain period. Maybe the 10% irregular variables in ω Cen are in the stage of transition from one mode to the other.

V. Coronae Borealis Stars

Progress has been made in understanding R CrB stars during the last years. It has been shown that the light which is missing in the visual during faint stages can be found in the infrared as 1000 K black-body radiation (STEIN *et al.*, 1969). This favours the Loreta hypothesis according to which carbon clouds obscure the star. Models have been constructed (PACZYNSKI, 1971; BIERMANN and KIPPENHAHN, 1971) indicating that these stars consist of a helium envelope and a degenerate carbon-oxygen core. The core masses are in the order of 0.6 to 0.7 $M_\odot$. The total mass is in the order of one solar mass – certainly not more than 2.5 $M_\odot$.

How can stars of this type be formed during their post-main-sequence evolution? How did the star get rid of its outer hydrogen envelope? The simplest explanation would be that the hydrogen has been blown into space. This is indicated in the upper part of Fig. 13. A star, say of 7 solar masses, forms a helium core and in the helium a carbon-oxygen core. Then the hydrogen-rich envelope is blown into space and within some 10^7 years a star of the type of the models for R CrB stars has been formed. But there is a serious difficulty. According to WARNER (1967) the spacial distribution as well as the velocity distribution of the R CrB stars suggest that the ages of R CrB stars are some 10^9 years – hundred times longer than the time of the evolution in the upper part of the Fig. 13! Therefore PACZYNSKI (1971) suggests another type of evolution (indicated schematically in the lower

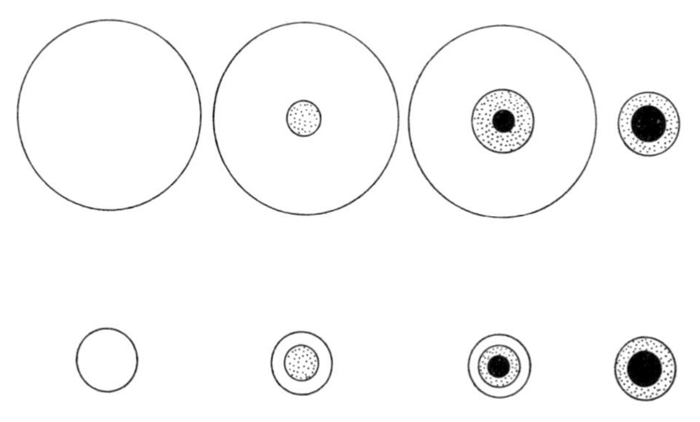

Fig. 13. Two schemes of evolution which lead to the same model of a star with a carbon-oxygen core (black) and a helium envelope (dotted). In the upper line one starts with a main-sequence star of about 7 solar masses, in the second line with a model of about one solar mass. White areas give the original hydrogen-rich composition. The diagrams do not give true geometrical relations

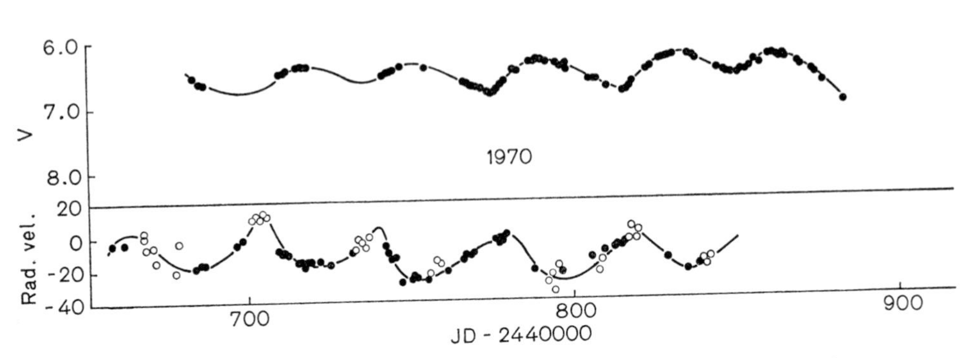

Fig. 14. Light curve and radial velocity of the pulsations of RY Sag according to ALEXANDER *et al.* (1972)

part of Fig. 13) which starts already with a star of the order of one solar mass. The star again forms a helium core in which a carbon core is built up but then the hydrogen rich envelope somehow is mixed into the hydrogen burning shell and the hydrogen is completely used up in this star. The final product is a star which again resembles our R CrB models but this time the time scale of evolution is of the order of some 10^9 years. Certainly here we see very late stages of stellar evolution which we haven't yet been able to understand from stellar evolution theory.

But then R CrB stars are even more exciting. It has been known already that superimposed to the light curve of R CrB there is a small periodic oscillation of the order of 40 days. Recently it has also been found by ALEXANDER *et al.* (1972) that RY Tau is pulsating as can be seen in Fig. 14. It seems that these stars happen to be accidentally in the Cepheid instability strip and, therefore, show pulsations driven by the κ-mechanism. TRIMBLE (1972) investigated the pulsation properties of models for R CrB stars. She indeed found that the models were pulsating. She also obtained the right periods. But the non-linear pulsation calculations gave wrong amplitudes and wrong radial velocities. I think this is not serious since it seems that convection (which until now cannot be included properly into pulsation calculations) is more important for pulsating R CrB stars than it is for classical cepheids.

The stages of evolution where theorists believe to be able to produce the best stellar models are also those for which we have the best observational material. The more scarce the observations the poorer also the theoretical models. This tells us that observation has been leading the theorist that it had helped him to avoid mistakes and to find the more or less correct procedure to construct models. In the stages of late stellar evolution as well as for the very early phases the observational material is not very plentiful since the evolution is fast and the chance to observe such an object is much smaller. The theorist no longer has the helpful guidance of the observer. But it seems to me that especially the variable stars can at least partially fill in these gaps. Their variability makes them easily detectable and in some cases—like in the R CrB stars as well as for instance in RR Lyrae stars—and even more in the case of the W Vir stars they give us information which may guide us into these unexplored stages of stellar evolution.

References

ALBADA, T. S., VAN, BAKER, N.: Preprint (1972).
ALEXANDER, J. B., ANDREWS, P. J., CATCHPOLE, R. M., FEAST, M. W., LLOYD EVANS, T., MENZIES, J. W., WISSE, P. N. J., WISSE, M.: Monthly Notices Roy. Astron. Soc. **158**, 305 (1972).
BAGLIN, A.: Astron. Astrophys. **19**, 45 (1972).
BAKER, N., SENGBUSCH, K. VON: Private communication (1972).
BIERMANN, P., KIPPENHAHN, R.: Astron. Astrophys. **14**, 32 (1971).
BREGER, M.: Astrophys. J. **171**, 539 (1972a).
BREGER, M.: Astrophys. J. **176**, 373 (1972b).
BREGER, M., DYCK, H. M.: Astrophys. J. **175**, 127 (1972).
CHEVALIER, C.: Astron. Astrophys. **14**, 24 (1971).
CHRISTY, R. F.: Astrophys. J. **144**, 108 (1966).
FRICKE, K.: Z. Astrophys. **68**, 317 (1968).
GAHM, G. F.: Astron. Astrophys. **8**, 73 (1970).
GOLDREICH, P., SCHUBERT, G.: Astrophys. J. **150**, 571 (1967).
IBEN, I., TALBOT, R. J.: Astrophys. J. **144**, 968 (1966).
KIPPENHAHN, R.: In L. GRATTON (ed.), Star Evolution, p. 330. New York: Acad. Press 1963.
KIPPENHAHN, R.: In The Position of Variable Stars in the Hertzsprung-Russell-Diagram. Third Colloquium on Variable Stars. Bamberg, p. 7 (1965).
KUHI, L. V.: Astrophys. J. **140**, 1409 (1964).
KUHI, L. V.: In Évolution stellaire avant la séquence principale. Liège Symp. 1969, p. 295 (1970).
LARSON, R. B.: Monthly Notices Roy. Astron. Soc. **145**, 271 (1969a).
LARSON, R. B.: Monthly Notices Roy. Astron. Soc. **145**, 297 (1969b).
LARSON, R. B.: Monthly Notices Roy. Astron. Soc. **156**, 437 (1972).
LOW, F. J., JOHNSON, H. L., KLEINMANN, D. E., LATHAM, A. S., GEISEL, S. L.: Astrophys. J. **160**, 531 (1970).
PACZYNSKI, B.: Acta Astron. **21**, 1 (1971).
ROXBURGH, I. W.: Monthly Notices Roy. Astron. Soc. **128**, 157 (1964).
SCHWARZSCHILD, M.: Astrophys. J. **95**, 441 (1942).
SERKOWSKI, K.: Astrophys. J. **156**, L55 (1969).
STEIN, W. A., GAUSTADT, J. E., GILLETT, F. C., KNACKE, R. F.: Astrophys. J. **155**, L3 (1969).
STOBIE, R. S.: Astrophys. J. **168**, 381 (1971).
STROM, K. M., STROM, S. E., YOST, J.: Astrophys. J. **165**, 479 (1971).
STROM, S. E., STROM, K. M., BROOKE, A. L., BREGMAN, J., YOST, J.: Astrophys. J. **171**, 267 (1972).
TRIMBLE, V.: Monthly Notices Roy. Astron. Soc. **156**, 411 (1972).
WALKER, M. F.: Astrophys. J. Suppl. **2**, 365 (1956).
WALKER, M. F.: In R. F. STEIN, and A. G. W. CAMEROH (eds.), Stellar Evolution, p. 405. New York: Plenum Press 1966.
WALKER, M. F.: In L. DETRE (ed.), Non-Periodic Phenomena in Variable Stars. Fourth Colloquium on Variable Stars, p. 102. Dordrecht-Holland: D. Reidel 1969.
WARNER, B.: Monthly Notices Roy. Astron. Soc. **137**, 119 (1967).

Discussion

STEPIEN, K.:

CHRISTY has found from his calculations relations connecting the transition period and luminosity, and the transition temperature and mass of variables. Do the results of BAKER and VON SENGBUSCH indicate that these relations are wrong?

KIPPENHAHN, R.:

The present results cover only pulsations of one mass at a certain luminosity, for which CHRISTY gives a broader overlap of fundamental and first overtone pulsations in $T_{\rm eff}$. No statement can yet be made about the relations in question.

GRATTON, L.:
Mention two facts: Peculiar nebulocities around T Tauri stars. Some δ Sct stars have very rich line spectrum (more than it might be expected from star of the same temperature).

DE JAGER, C.:
Could you explain how element diffusion in slow rotators produces an overabundance of metals in the stellar photospheres?

KIPPENHAHN, R.:
It is not only the separation due to gravity but also the effect of radiation pressure which acts differently on different atoms.

On the Stability of the Light Curves of Galactic Cepheids

By G. Asteriadis, L. N. Mavridis, and A. Tsioumis
Department of Geodetic Astronomy, University of Thessaloniki
Thessaloniki, Greece

With 16 Figures

Abstract

High accuracy photoelectric three-color (U, B, V) observations of the galactic cepheids CD Cyg; X, Z, RR Lac and U Vul have been carried out during the years 1967–1970 with the 38-cm reflector of the Hamburg Observatory installed at the Stephanion Observatory in Greece. A comparison of the results with the photoelectric two-color (B, V) observations of comparable accuracy of the same cepheids carried out in the years 1956 – 1959 with the 72-cm reflector of the Landessternwarte Heidelberg shows that no changes greater than the observational errors have occured in the form of the light and color curves of these cepheids between the two observational series.

I. Introduction

Starting with the pioneer work of Eggen (1951) a great number of photoelectric observations of cepheid variables have been carried out during the last twenty years. In most of these programs, however, an effort was made to increase the number of the stars observed by measuring only the points considered absolutely necessary for a relatively good determination of the light and color curves. In this way rich material for the study of the structure and rotation of the Galaxy with the help of cepheid variables has been obtained (Fernie and Hube, 1968).

The determination of as complete and accurate light and color curves of cepheid variables, as possible, on the other hand, could also be of considerable interest, for example from the following points of view:

a) For a more thorough study of the relations existing between the different parameters characterizing the light and color curves and the period. These relations could then be used for a better separation between the population I and population II cepheids in the disk of the Galaxy as well as for an eventual subdivision of these two groups of cepheids into further sub-groups.

b) For a control of the stability of the periods and, after reobservation of the same stars at a later time, of the form of the light and color curves. This information could be of great value for the determination of the time scale of the cepheid phaenomenon.

II. The Heidelberg Observations

As a contribution to this second approach an effort has been made by BAHNER and MAVRIDIS (1971) to determine light and color curves (B, V) as completely and accurately as possible for the following 18 galactic cepheids with $2^d < P < 17^d$: RT, RX, SY, Aur; RW Cam; SU Cas; VZ, CD Cyg; V, X, Y, Z, RR, BG Lac; RS Ori; SV, AW Per; U Vul and TU Cas (the cepheid shows beat phaenomena). The observations have been obtained in 1956–1959 with the reflector of the Landessternwarte Heidelberg-Königstuhl (aperture 72 cm, $f/17$). In order to achieve greater accuracy a strictly differential method has been used. Close to each variable V two comparison stars A, B were selected. A symmetrical measuring sequence was adopted which within 25 minutes gives two essentially independent differences V−A, V−B for the same instant and a difference A−B of half weight both in v and $b-v$, where v and b are the visual and blue magnitudes in the instrumental system.

The mean errors for one observation of a comparison pair v difference were between $0^m_{\cdot}007$ and $0^m_{\cdot}012$. The typical mean error for one observation of a comparison pair color difference in $b-v$ was $0^m_{\cdot}009$. No observation showing reasonable deflections was excluded. If the deviations from the mean comparison star difference was more than $0^m_{\cdot}04$ or the difference between the symmetrical halves of a cepheid observation surpassed $0^m_{\cdot}025$ the observation was marked uncertain.

The results obtained in this way for the cepheids TU Cas; CD Cyg; X, Z, RR Lac and U Vul have been published already (BAHNER and MAVRIDIS, 1971).

III. The Stephanion Observations

In the year 1967–1970 a new series of photoelectric three-color (U, B, V) observations of the cepheids CD Cyg; X, Z, RR Lac and U Vul has been carried out with the 38-cm reflector ($f/21$) of the Hamburg Observatory installed at the Stephanion Observatory in Greece ($\lambda = -22° 49' 44''$, $\varphi = +37° 45' 15''$, $H = 800$ m above see level).

In order to make the two observational series comparable the same differential method of observations as well as the same method of reductions have been used at Stephanion as in Heidelberg. That is close to each variable V two comparison stars A, B (the same as in Heidelberg) were selected and a symmetrical measuring sequence Au, b, v Vv, b, u, u, b, v Bv, b, u, u, b, v Vv, b, u, u, b, v Av, b, u was adopted, which within 45 minutes gives two essentially independent differences V−A, V−B for the same instant and a difference A−B of half weight both in v and $b-v$, $u-b$, where v, b and u are the visual, blue and ultraviolet magnitudes in our instrumental system.

The differences in air mass between variable and comparison stars were always quite small. The v magnitudes were reduced assuming that the second order term in the v extinction coefficient is zero. The mean errors for one observation of a comparison pair v difference were between $0^m_{\cdot}014$ and $0^m_{\cdot}023$.

For the $(b-v)$ and $(u-b)$ colors, the observations were reduced with the relations

$$\Delta(b-v)_0=\Delta(b-v)_X-K''_{b-v}X\Delta(b-v)_0\simeq\Delta(b-v)_X[1-K''_{b-v}X], \quad (1)$$

$$\Delta(u-b)_0=\Delta(u-b)_X-K''_{u-b}X\Delta(u-b)_0\simeq\Delta(u-b)_X[1-K''_{u-b}X], \quad (2)$$

where X is the air mass. The second order coefficients K''_{b-v} and K''_{u-b} were determined for each night using all the comparison star observations; the resulting values were between $0^m.00$ and $-0^m.08$ for K''_{b-v} and $+0^m.04$ and $+0^m.12$ for K''_{u-b}. The mean errors for one observation of a comparison pair color difference were between $0^m.008$ and $0^m.022$ for $b-v$ and $0^m.03$ and $0^m.06$ for $u-b$.

As in Heidelberg, no observation showing reasonable deflections was excluded. If the deviation from the mean comparison star difference was more than $0^m.04$ or the difference between the symmetrical halves of a cepheid observation surpassed $0^m.025$ the observation was marked uncertain. These limits were used for v, and $b-v$. The corresponding limits for $u-b$ were taken correspondingly equal to $0^m.065$ and $0^m.04$.

On a few good nights the comparison stars were observed together with stars which have high weight in Table 9 of JOHNSON *et al.* (1966). The stars representing the *U*, *B*, *V*-system cover an adequate range in color and position in the sky. From these observations the transformations of our instrumental system to the international *U*, *B*, *V*-system were found.

As a consequence of the differential method used the light and color curves obtained for each variable relative to each comparison star both in Heidelberg and at Stephanion are better defined than the magnitudes and colors of the comparison stars in a common system or this system in the *U*, *B*, *V* frame.

More details about the observations carried out at Stephanion and their reduction as well as the values of the magnitudes and colors obtained with the help of these observations will be published elsewhere (ASTERIADIS *et al.*, 1973). The corresponding light and color curves are given in Figs. 1–5. The phases given both in the paper of BAHNER and MAVRIDIS (1971) and in the present paper were computed with the help of the epochs and periods given in KUKARKIN *et al.* (1969). Open circles represent in both cases values of lower weight.

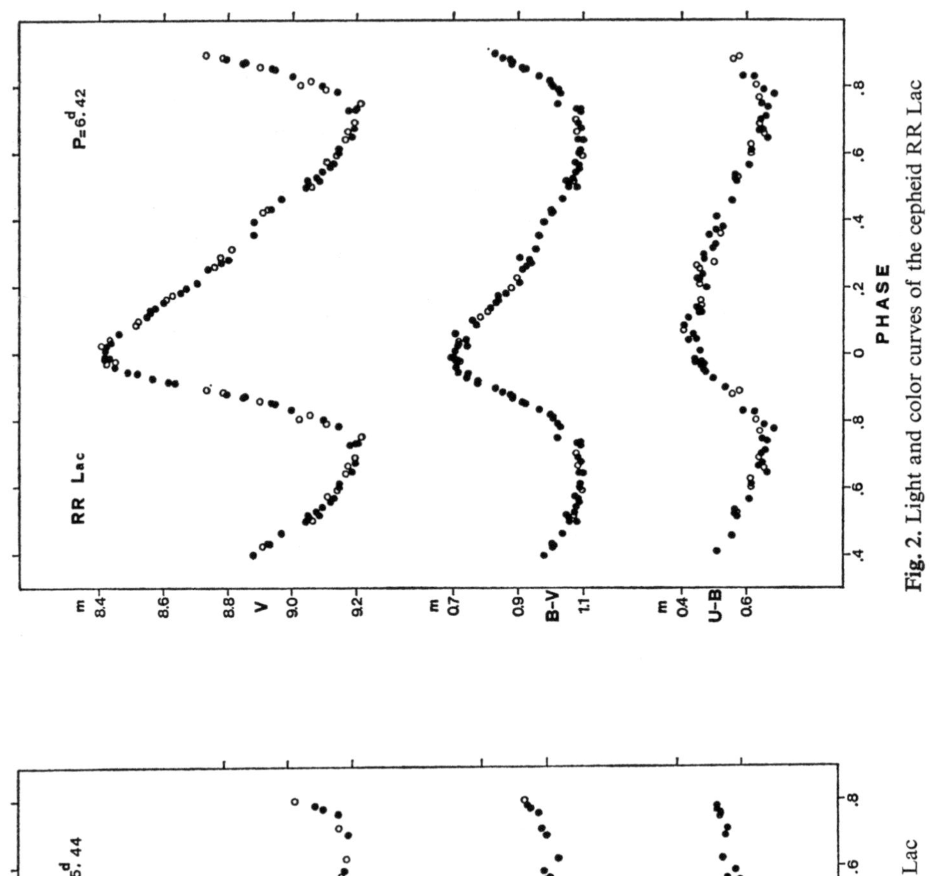

Fig. 2. Light and color curves of the cepheid RR Lac

Fig. 1. Light and color curves of the cepheid X Lac

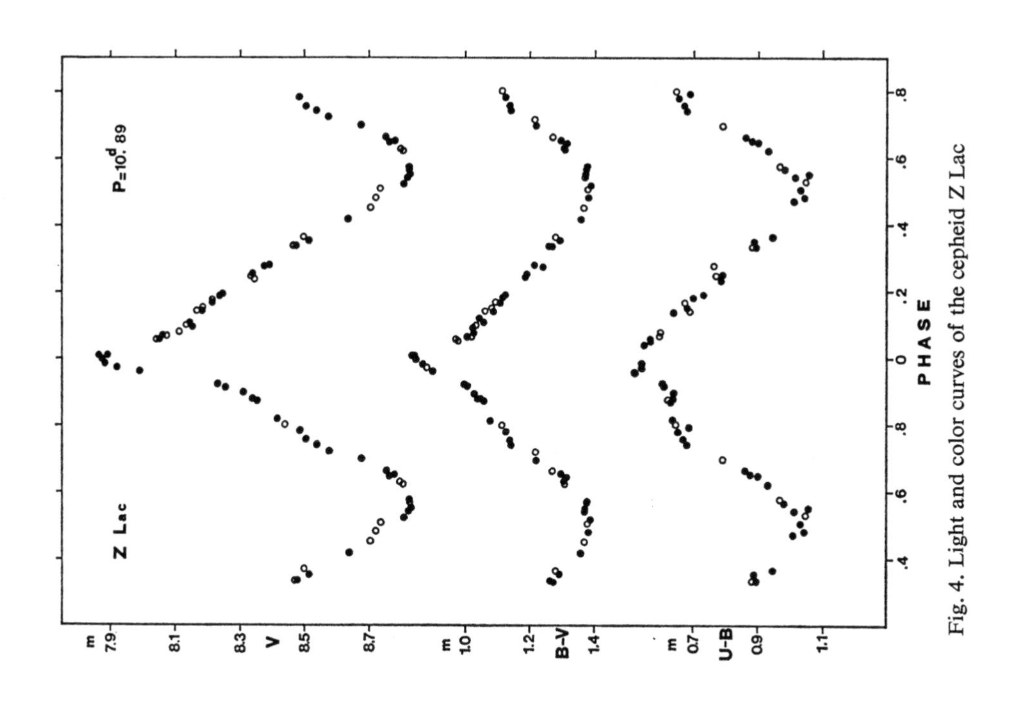

Fig. 4. Light and color curves of the cepheid Z Lac

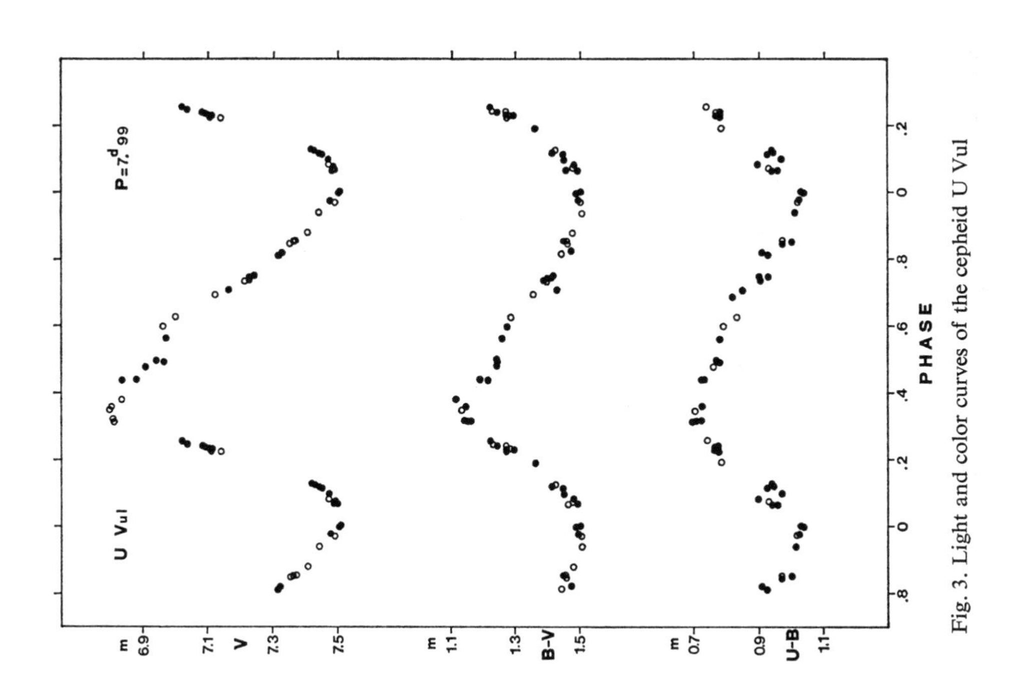

Fig. 3. Light and color curves of the cepheid U Vul

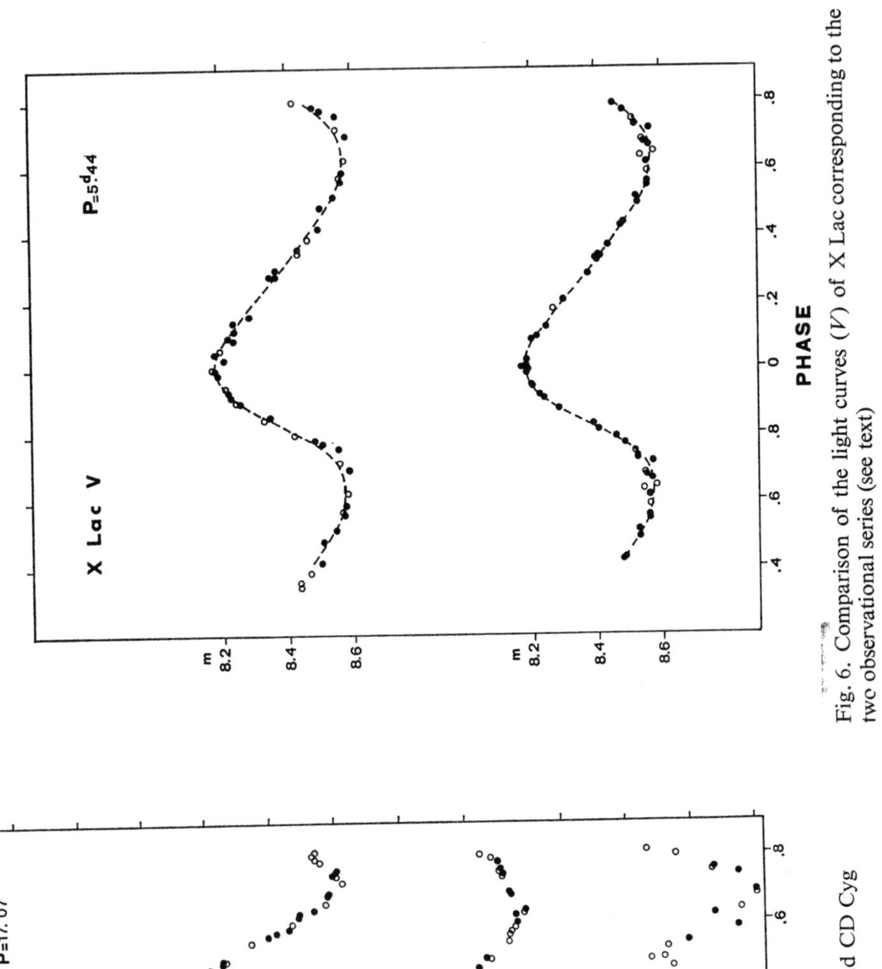

Fig. 5. Light and color curves of the cepheid CD Cyg

Fig. 6. Comparison of the light curves (V) of X Lac corresponding to the two observational series (see text)

IV. Comparison of the Two Observational Series

One of the main objectives of the reobservation of the five cepheids mentioned above was the control of the stability of the form of the light and color curves in the eleven-year time interval elapsed between the Heidelberg and Stephanion observational series.

In order to carry out a detailed comparison of the light and color curves of the same cepheid corresponding to two different observational series one has first to determine the phase shift ΔP which has to be applied to the earlier light curve in order to bring it into agreement with the later one because of an eventual inaccuracy in the value of the period used. The values of ΔP were obtained with the help of the method of HERTZSPRUNG (1919). The results found are given in Table 1,

Table 1. Phase shift between the Heidelberg and Stephanion observational series

Cepheid	Period	$\overline{JD_S} - \overline{JD_H}$	ΔP
X Lac	$5^d.44$	+4432.049	$-0^P.016$
RR Lac	6.42	4614.602	+0.006
U Vul	7.99	4204.953	−0.001
Z Lac	10.89	4462.891	−0.004
CD Cyg	17.07	4363.570	+0.017

where $\overline{JD_S}$ and $\overline{JD_H}$ represent respectively the mean Julian Date for the Stephanion and the Heidelberg observations.

From the values given in Table 1 one can see that while for the cepheids RR Lac, U Vul and Z Lac both the Heidelberg and the Stephanion observations can be described satisfactorily with the epochs and periods given in the third edition of the General Catalogue of Variable Stars (KUKARKIN *et al.*, 1969) the same is not valid for the cepheids X Lac and CD Cyg.

The ΔP shifts given in Table 1 were then applied to the Heidelberg light and color curves and at this stage a detailed comparison of the Heidelberg and the Stephanion photometric systems was carried out. Following transformations were found:

$$\begin{aligned} V_S &= +0.054 + 0.987\, V_H + 0.069\, (B-V)_H, \\ &\quad \pm 0.014 \pm 0.002 \quad\; \pm 0.004 \end{aligned} \tag{3}$$

$$\begin{aligned} (B-V)_S &= -0.022 + 0.990\, (B-V)_H. \\ &\quad \pm 0.006 \pm 0.005 \end{aligned} \tag{4}$$

A more detailed study of the transformation of the Heidelberg photometric system to the Stephanion system will be given elsewhere (ASTERIADIS *et al.*, 1973).

In Figs. 6–15 the lower part gives for each of the cepheids studied here the light and color curves corresponding to the Heidelberg observational series shifted in phase by the ΔP values given in Table 1 and transformed to the Stephanion photo-

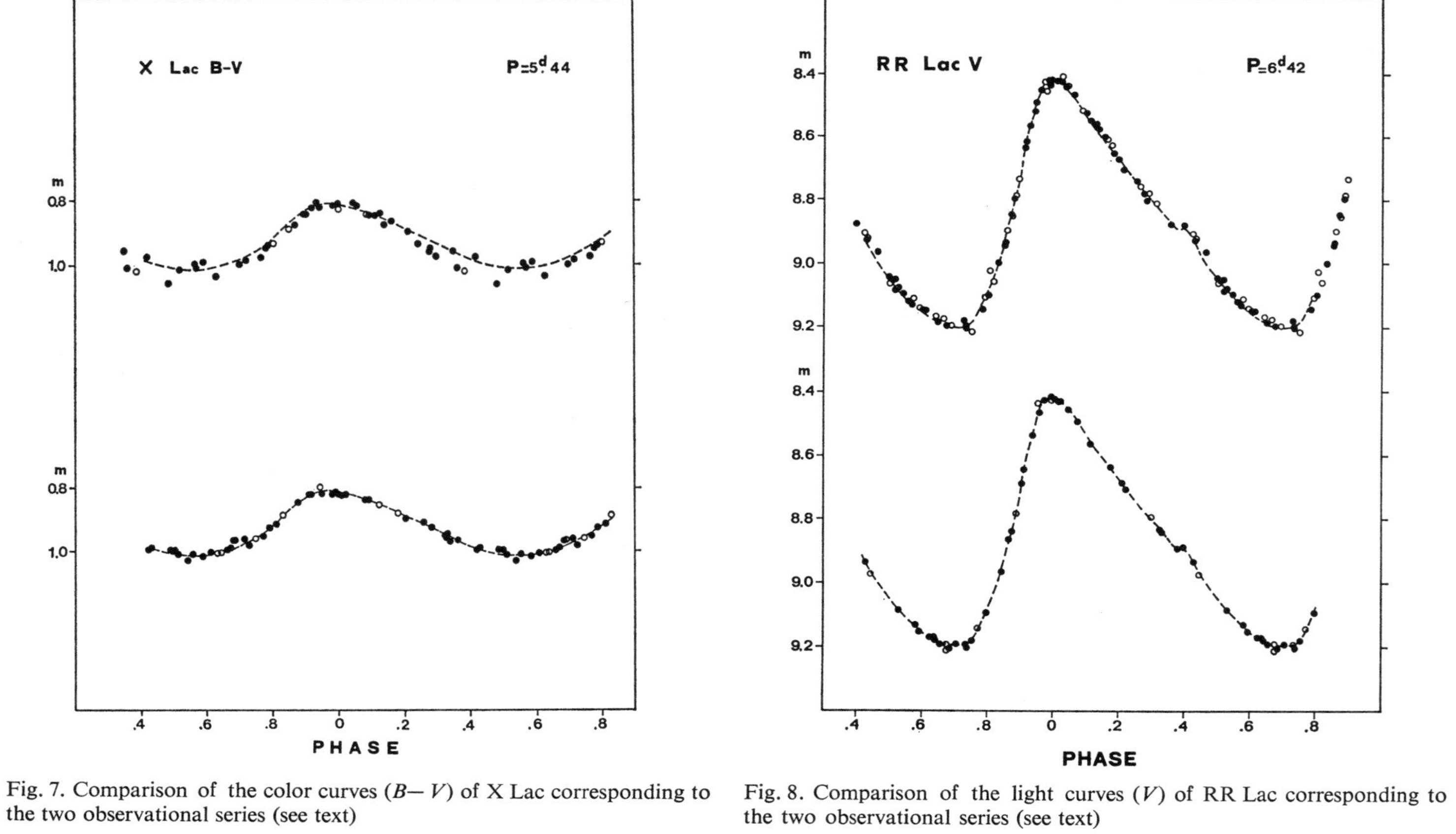

Fig. 7. Comparison of the color curves $(B-V)$ of X Lac corresponding to the two observational series (see text)

Fig. 8. Comparison of the light curves (V) of RR Lac corresponding to the two observational series (see text)

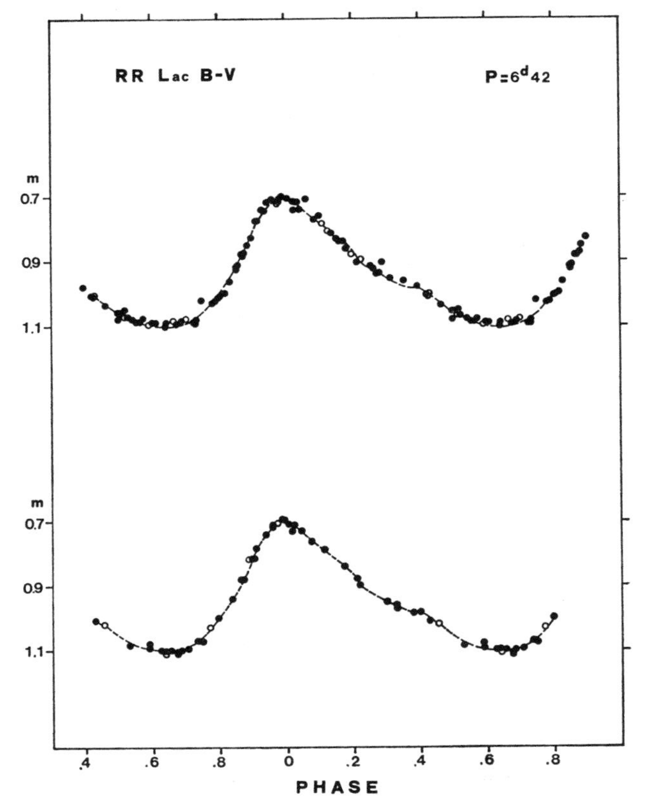

Fig. 9. Comparison of the color curves $(B-V)$ of RR Lac corresponding to the two observational series (see text)

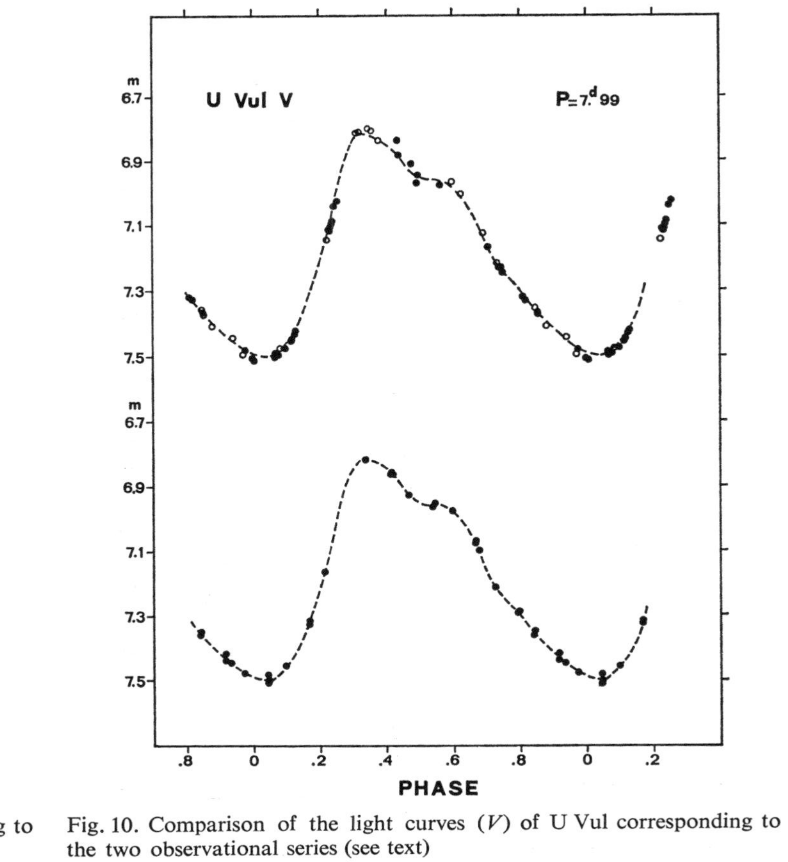

Fig. 10. Comparison of the light curves (V) of U Vul corresponding to the two observational series (see text)

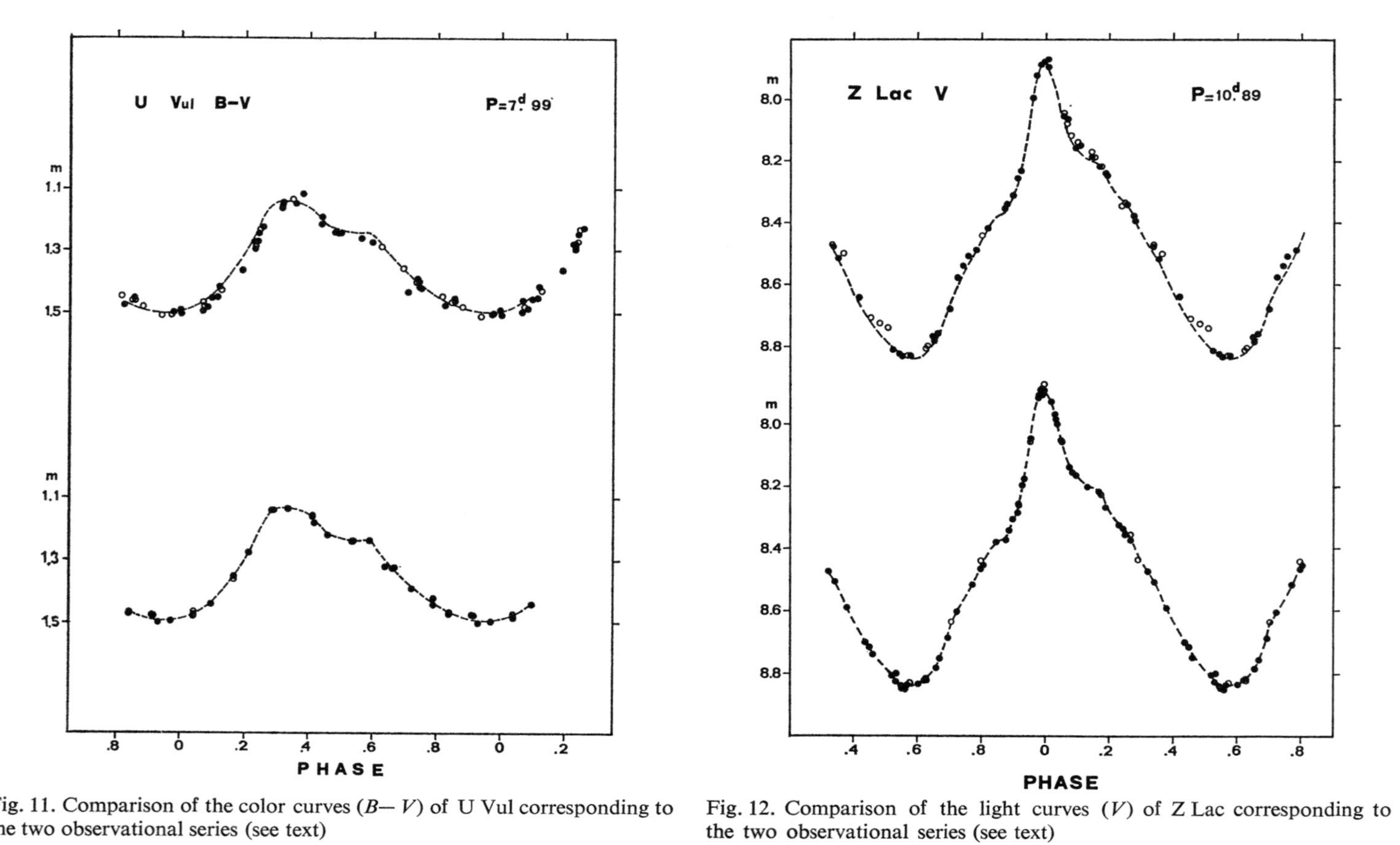

Fig. 11. Comparison of the color curves ($B-V$) of U Vul corresponding to the two observational series (see text)

Fig. 12. Comparison of the light curves (V) of Z Lac corresponding to the two observational series (see text)

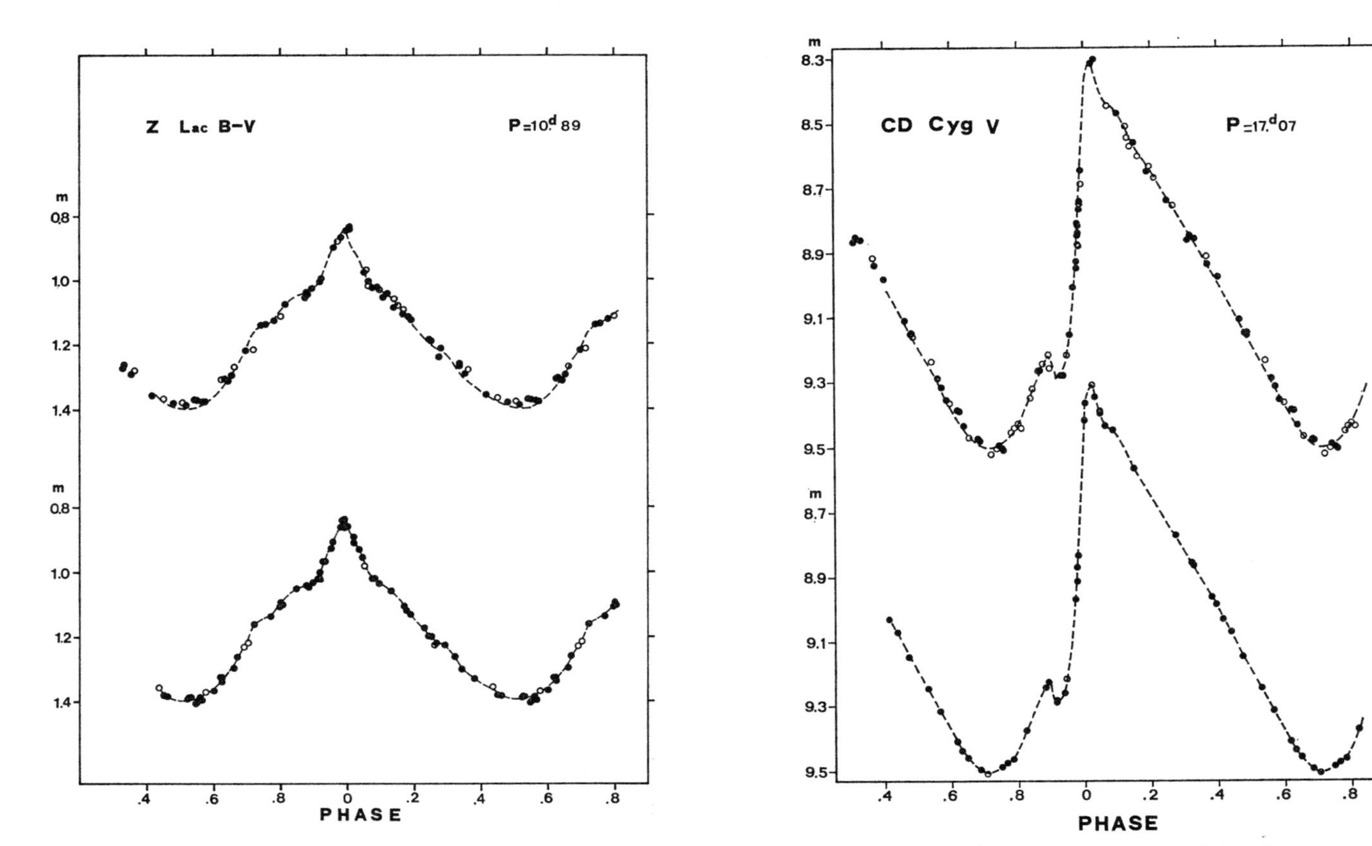

Fig. 13. Comparison of the color curves ($B-V$) of Z Lac corresponding to the two observational series (see text)

Fig. 14. Comparison of the light curves (V) of CD Cyg corresponding to the two observational series (see text)

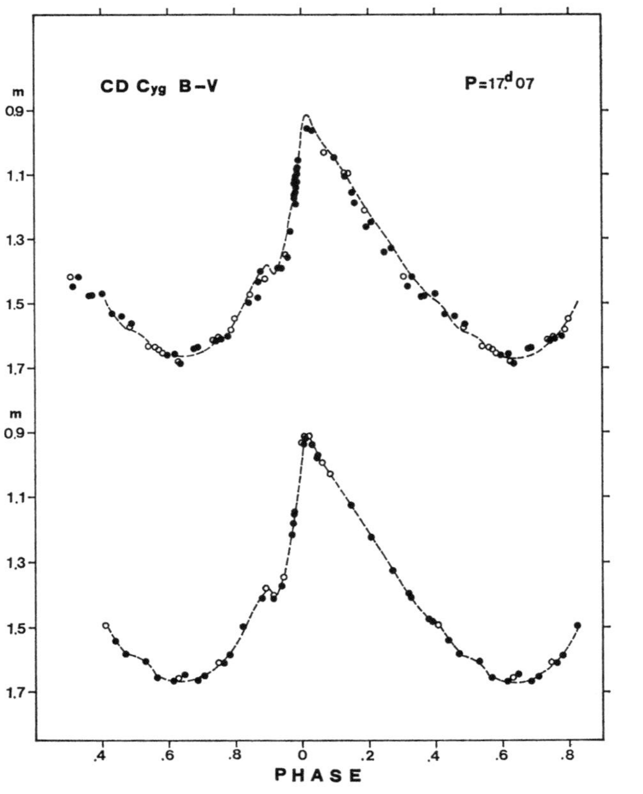

Fig. 15. Comparison of the color curves $(B-V)$ of CD Cyg corresponding to the two observational series (see text)

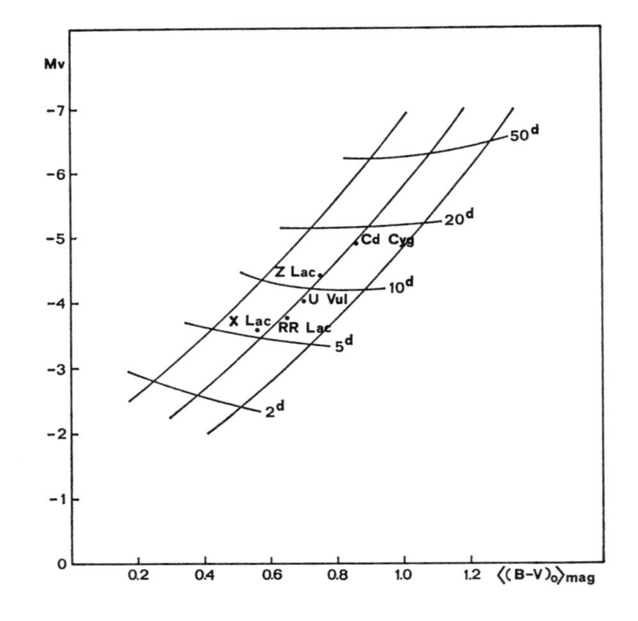

Fig. 16. Position of the five cepheids studied in the present paper in the instability strip

metric system with the help of Eqs. (3) and (4). A dashed-line has been drawn which is considered to give the best representation of these observations.

In the upper part of these figures the open and full circles represent the magnitudes and colors of the same cepheids obtained at Stephanion. The original phases were used for these observations which are also expressed in the original (Stephanion) photometric system. The dashed line drawn in the upper part of these figures is the same as the one defined above, i.e. the line representing the Heidelberg observations, drawn in the scale of the upper part of each figure.

From Figs. 6–15 one can see that no changes greater than the observational errors have occured in the form of the light and color curves of the five cepheids studied here in the eleven-year time interval elapsed between the Heidelberg and the Stephanion observational series.

Fig. 16 represents the position of the five cepheids in the instability strip as drawn by Fernie (1967b). The absolute visual magnitudes M_V have been computed with the help of the period-luminosity relation given by Fernie (1967b, equation 25). The intrinsic colors $\langle (B-V)_0 \rangle_{\mathrm{mag}}$ have been computed with the help of the relation

$$\langle (B-V)_0 \rangle_{\mathrm{mag}} = \langle (B-V) \rangle_{\mathrm{mag}} - E_{B-V}, \tag{5}$$

where $\langle (B-V) \rangle_{\mathrm{mag}}$ is the average of the observed $B-V$ color (Stephanion) over the cycle and E_{B-V} is the color excess in $B-V$ as given by Fernie (1967a). From this figure one can see that the cepheids RR Lac, U Vul, Z Lac, and CD Cyg are very close to the central line of the instability strip while X Lac is at a relatively greater distance from this line.

Acknowledgement

The second of the authors would like to express his gratitude to the following Organizations: 1) to the Science Committee, North Atlantic Treaty Organization for a research grant which made possible the establishment of the Stephanion Observatory where the observations discussed in this paper have been carried out and 2) to the National Hellenic Research Foundation for a research grant supporting the present research program.

References

Asteriadis, G., Mavridis, L. N., Tsioumis, A.: In preparation (1973).
Bahner, K., Mavridis, L. N.: Ann. Faculty of Technology, Univ. of Thessaloniki **5**, 65 (1971). Contr. Department Geodetic Astronomy, Univ. Thessaloniki No. 3.
Eggen, O. J.: Astrophys. J. **113**, 367 (1951).
Fernie, J. D.: Astron. J. **72**, 422 (1967a).
Fernie, J. D.: Astron. J. **72**, 1327 (1967b).
Fernie, J. D., Hube, J. O.: Astron. J. **73**, 492 (1968).
Hertzsprung, E.: Astron. Nachr. **210**, 5018 (1919).
Johnson, H. L., Mitchell, R. I., Iriarte, B., Wisniewski, W. Z.: Commun. Lunar Planet. Lab. **4**, p. 99 (1966).
Kukarkin, B. V., Kholopov, P. N., Efremov, Yu. N., Kukarkina, N. P., Kurochkin, N. E., Medvedeva, G. I., Perova, N. B., Fedorovich, V. P., Frolov, M. S.: General Catologue of Variable Stars. Third Edition, Moscow (1969).

Discussion

Todoran, J.:
Is it not useful to determine some periodic functions to see if their parameters are variable?

Mavridis, L. N.:
We have not tried that so far.

On the Pulsation Amplitude of Cepheid Variables

By Nikola Nikolov and Tsvetan Tsvetkov
Department of Astronomy, Faculty of Physics, University of Sofia, Bulgaria

With 3 Figures

Abstract

On the basis of more numerous data a relation $\log P\Delta V/\log \Delta R$ (P is the period, ΔR is the amplitude of the radius variation and ΔV is the amplitude of the V-magnitude variation) for cepheid variables (Fernie, 1965) has been established. A linear relation between $\log P\Delta V$ and $\log \Delta R$ for classical cepheids is found, which perhaps has a break at $\Delta R = 10\ R_\odot$.

In the $\log \Delta R/\log P$ diagram the s-cepheids (Efremov, 1968) form a distinct sequence. All s-cepheids present a relative variation of the radii $\Delta R/R \leqq 0.075$. Very probably this is in accordance with Efremov's (1968) hypothesis that these cepheids cross for the first time the instability region in the HR-diagram.

On the basis of the $\log \Delta R/\log P$ diagram one could suppose that cepheids with $\log P \geqq 1.1$ pulsate in the first overtone whereas those with $\log P \leqq 1.1$ pulsate in the fundamental one. If these two groups of cepheids pulsate really differently, they would have a distinct $P-L$ relation, as some investigators have found a break in the $P-L$ relation of cepheids near $\log P \approx 1$.

The investigation of the amplitude of the radius variation and its relation to other characteristics of the variable stars is important for many reasons. The main one is that this amplitude refers directly to the pulsation of the variables. But it is well known that the pulsation of stars is closely connected with their evolution. Thus, the investigation of the radius amplitude can throw light into the evolution of the stars.

The cepheids are one of the most regularly pulsating variable star groups and are very suitable for such an investigation. The present paper summarizes some investigations on the amplitude of the radius variation of cepheids and its relation to the amplitude of the light variation and to the period.

The amplitude of the radius variation ΔR can be found directly from the velocity curve for a relatively small number of variables. Fernie (1965) has shown that for cepheids ΔR is nearly proportional to $P\Delta V$, where P is the period and ΔV is the amplitude of the light variation. From 16 stars with sufficiently accurate velocity curves and hence a well determined ΔR Fernie has found an approximately linear relation between $\log \Delta R$ and $\log P\Delta V$.

Since the stars used by Fernie are not very numerous, it would be interesting to see whether Fernie's relation is applicable to a greater number of cepheid

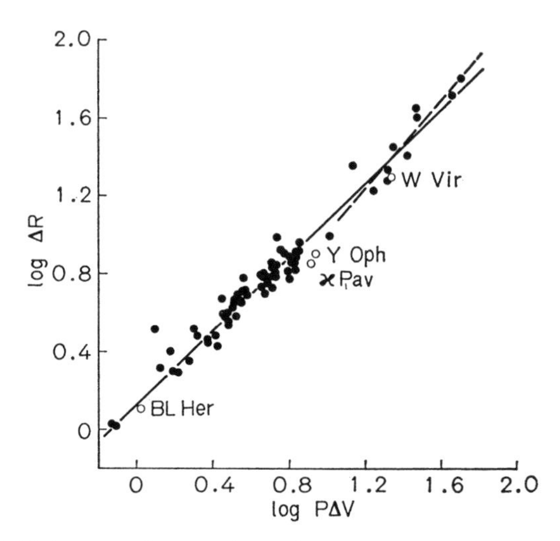

Fig. 1. $\log \Delta R/\log P\Delta V$ diagram. Open circles indicate population II cepheids. The solid line and the dashed line represent equations (1) and (2) respectively

variables. We have made use of the amplitudes ΔR found by KUROCHKIN (1966) on the assumption that the cepheids irradiate as black bodies on the one hand and of the amplitudes ΔV taken from a list of photometric characteristics of cepheids (NIKOLOV, 1968) on the other hand.

In Fig. 1 $\log \Delta R$ is plotted against $\log P\Delta V$ for 76 stars from the General Catalogue of Variable Stars (KUKARKIN *et al.*, 1969). ΔR is expressed in solar radii, P in days and ΔV in magnitudes. Open circles indicate Population II cepheids. It is evident that FERNIE'S relation is well satisfied by a larger number of cepheids and is probably applicable to all cepheids.

For 71 classical cepheids (three Population II cepheids as well as BY Cas and Y Oph, which deviate considerably from the relation, have been omitted) we have found by least squares the following linear relation

$$\log \Delta R = (0.944 \pm 0.020) \log P\Delta V + (0.134 \pm 0.015) \tag{1}$$

with a mean square error equal to ± 0.063.

BY Cas and Y Oph are so-called *s*-cepheids (EFREMOV, 1968) with a small light amplitude and a symmetrical light curve. These stars show peculiarities in many characteristics and especially they are very blue (EFREMOV, 1968). BY Cas is the bluest of all cepheids (NIKOLOV, 1966a, b) and has a rapidly increasing period (MALIK, 1965).

In Fig. 1 one can notice a gap after $\log \Delta R \approx 1.0$. One can suppose that cepheids with $\log \Delta R > 1.0$ deviate in our as well as in FERNIE'S $\log \Delta R/\log P\Delta V$ plot. For classical cepheids with $\log \Delta R > 1$ we have found by least squares

$$\log \Delta R = (1.140 \pm 0.107) \log P\Delta V - (0.122 \pm 0.150) \tag{2}$$

with a mean square error equal to ± 0.055. It would be interesting to verify by means of more complete data whether these cepheids have a distinct relationship.

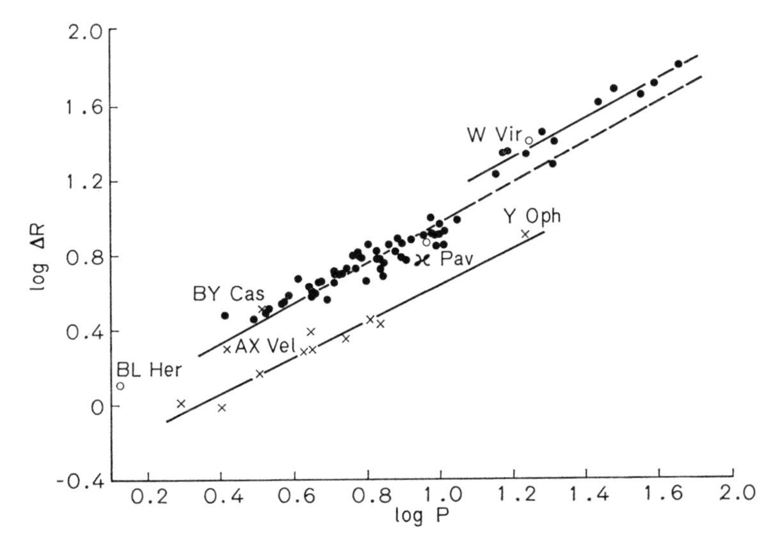

Fig. 2. Log$\varDelta R$/log$P$ diagram. Open circles indicate population II cepheids and crosses *s*-cepheids. The straight lines represent equations (3), (4) and (5)

Fig. 2 shows the log$\varDelta R$/log$P$ diagram for 79 stars. It is evident that the *s*-cepheids (denoted by crosses) form a distinct sequence which is approximately parallel to the sequence formed by the other cepheids. The fact that the *s*-cepheids satisfy the relation log$P\varDelta V$/log$\varDelta R$ which is the same for all stars shows that a factor should exist which changes the amplitudes $\varDelta V$ of these cepheids in a manner compensating exactly their different (for a given $\varDelta R$) periods. AX Vel is a doubtful *s*-cepheid (Efremov, 1968), though it has an instable period like a great percentage of these cepheids (Petit, 1970). Omitting AX Vel and BY Cas following relation has been found

$$\log \varDelta R = (0.957 \pm 0.069) \log P - (0.328 \pm 0.049) \tag{3}$$

with a mean square error equal to ± 0.054. The rest of the classical cepheids cannot be represented by an unique relation. They have been separated into two groups according to their period. For 52 stars with $\log P < 1.1$ we get

$$\log \varDelta R = (1.066 \pm 0.055) \log P - (0.106 \pm 0.044) \tag{4}$$

with a mean square error equal to ± 0.058, and for 12 stars with $\log P > 1.1$

$$\log \varDelta R = (1.037 \pm 0.117) \log P + (0.059 \pm 0.160) \tag{5}$$

with a mean square error equal to ± 0.068.

It is interesting to mention here that if we assume that cepheids with $\log P > 1.1$ are pulsating in the first overtone according to relations (4) and (5) we obtain a ratio in period of the first overtone to the fundamental $P_1/P_0 = 0.74$ and 0.78 for $\log \varDelta R = 1.0$ and $\log \varDelta R = 1.8$ respectively. For some cepheids it is known that this ratio is 0.70–0.71 (Christy, 1966; Efremov, 1970; Fitch, 1970). Assuming that

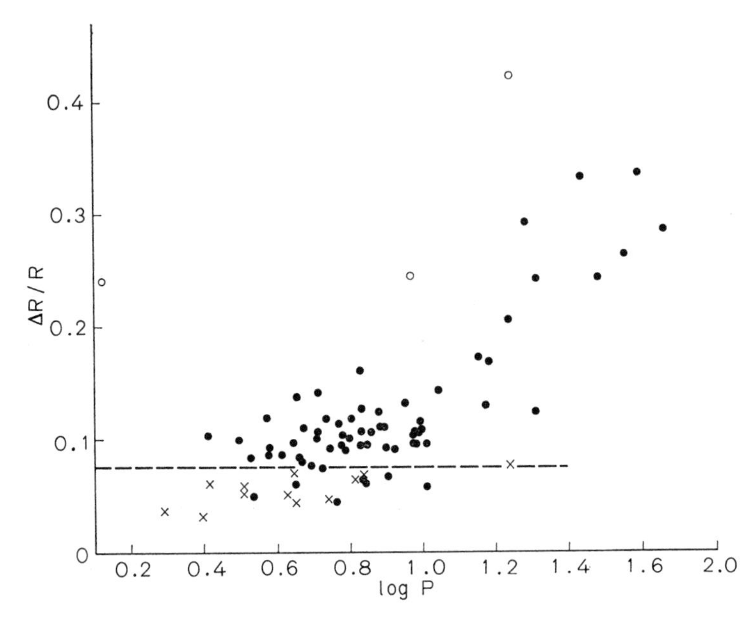

Fig. 3. $\Delta R/R$ versus $\log P$ diagram. The symbols are the same as those in Fig. 2. The dashed line is the upper limit of *s*-cepheids

η Aql, W Sgr, β Dor and X Cyg are pulsating in the first overtone and accepting $P_1/P_0 = 0.71$ for 15 stars with well determined radii FERNIE (1968) has found a linear relation between $\log R/R_\odot$ and $\log P$. Both our P_1/P_0 fit well this relation.

Here there are two interesting points:

1) If the two groups of cepheids according to P pulsate really differently, they would have a distinct $P-L$ relation. But it is well known that some investigators (KUKARKIN, 1949; PAYNE-GAPOSCHKIN and GAPOSCHKIN, 1966; EFREMOV, 1967) have found a break in the $P-L$ relation of cepheids near $\log P \approx 1$.

2) Recently EFREMOV (1972) has constructed a composite period-age relation for cepheid variables in the Magellanic Clouds, M 31 and the Galaxy. Quite probably this relation has a break again near $\log P = 1.1$. This break is due to the fact that the cepheids with $\log P \geqq 1.1$ are younger than the rest. Our result combined with EFREMOV's one shows that the mode of pulsation of the cepheids is in close connection with their age.

The relative variations of the radii $\Delta R/R$ are plotted against $\log P$ in Fig. 3. It is evident that $\Delta R/R$ of all *s*-cepheids does not exceed 0.075. This means that these stars pulsate not only with absolutely but also with relatively small variation of the radius for their period. This can be put in accordance with the hypothesis of EFREMOV (1968) that *s*-cepheids cross for the first time the instability region in the HR-diagram. This raises a point about the evolutionary place of non *s*-cepheids, which also pulsate with $\Delta R/R \leqq 0.075$.

$\Delta R/R$ for the remaining stars rapidly increases for $P > 10$ days. One can assume that Population II cepheids form probably in Fig. 3 a different sequence than the rest of the cepheids due to the fact that the relative variation of the radius of these stars contributes considerably more in the light variation in comparison with the classical cepheids.

References

Christy, R. F.: Ann. Rev. Astron. Astrophys. **4**, 353 (1966).
Efremov, Yu. N.: Astron. Circ. No. 443 (1967).
Efremov, Yu. N.: Perem. Zvezdy **16**, 365 (1968).
Efremov, Yu. N.: In "Pulsirujuscije Zvezdy". Moscow 1970.
Efremov, Yu. N.: Astron. Circ. No. 671 (1972).
Fernie, J. D.: Observatory **85**, 185 (1965).
Fernie, J. D.: Astrophys. J. **151**, 197 (1968).
Fitch, W. S.: Astrophys. J. **161**, 669 (1970).
Kukarkin, B. V.: Isledovanije stroenija i razvitija zvezdnih sistem. Moscow 1949.
Kukarkin, B. V., Kholopov, P. N., Efremov, Yu. N., Kukarkina, N. P., Kurochkin, N. E., Medvedeva, G. I., Perova, N. B., Fedorovich, V. P., Frolov, M. S.: General Catalogue of Variable Stars, Third Edition. Moscow 1969.
Kurochkin, N. E.: Perem. Zvezdy **16**, 10 (1966).
Malik, G. M.: Astron. J. **70**, 94 (1965).
Nikolov, N. S.: Astron. Z. **43**, 783 (1966a).
Nikolov, N. S.: Soviet Astron. **10**, 623 (1966b).
Nikolov, N. S.: Perem. Zvezdy **16**, 312 (1968).
Payne-Gaposchkin, C., Gaposchkin, S.: Smithsonian Contr. Astrophys. **9** (1966).
Petit, M.: Comm. 27 IAU Inf. Bull. Var. Stars 455 (1970).

The Influence of an Axial Rotation and a Tidal Action on the Non-Radial Oscillations of a Star, with Reference to the β Canis Majoris Stars

By J. DENIS and P. SMEYERS
Astronomisch Instituut, Katholieke Universiteit Leuven, Leuven, Belgium

Abstract

The perturbation method has been further developed in order to study the influence of an axial rotation and a tidal action on the non-radial oscillations of a star. Results are obtained for the *R*-mode and the *S*-mode in the case of an homogeneous model and their implication for the interpretation of the beat phenomena observed in *β* Canis Majoris stars is discussed.

We are going to report upon theoretical work which we have done recently on the influence of axial rotation and tidal action on the non-radial oscillations of a gaseous star. This work is still in progress. We believe it may be of some interest for the interpretation of the beat phenomena observed in *β* Canis Majoris stars.

The equations which govern the adiabatic non-radial oscillations of a gaseous star in hydrostatic equilibrium under its own gravity, together with the boundary conditions, give rise to a fourth order eigenvalue problem for the frequencies. When we introduce an axial rotation or a tidal action in the problem, we distinguish two kinds of perturbations. First, we see that the pressure and the density change at every point of the equilibrium configuration and that, in the case of an axial rotation, the Coriolis force appears as a supplementary term in the equations of motion. These are *volume perturbations* and they can be treated by the usual perturbation method. On the other hand, the equilibrium configuration is distorded and is no longer a sphere. Therefore, we have also to deal with a *surface perturbation*. This perturbation makes the treatment of the problem complex from a mathematical point of view.

To take account of the surface perturbation we proceed as follows. In a first step, we introduce an appropriate coordinate transformation, so that we associate to every point of the distorted equilibrium configuration the coordinates of a point in a reference sphere which is on the same radius. It is assumed that the reference sphere is close to the distorted configuration and that it is in equilibrium. The coordinate transformation produces an one-to-one mapping of the distorted configuration upon the reference sphere. Its introduction is equivalent to the procedure used by BRILLOUIN (1937) to reduce surface perturbations to volume perturbations in the case of eigenvalue problems for purely scalar functions, as for example wave functions.

However, in the eigenvalue problem of stellar oscillations the eigenfunctions are the displacement vectors of the material points and we must make one more step: we have to map the displacement vector at a point of the distorted configuration on an image vector at the associated point in the reference sphere. In terms of Differential Geometry it corresponds to accomplishing an infinitesimal parallel displacement along the radius.

In this way, we are able to treat the problem as one of a volume perturbation of the oscillations of a reference sphere. The image vector can be expanded in terms of the zero order oscillations of the sphere, we get in absence of any perturbing term. As these solutions satisfy the boundary condition that the Lagrangian perturbation of the pressure has to vanish at the surface, the solutions of the distorted configuration will also satisfy that condition at the distorted surface.

Our method corresponds to a geometrical generalization of the method used by SIMON (1969) in order to determine the second order rotational perturbation of the radial oscillations of a star. The method is accurate up to the second order in the angular velocity in the case of a rotational perturbation, and to the third order in the ratio of the radius of the primary to the distance to the secondary in the case of a tidal perturbation. Although the method is complex, it can be used in principle for any mode. We have applied it to determine the rotational and the tidal perturbation of non-radial oscillations of a star (SMEYERS and DENIS, 1971; DENIS 1972). Here we will only comment on one of the results we obtained.

CHANDRASEKHAR and LEBOVITZ (1962, 1964) have drawn attention to the fact that for a polytropic model a critical value of Γ_1 can be found for which an accidental degeneracy occurs between the fundamental radial mode and the *f*-mode or pseudo-Kelvin mode belonging to the spherical harmonic of degree 2 (the azimuthal parameter $m=0$). This critical value of Γ_1 depends on the central concentration of the model, but not in a very sensitive manner: for the homogeneous model the degeneracy occurs for $\Gamma_1=1.6$, while for the polytrope $n=3$ it occurs for $\Gamma_1=1.581$ (HURLEY *et al.*, 1966). Rotation and tidal action remove this degeneracy. They couple the two modes in a so-called *R*-mode and a *S*-mode, both of which are non-radial and have slightly different frequencies. We show here the frequencies we obtain for the *R*-mode and the *S*-mode in the case of an homogeneous model (CHANDRASEKHAR, 1933).

Table 1. Characteristic frequencies of *R*- and *S*-modes

M/M'	σ_R^2	σ_S^2
∞ (purely rotational case)	$1.0667-0.0216\,e^2$	$1.0667+0.7530\,e^2$
5	$1.0667-0.2316\,e^2$	$1.0667+1.1281\,e^2$
2	$1.0667-0.4620\,e^2$	$1.0667+1.5236\,e^2$
1	$1.0667-0.6982\,e^2$	$1.0667+1.9249\,e^2$
1/2	$1.0667-0.9369\,e^2$	$1.0667+2.3286\,e^2$
1/5	$1.0667-1.1768\,e^2$	$1.0667+2.7336\,e^2$

In these expressions e represents the eccentricity of the configuration of the primary and it is related to the angular velocity Ω by

$$\Omega^2=\frac{2}{5}\,\frac{GM}{R^3}\,e^2,$$

where M and R are respectively the mass and the radius of the primary. The angular velocity corresponds to its Keplerian value

$$\Omega^2 = \left(1 + \frac{M}{M'}\right) \frac{GM'}{d^3},$$

where M' denotes the mass of the secondary and d the distance between the primary and the secondary.

The calculations have been made for several values of the ratio M/M'. We find that the results are quite sensitive to this quantity. We can compare these results with the one we obtain when we do not take into account tidal action ($M/M' \to \infty$). This comparison shows that the R-mode and the S-mode are splitted by a much larger amount under the combined influence of rotation and tidal action than under the influence of rotation alone.

In the future we shall make similar calculations for the case of the polytrope $n=3$. We expect that the effect of tidal action will be somewhat less in a polytrope $n=3$ through the higher central concentration of the mass.

Our results may have some implication for the interpretation of the beat phenomena observed in β Canis Majoris stars. In the past, CHANDRASEKHAR and LEBOVITZ (1962) have proposed the excitation of the R-mode and the S-mode as a possible explanation for the occurence of double periods in these stars. Some authors (BÖHM-VITENSE, 1963; VAN HOOF, 1964) have objected that the rotational velocities which are required to give a sufficient splitting of the characteristic frequencies, are much higher than can be accounted for by observations: the β Canis Majoris stars were known to have generally narrow spectral lines, and it is only recently that the existence of two broad-lined β Canis Majoris stars has been reported (SHOBBROOK and LOMB, 1972). The results of our theoretical work show, however, that tidal action tends to reduce the required rotational velocities by a certain amount. We conclude by saying that the difficulty mentioned above may be less for β Canis Majoris stars which are components of binaries, but we must add that there is no observational evidence at the present time that all β Canis Majoris stars belong to binary systems.

References

BÖHM-VITENSE, E.: Publ. Astron. Soc. Pacific **75**, 154 (1963).
BRILLOUIN, L.: Compt. Rend. Acad. Sci. Paris **204** A, 1863 (1937).
CHANDRASEKHAR, S.: Monthly Notices Roy. Astron. Soc. **93**, 539 (1933).
CHANDRASEKHAR, S., LEBOVITZ, N. R.: Astrophys. J. **136**, 1105 (1962).
CHANDRASEKHAR, S., LEBOVITZ, N. R.: Astrophys. J. **140**, 1517 (1964).
DENIS, J.: Astron. Astrophys. **20**, 151 (1972).
HOOF, A. VAN: Z. Astrophys. **60**, 184 (1964).
HURLEY, M., ROBERTS, P. H., WRIGHT, K.: Astrophys. J. **143**, 535 (1966).
SHOBBROOK, R. R., LOMB, N. R.: Monthly Notices Roy. Astron. Soc. **156**, 181 (1972).
SIMON, R.: Astron. Astrophys. **2**, 390 (1969).
SMEYERS, P., DENIS, J.: Astron. Astrophys. **14**, 311 (1971).

An Atlas of Nova Delphini 1967

By W. SEITTER
Observatorium Hoher List, Daun, Eifel, F.R.G.

With 1 Figure

Abstract

An atlas of objective prism spectra of Nova Delphini 1967 is being prepared at Bonn Observatory. Some of its major features are outlined with special emphasis on the wealth of information obtainable from objective prism spectrograms of novae if careful reduction methods are used.

With several new large Schmidt telescopes, equipped with objective prisms, going into operation in the near future, the number of novae discovered and observed on low-resolution spectral plates is likely to increase.

While STRATTON's and MANNING's "Atlas of Nova Herculis (1934)" gives a very detailed survey over the spectral evolution of a fast nova—largely illustrated on higher resolution plates—samples of low-dispersion spectra useful for classification purposes are rarely found in the literature.

The "Atlas of Nova Delphini 1967" presents material much less complete than STRATTON'S and MANNING'S, yet it is hoped that several among the following features will make it more than a lower grade duplication of the first nova atlas.

1. Nova Delphini belongs to the rare class of very slow novae and shows a particularly large variety of spectral changes.

2. The spectra represent a uniform sample obtained with the 340/500/1375 Schmidt telescope of the Hoher List Observatory of Bonn University, equipped with prism I, giving a reciprocal linear dispersion of 240 Å/mm at $H\gamma$ and a wavelength resolution of about 2 Å at $H\gamma$.

3. The spectra are distributed over a period from 7 days after discovery until November 1968, that is way into the nebular stage which began in late July 1968.

4. The spectra cover the wavelength region from 3400 Å to 9000 Å. Though the long-wavelength parts show extremely low dispersion and the short-wavelength parts are frequently underexposed, major features are visible in all regions of the spectrum.

5. Identifications of all emission and absorption lines are attempted and indicated on the reproductions. It is hoped that the large number of identifiable lines (as well as the very good radial velocity determinations obtainable from objective prism spectrograms as discussed in 7.) will encourage observers to evaluate their low-dispersion spectrograms in detail.

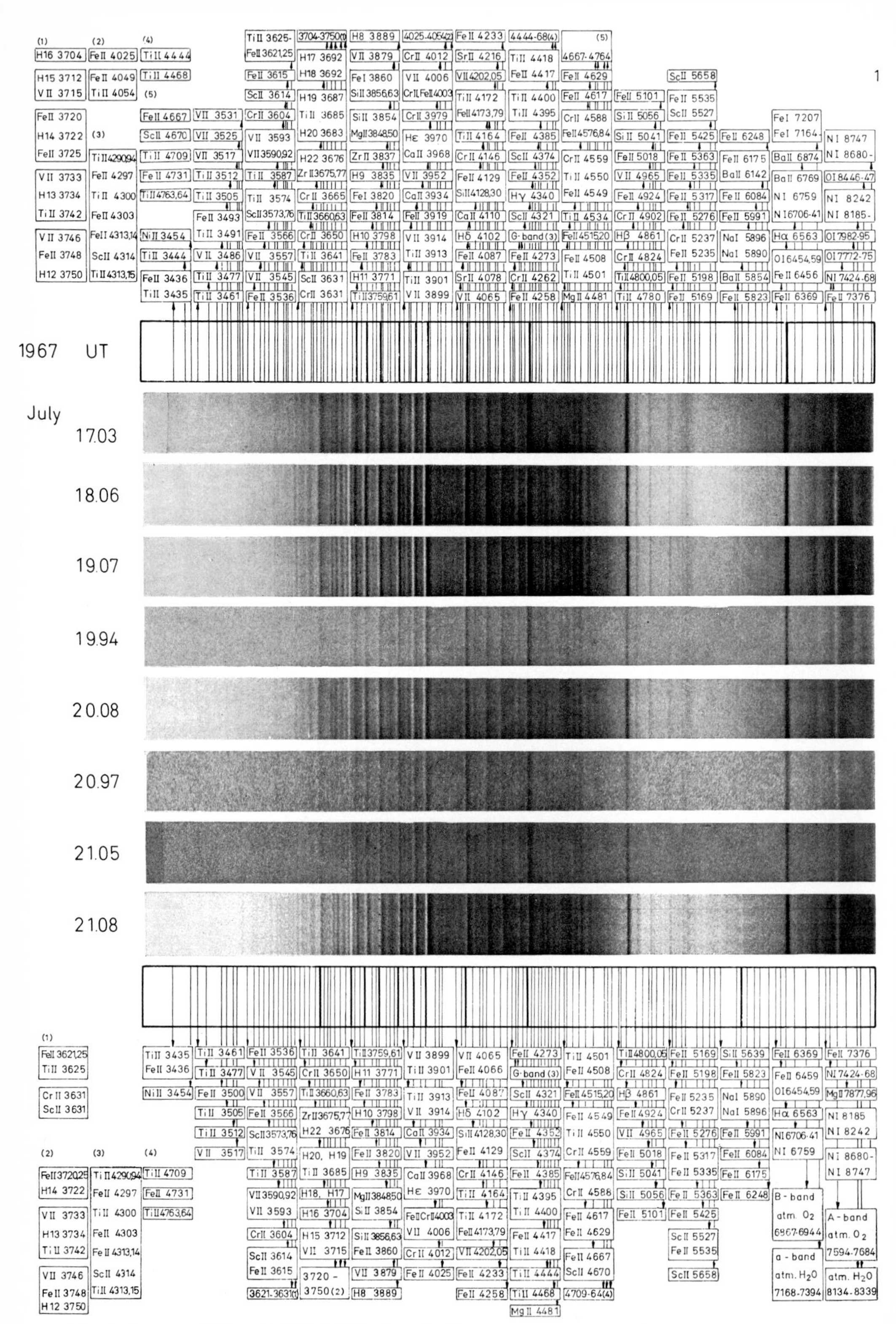

Fig. 1. Spectra of Nova Delphini 1967 after the first outburst. (Plate 1 of the Nova Atlas)

6. A complete light curve assembled by HILMAR DÜRBECK of the Bonn Observatory from nearly all available data (published and unpublished) will be given on overlay sheets such that the spectral and light variations can be followed simultaneously.

7. A second set of overlays shows the radial velocity variations of the major absorption lines as determined from our own spectra. Table 1 gives a sample of mean absorption line velocities of different ions on a given date (B) in comparison with data obtained at Victoria (V). The agreement is extremely good. The mean error of the Victoria values is $\pm 10\ \mathrm{km \cdot s^{-1}}$, of the Bonn values for the H-lines $\pm 50\ \mathrm{km \cdot s^{-1}}$. It must be remembered, however, that the Victoria dispersion is generally 15 Å/mm and the Bonn dispersion 240 Å/mm.

8. Samples of spectra of a few other novae (Nova Herculis 1963, Nova Vulpeculae 1968 I, Nova Cephei 1971) are included.

9. Several spectra obtained with dispersions 645 and 1280 Å/mm at $H\gamma$ are also shown.

10. The atlas is accompanied by a brief introduction and explanation. A catalogue lists the major multiplets of elements and ions present in the nova spectra and the times of their appearance and disappearance. This part is followed by an evaluation of our data.

Samples of atlas sheets were shown to the audience, one page is reproduced in Fig. 1.

Table 1. Radial velocities of Nova Delphini 1967 in July 1967 given in $\mathrm{km \cdot s^{-1}}$

	July 12, (V)	July 19, (B)	July 31, (V)
H	−633	−538	−468
Fe II	−645	−510	−433
Ti II	−627	−436	−415
Cr II	−650	−471	−400

Discussion

MCCARTHY, M. F.:
Can you tell us how the radial velocities are determined on your objective prism plates?

SEITTER, W.:
It is assumed that the centers of the emission lines are not displaced from the rest position within the error of measurement. The difference between the emission line center and the absorption line center both measured graphically on a dispersion curve to the nearest full Ångström unit is used for the radial velocity determination. The dispersion curve is derived from the major emission lines under the above stated assumption.

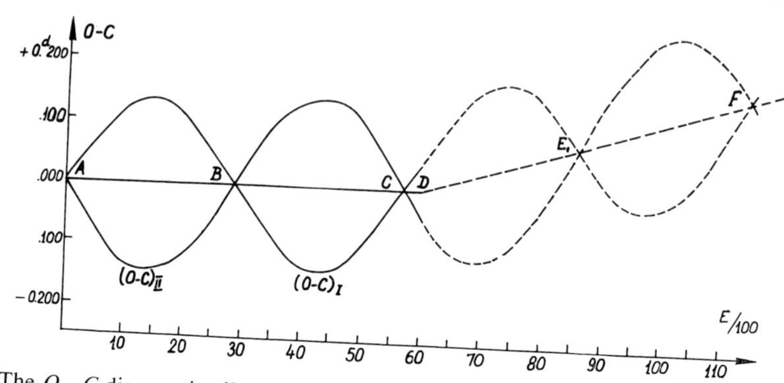

Fig. 1. The $O-C$ diagram is affected by an abrupt change in the orbital period

to $E \leqq E_j$ and $E \geqq E_j$ respectively. After taking the jump of the orbital period into account and after having superposed the straight line DF over AD, we may distinguish the following three cases:

i) The $O-C$ differences from D to F may be exactly superposed over the corresponding differences from A to D. In this case the jump of the orbital period must be considered as a result of an external cause, the apsidal period and orbital parameters having their constant size in all this interval of time.

ii) The apsidal period remains constant but the shape of the $O-C$ diagrams is altered. In this case the orbital parameters have to change during the jump of the orbital period.

iii) Together with the jump of the orbital period, the apsidal period and the orbital parameters underwent corresponding changes. In this case a change of the internal structure of the two stars must be the cause of the observed variations and this situation may be very important for the study of stellar evolution.

If we discover a case like (ii), we can write for the two straight lines AD and DF and for the two constants of internal structure k_1 and k_2 (KOPAL, 1965)

$$\left.\begin{aligned} \left(\frac{P}{U}\right)_{AD} &= (c_1)_{AD} k_1 + (c_2)_{AD} k_2, \\ \left(\frac{P}{U}\right)_{DF} &= (c_1)_{DF} k_1 + (c_2)_{DF} k_2, \end{aligned}\right\} \tag{9}$$

with

$$c_i = \left\{ \left(\frac{w_i}{w_k}\right)^2 \left(1 + \frac{m_{3-i}}{m_i}\right) \frac{1}{(1-e^2)^2} + 15 \frac{m_{3-i}}{m_i} \frac{8 + 12e^2 + e^4}{8(1-e^2)^5} \right\} \left(\frac{a_i}{a}\right)^5, \tag{10}$$

where w_i is the angular velocity of axial rotation, w_k the Keplerian angular velocity, m_{3-i}, m_i and a_{3-i}, a_i are the masses and the radii of the two components and a is the semi-major axis of the relative orbit.

In Eq. (9) k_1 and k_2 are the only unknown parameters and the importance of such a system is evident.

b) *The $O-C$ Diagrams of the Apsidal Motion are Affected by a Continuous Variation of the Orbital Period*

If the orbital period undergoes a continuous variation i.e.

$$\Delta P = \text{const. } E, \tag{11}$$

the $O-C$ diagrams of the apsidal motion will be affected by the term

$$\text{const. } E^2.$$

In this case the points where $\cos\omega=0$ will be situated on a vertical parabola which may be determined from at least three such points.

In Fig. 2 we have computed the $O-C$ diagrams by the ephemeris formulae (8) and with the arbitrary square term $+0\overset{d}{.}3 \cdot 10^{-8} E^2$. Here the points A, B, C, D, and E_1 are points where $\cos\omega=0$.

If we dispose only of one observed cycle or less, we do not know the curve on which the points of $\cos\omega=0$ are situated. In this case some distortion of the $O-C$ diagrams caused by a square term, may be attributed to the orbital eccentricity and, therefore, we shall obtain a spurious value of this parameter. As a matter of fact, in literature there are some cases where the orbital eccentricity obtained from a study of apsidal motion is different from its value obtained by other methods.

The examination of the three cases (i), (ii), or (iii) of the previous section may be resumed, but now we must consider each cycle as an independent series of observations and mean values may be determined for apsidal motion and orbital parameters. If we meet two successive cycles like (ii) we may use Eq. (9) for the determination of the mean constants k_1 and k_2 of the internal structure.

c) *Periodic Change of the Orbital Period*

If the $O-C$ diagrams besides apsidal motion are also affected by another periodic effect e.g. the presence of a third component, their appearance becomes very strange because the result of the superposition of the two periodic functions depends on a great number of parameters. But in this case the position of the points where $\cos\omega=0$ will determine the second periodic function because these points are not affected by apsidal motion as it has already been stressed.

After having succeeded in removing the influence of the second periodic function, we may determine the apsidal motion by a suitable method.

If the period of the second function is long and we dispose of observed minima only in a short interval of time, the corresponding curve may be confused with one caused by a square term. We may decide on the true situation only if we can study one of the three cases (i), (ii), (iii) as they have been presented before.

III. Remarks on the Determination of the Apsidal Motion

Keeping in mind that the apsidal motion may be determined both from the differences T_2-T_1 and from $O-C$ diagrams, we shall examine below the corresponding methods for its determination.

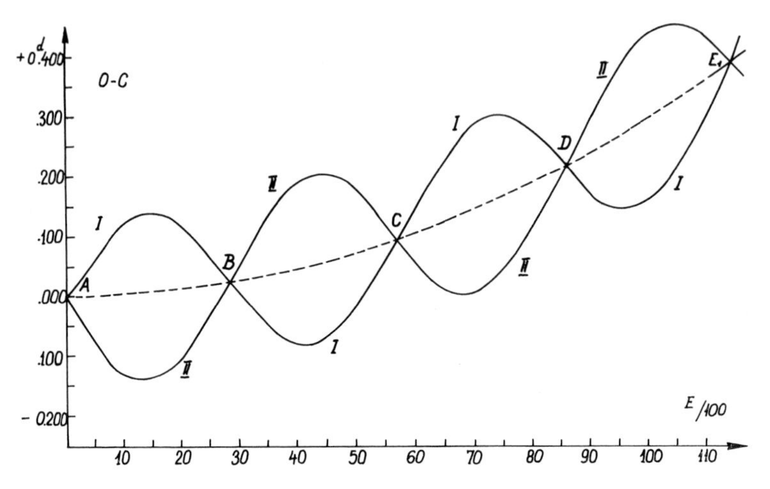

Fig. 2. The $O-C$ diagram is affected by a continuous variation of the orbital period

a) *The Use of T_2-T_1 Differences for the Determination of the Apsidal Motion*
From (1) and (2) we may write

$$\begin{aligned}T_2-T_1=&\frac{P}{2}+\frac{Pe}{\pi}\{\operatorname{cotg}^2 i+2+\\&+e^2[\tfrac{3}{2}\operatorname{cotg}^2 i+(2+\operatorname{cosec}^2 i)\operatorname{cotg}^2 i\operatorname{cosec}^2 i+1]\}\cos\omega-\\&-\frac{Pe^3}{\pi}[3\operatorname{cotg}^2 i+(2+\operatorname{cosec}^2 i)\operatorname{cotg}^2 i\operatorname{cosec}^2 i+\tfrac{4}{3}]\cos^3\omega+\cdots\end{aligned}\tag{12}$$

If the orbital eccentricity is not too large, and we can make this assumption at least in the first approximation, it follows from (12) that the extreme values of $T_2-T_1=f(E)$ have the same abscissae as those of the function $\cos\omega$. Thus, for the apsidal period U, we may write

$$U=2\,|E_{\max}-E_{\min}|\,P,\tag{13}$$

where the abscissae $E_{\max}$ and $E_{\min}$ of the extreme values of T_2-T_1 may be determined like epochs of minima for light curves.

In agreement with the above remark, we may write

$$\cos\omega_{\max,\,\min}=\pm 1,\tag{14}$$

and in accordance with (4), we have

$$\omega_0+\omega_1 E_{\max}=0^\circ \quad\text{or}\quad 360^\circ,\tag{15}$$

$$\omega_0+\omega_1 E_{\min}=180^\circ,\tag{16}$$

or

$$\omega_1=\frac{360^\circ}{U}P=\frac{180^\circ}{|E_{\max}-E_{\min}|},\tag{17}$$

$$\omega_0 = 90^\circ \left(1 - \frac{E_{max} + E_{min}}{|E_{max} - E_{min}|}\right), \tag{18}$$

$$\omega_0 = 90^\circ \left(3 - \frac{E_{max} + E_{min}}{|E_{max} - E_{min}|}\right). \tag{19}$$

The formulae (18) and (19) are a consequence of the two values 0° and 360° in (15) and from the observed curve $T_2 - T_1 = f(E)$ we may decide which one is more suitable for a certain case.

Now, by using (4) we will compute the values of ω for those values of E for which we have observed $T_2 - T_1$ differences and neglecting the term in e^3, we may write

$$e = \frac{\pi}{P(\operatorname{cotg}^2 i + 2)} \, \frac{\sum_{j=1}^{n} \left(\frac{T_2 - T_1 - \frac{P}{2}}{\cos \omega}\right)_j w_j}{\sum_{j=1}^{n} w_j}, \tag{20}$$

where n is the number of observed values $T_2 - T_1$ for $\cos \omega \neq 0$ and w_j is the weight of the corresponding observation.

If the orbital eccentricity e is large and terms with e^3 cannot be ignored, we may consider the above determined values as preliminary ones $\tilde{\omega}_0$, $\tilde{\omega}_1$ and $\tilde{e}$ and we must proceed to their improvement. With that end in view, from (12) we obtain

$$(T_2 - T_1)_0 - (T_2 - T_1)_c = (O - C)_j = \mathcal{A}_j \, \delta e + \mathcal{B}_j \, \delta \omega_0 + \mathcal{C}_j \, \delta \omega_1, \tag{21}$$

with

$$\mathcal{A}_j = \frac{P}{\pi} \{\operatorname{cotg}^2 i + 2 + 3e^2 [\tfrac{3}{2} \operatorname{cotg}^2 i + (2 + \operatorname{cosec}^2 i) \operatorname{cotg}^2 i \operatorname{cosec}^2 i + 1] - \\ - 3e^2 [3 \operatorname{cotg}^2 i + (2 + \operatorname{cosec}^2 i) \operatorname{cotg}^2 i \operatorname{cosec}^2 i + \tfrac{4}{3}] \cos^2 \omega_j\} \cos \omega_j, \tag{22}$$

$$\mathcal{B}_j = -\frac{Pe}{\pi} \left\{\operatorname{cotg}^2 i + 2 + e^2 [\tfrac{3}{2} \operatorname{cotg}^2 i + (2 + \operatorname{cosec}^2 i) \operatorname{cotg}^2 i \operatorname{cosec}^2 i + 1] - \\ - \frac{3Pe^3}{\pi} [3 \operatorname{cotg}^2 i + (2 + \operatorname{cosec}^2 i) \operatorname{cotg}^2 i \operatorname{cosec}^2 i + \tfrac{4}{3}] \cos^2 \omega_j \right\} \sin \omega_j, \tag{23}$$

$$\mathcal{C}_j = \mathcal{B}_j \, E_j. \tag{24}$$

A least-square solution of Eq. (21) determines the corrections $\delta \omega_0$, $\delta \omega_1$ and δe. Thus, we have

$$\omega_0 = \tilde{\omega}_0 + \delta \omega_0, \quad \omega_1 = \tilde{\omega}_1 + \delta \omega_1, \quad e = \tilde{e} + \delta e. \tag{25}$$

If e is very large, the corresponding improvement will be made by the method of successive approximations.

b) *Numerical Application for Y Cygni*

Having in view that apsidal motion for Y Cygni was well studied by DUGAN (1931), we have used his tables of supernormal minima (p. 27) and we have drawn up Table 1.

Table 1. Values of T_1, T_2, w, E and T_2-T_1 for Y

T_1	T_2	w	E	T_2-T_1
10271.4373	10272.8596	2	246	1ᵈ4223
10879.7582	10881.1152	2	449	1.3572
11314.2265	11315.5581	2	594	1.3316
12012.3783	12013.6626	3	827	1.2843
12716.5558	12717.7802	2	1062	1.2244
13567.5183	13568.7349	2	1346	1.2166
14523.3510	14524.5818	3	1665	1.2308
15350.3220	15351.5678	2	1941	1.2458
15491.1497	15492.4313	1	1988	1.2816
15976.5626	15977.8373	1	2150	1.2747
16710.6623	16711.9928	1	2395	1.3305
17444.6543	17446.0944	1	2640	1.4401
17684.3609	17685.8208	1	2720	1.4599
17708.3226	17709.7915	2	2728	1.4689
20036.3725	20038.0450	2	3505	1.6725
20333.0095	20334.7200	2	3604	1.7105
20716.5090	20718.2506	2	3732	1.7416
20785.4425	20787.1663	2	3755	1.7238
21192.9438	21194.6766	1	3891	1.7328
21471.6028	21473.3729	1	3984	1.7701
21843.1227	21844.8860	1	4108	1.7633
22909.7803	22911.5893	2	4464	1.8090
23002.7144	23004.4757	2	4495	1.7613
23410.2240	23411.9770	2	4631	1.7530
23745.8133	23747.5506	2	4743	1.7373
24078.4186	24080.1436	1	4854	1.7250
24336.1032	24337.7669	1	4940	1.6637
24824.5221	24826.1958	1	5103	1.6737

Since the orbital period of Y Cygni is about 3^d, we were not able to dispose of successive principal and secondary minima; that is why we are forced to shift some of the observed supernormal minima by the term $\pm P\,\Delta E$. This is always correct for that range ΔE in which we may write $(T_2)_j$ or $(T_1)_j = \text{const.} + (\Delta E)_j\,P$ (here P may be taken instead of $\mathbb{P} = 2\overset{d}{.}9963331$).

By the method of the bisection curve (TODORAN, 1968) we have determined

$$E_{\min} = 1453\,P \pm 14\,P, \qquad E_{\max} = 4300\,P \pm 20\,P.$$

Taking $i = 90°$ $\{i = 88° \pm 1\overset{\circ}{.}5$ (KOPAL and SHAPLEY, 1956)$\}$ and using the above presented formulae, we have determined the final values which are compared with DUGAN's results in Table 2.

Table 2. Values of U, ω_0, ω_1 and e for Y

	DUGAN	TODORAN
U	$5745\,P \pm 22\,P$	$5694\,P \pm 20\,P$
ω_0	$90°$	$88\overset{\circ}{.}2 \pm 1\overset{\circ}{.}8\,P$
ω_1	$0\overset{\circ}{.}06266 \pm 0\overset{\circ}{.}00024$	$0\overset{\circ}{.}0632 \pm 0\overset{\circ}{.}00022$
e	0.144 ± 0.0016	0.156 ± 0.0024

Finally, comparing the two sets of results listed in Table 2, we see that within the limits of their errors, these results agree with each other.

By using formula (12) we get

$$(T_2 - T_1)_c = 1\overset{\mathrm{d}}{.}4982 + 0\overset{\mathrm{d}}{.}2777 \cos\omega, \tag{26}$$

the term with e^3 being smaller than $\pm 0\overset{\mathrm{d}}{.}0009$.

The solid curve in Fig. 3 is given by (26) and the observed points of the corresponding curve are listed in Table 1.

From (1) and (2) and by using the corresponding values from Table 2, we get formulae (8) which were used before, and the corresponding observed $O-C$ differences are computed by using DUGAN's linear ephemeris

$$\left.\begin{matrix} 2409534.3195 \\ 35.8175 \end{matrix}\right\} + 2\overset{\mathrm{d}}{.}9963331\, E.$$

The comparison between observed points and computed $O-C$ differences is illustrated in Fig. 4, where the minima for $E > 7000\,P$ are given by the following observers:

AHNERT (1963), DIETHELM *et al.* (1972), DIETHELM *et al.* (1971), FLIN (1969), KIZILIRMAK and POHL (1966), KIZILIRMAK (1970), KORDYLEWSKI (1957), MAGALASCHWILI and KUMSISCHWILI (1959), MARKS (1962), OBURKA (1965), O'CONNELL (1971), PLAVEC and MAYER (1962), POPOVICI (1970), ROBINSON (1966, 1967), ZAITSEVA *et al.* (1971).

c) *The Use of $O-C$ Diagrams for the Determination of the Apsidal Motion*

Obviously, the determination of the apsidal motion only from $T_2 - T_1$ differences is not affected by a change of the orbital period, but the number of the close binary systems with both kinds of observable minima is very limited and the apsidal period is often very large. That is why, the apsidal motion determined from $O-C$ differences may be useful for the determination of the internal structure of the stars. But, in this case, the secondary minima have to be observed in order to confirm or to refute the hypothesis of the apsidal motion.

Of course, if we dispose only of one $O-C$ diagram, we are not able to determine the position of the points where $\cos\omega = 0$. But this drawback may be avoided by taking, in the first approximation, the points situated at the middle of the curve between two successive extreme values.

The parameters of the apsidal motion may be determined by an iterative method. With this purpose in view, from (1) (TODORAN, 1971), for the case (*b*), we have

$$\Delta(O-C)_j = \Delta T_0 + \Delta \mathbb{P}\, E_j + \Delta q\, E_j^2 + \mathscr{A}_j\, \delta e + \mathscr{B}_j\, \delta\omega_0 + \mathscr{C}_j\, \delta\omega_1, \tag{27}$$

where

$$\begin{aligned} \mathscr{A}_j = & -\frac{P}{2\pi}\{\operatorname{cotg}^2 i + 2 - 3e^2[\tfrac{3}{4}\operatorname{cotg}^2 i - \tfrac{1}{4}(2+\operatorname{cosec}^2 i)\operatorname{cotg}^2 i \operatorname{cosec}^2 i]\}\cos\omega_j + \\ & + \frac{Pe}{2\pi}[\operatorname{cotg}^2 i \operatorname{cosec}^2 i + 2\operatorname{cotg}^2 i + \tfrac{3}{2}]\sin 2\omega_j + \\ & + \frac{3Pe^2}{8\pi}[(2+\operatorname{cosec}^2 i)\operatorname{cotg}^2 i \operatorname{cosec}^2 i + 3\operatorname{cotg}^2 i + \tfrac{4}{3}]\cos 3\omega_j, \end{aligned} \tag{28}$$

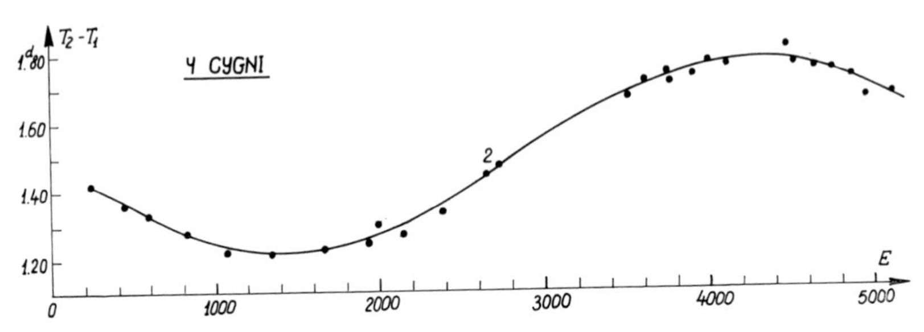

Fig. 3. The apsidal motion is determined from $T_2 - T_1$ differences

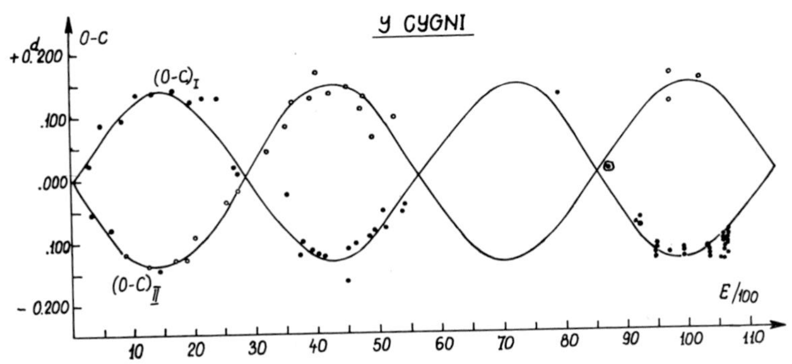

Fig. 4. Comparison between observed points and computed $O-C$ differences for Y Cygni

$$\begin{aligned}\mathscr{B}_j = & \frac{Pe}{2\pi}\{\operatorname{cotg}^2 i + 2 - e^2[\tfrac{3}{4}\operatorname{cotg}^2 i - \tfrac{1}{4}(2+\operatorname{cosec}^2 i)\operatorname{cotg}^2 i \operatorname{cosec}^2 i]\}\sin\omega_j + \\ & + \frac{Pe^2}{2\pi}[\operatorname{cotg}^2 i \operatorname{cosec}^2 i + 2\operatorname{cotg}^2 i + \tfrac{3}{2}]\cos 2\omega_j - \\ & - \frac{3Pe^3}{8\pi}[(2+\operatorname{cosec}^2 i)\operatorname{cotg}^2 i \operatorname{cosec}^2 i + 3\operatorname{cotg}^2 i + \tfrac{4}{3}]\sin 3\omega_j,\end{aligned} \tag{29}$$

$$\mathscr{C}_j = \mathscr{B}_j E_j. \tag{30}$$

The differences $O-C$ must be computed by using a square ephemeris formula

$$T_c = T_0 + \mathbb{P} E + \tilde{q} E^2, \tag{31}$$

where $\tilde{q}$ is determined by the points situated at the middle of the curves between the two successive extreme values. If $q=0$, there is the case (a); for the case (c) instead of $\Delta q E^2$ and $\tilde{q} E^2$, there will be a number of corresponding terms.

Unfortunately, we could not do a numerical application because there are no $O-C$ diagrams (Kreiner, 1971) which permit us to determine the three points required for the determination of the preliminary value of q.

In Fig. 5 we have represented the $O-C$ diagram of W Delphini and it is clear that for $E \leqq 3600 P$ the observed principal minima may be represented by a si-

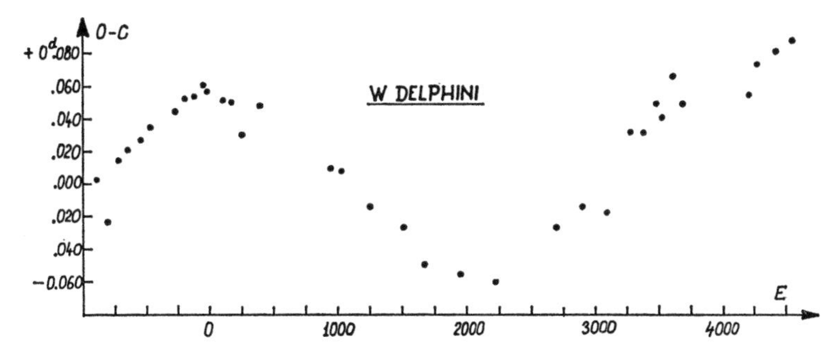

Fig. 5. The $O-C$ diagram of W Delphini

nusoidal function as it was determined by PLAVEC (1959). After $E \geqq 3600P$, the corresponding $O-C$ differences are increasing outgrowing the values of PLAVEC's sinusoidal curve. But, having in view Figs. 1 and 2, such an increase could have been explained by the apsidal motion if WALTER's (1970) observations had not pointed out a circular photometric orbit.

Probably, the $O-C$ differences of W Delphini could be explained by other causes described by periodic and parabolic or only by periodic functions.

IV. Stars Proposed for New Observations

In order to put into practice all the above exposed ideas, it is very important to obtain new observed epochs of minima. We are transcribing below the stars which can be very useful after a new series of observed minima.

a) *Systematic Observations of Primary and Secondary Minima*

GL Car; Y, MR, V477 Cyg; HS Her; CO Lac; RU Mon; AG Per.

b) *Systematic Observations of Primary Minima and Sporadic Examination of the Secondary Minima*

TW, XZ, AB And; RY Aqr; XZ, KO Aql; RZ Aur; Y Cam; RS CVn; R CMa; AR, RZ Cas; XX Cep; SW, SY, WW Cyg; W Del; Z Dra; TX, CC Her; AR Lac; Y Leo; T LMi; RV Lyr; TY Peg; RT Per; Y Psc; RW Tau; X Tri; TX UMa.

V. Conclusions

From the above presented considerations, we can draw following conclusions:

If the apsidal motion is determined only from $T_2 - T_1$ differences, its period and orbital eccentricity will be not affected by any change in orbital period.

In the case when the apsidal motion is determined from $O-C$ diagrams, the obtained results may be altered by some variation in orbital period.

If we discover a jump in the orbital period and the apsidal period remains constant before and after this variation, we are able to determine the two constants k_1 and k_2 of the internal structure of the two components.

The existing $O-C$ diagrams are still not sufficient for a good separation of the apsidal motion from a parabolic or periodic variation of the orbital period.

An ever increasing attention should be paid to the observations of primary and secondary minima in the future.

References

AHNERT, P.: Mitt. Veränderl. Sterne No. 732 (1963).
DIETHELM, R., GERMANN, R., LOCHER, K., PETER, H.: BBSAC Bull. 1 (1972).
DIETHELM, R., ISLES, J., LOCHER, K.: Orion No. 126–128 (1971).
DUGAN, R. S.: Contr. Princeton Obs. No. 12 (1931).
FLIN, P.: Acta Astron. **19**, 173 (1969).
KIZILIRMAK, A.: Comm. 27 IAU Inf. Bull. Var. Stars, No. 456 (1970).
KIZILIRMAK, A., POHL, E.: Astron. Nachr. **289**, 191 (1966).
KOPAL, Z.: Adv. Astron. Astrophys. **3**, 89 (1965).
KOPAL, Z., SHAPLEY, M. B.: Ann. Jodrell Bank **1**, 205 (1956).
KORDYLEWSKI, K.: Rocznik Astron. **28**, 108 (1957).
KREINER, J. M.: Acta Astron. **21**, 365 (1971).
MAGALASCHWILI, N. L., KUMSISCHWILI, JA. J.: Bull. Abastumani Obs. **21**, 13 (1959).
MARKS, A.: Acta Astron. **12**, 138 (1962).
OBURKA, O.: Bull. Astron. Inst. Czech. **16**, 212 (1965).
O'CONNELL, D. J. K., S. J.: Comm. 27 IAU Inf. Bull. Var. Star., No. 542 (1971).
PLAVEC, M.: Bull. Astron. Inst. Czech. **10**, 185 (1959).
PLAVEC, M., MAYER, P.: Bull. Astron. Inst. Czech. **13**, 128 (1962).
POPOVICI, C.: Comm. 27 IAU Inf. Bull. Var. Stars No. 419 (1970).
ROBINSON, L. J.: Comm. 27 IAU Inf. Bull. Var. Stars No. 154 (1966).
ROBINSON, L. J.: Comm. 27 IAU Inf. Bull. Var. Stars No. 180 (1967).
TODORAN, I.: Acta Astron. **18**, 61 (1968).
TODORAN, I.: Studii Cerc. Astron. **16**, 177 (1971).
WALTER, K.: Astron. Nachr. **292**, 145 (1970).
ZAITSEVA, G. V., LYUTYI, V. M., MARTYNOV, D. YA.: Astron. Circ. No. 662 (1971).

Discussion

KREINER, M. J.:
The photoelectric observations of secondary minima are very important in this problem. This is connected with the necessity of developing a wide campaign of observing the times of minima (especially secondary ones) of eclipsing variables.

BAKOS, G. A.:
I have observations of times of minima of TX Her and will be published soon. Another star that shows period changes is AH Vir. It may be a good candidate for investigation of apsidal motion.

The Radius-Luminosity Relation in Eclipsing Binaries

By Călin Popovici and Al. Dumitrescu
Bucharest Observatory, Rumania

With 7 Figures

Abstract

The radius-luminosity relation in the form $M_b = x \log R + c$, is determined for main-sequence stars and subgiants in detached and semi-detached close binary systems. The $|x|$ value diminishes from $|x| = 12$ to $|x| = 6$ as we pass from the main-sequence stars to the subgiants. This is interpreted as an evolutionary effect and is compared with the evolutionary tracks by constant mass and by mass exchange.

The existence of a radius-luminosity relation was not generally considered, perhaps because it was regarded as a consequence of more fundamental relations, as the mass-luminosity relation and the mass-radius relation, or even only of the spectrum-luminosity relation.

The eclipsing binaries permit, however, to determine the radii of the stars – if they are also spectroscopic binaries – and their luminosities – if their parallaxes are known- or even better we can determine the difference of the magnitudes and the ratio of the radii of both stars in each pair, only by photometric means. In this way they can be used to determine directly the radius-luminosity relation in its final form ($M_b - \log R$), or in differential form ($\Delta M_b - \log k$), which gives us a new possibility to verify the stellar structure and stellar evolution theories. In a previous paper (Popovici, 1956) one of us has tried to determine such a relation with the observational data then available, and in 1971 we have published a paper about radius-luminosity relation based on better observational data (Popovici and Dumitrescu, 1971).

Z. Kopal (1959) has given mass-luminosity and mass-radius relations for detached binaries, and recently (Kopal, 1971) a diagram representing the correlation of the ratios of radii and the differences of magnitudes of pairs of stars in detached systems of eclipsing binaries with the ratios of their masses. Binnendijk (1965) has given diagrams for the relations ($\Delta M_b - \log k$) and ($m_2/m_1 - \log k$) for W Ursae Majoris stars. Some similar relations were given by Popper (1967).

From the mass-luminosity relation $L \sim m^p$ and the mass-radius relation $R \sim m^s$, or even from the spectrum-luminosity relation ($L \sim T_e^q$) one can find the radius-luminosity relation in the form

$$M_b = x \log R + c, \tag{1}$$

with

$$x = \frac{5s}{2p} = -\frac{5q}{q-4}, \tag{2}$$

or in differential form:

$$\Delta M_b = -x \log k, \tag{3}$$

with $k = \frac{R_2}{R_1}$, if the two stars of the pair follow the same relation (1).

Fig. 1 gives the $(M_b - \log R)$ relation from the more homogeneous sample of the KOPAL and SHAPLEY'S (1956) catalogue for 52 main-sequence stars in detached systems with $x = -12.55 \pm 0.45$ and $c = +4.80 \pm 0.20$. Fig. 2 gives the same relation from 81 similar stars in the SVETCHNIKOFF'S (1969) catalogue, with $x = -11.42 \pm 0.11$ and $c = +4.21 \pm 0.04$.

With the mass-luminosity relations and mass-radius relations from KOPAL'S (1959) paper for detached systems:

$$\begin{aligned} \log m &= 0.45 - 0.143\, M_b = 1.57 \log R - 0.1 \qquad m > 2\,\mathrm{m}_\odot \\ \log m &= 0.42 - 0.086\, M_b = 1.02 \log R \qquad\quad\;\; m \ll 2\,\mathrm{m}_\odot \end{aligned} \tag{4}$$

we obtain $x_1 = -10.99$, $c_1 = +4.2$; $x_2 = -11.86$, $c_2 = +4.9$.

From $x = -12.55$ we have for the slope of the $(M_b - \log T_e)$ or HR-relation, $-2.5q = -16.65$, with $q = \frac{4x}{x+5}$, $L \sim T^q$. From the x values from KOPAL we obtain $-2.5q = -18.35$ $(m > 2\mathrm{m}_\odot)$, and $-2.5q = -17.29$ $(m \ll 2\mathrm{m}_\odot)$.

We can try to determine x from pairs of stars in detached systems and we obtain $x = -9.86 \pm 0.64$ from 23 pairs of KOPAL and SHAPLEY'S catalogue (Fig. 3), and $x = -10.78 \pm 0.72$ from 44 pairs of the SVETCHNIKOFF'S catalogue (Fig. 4). The dispersion is greater in this case owing to more difficult computational conditions and perhaps due to some evolutionary effects. As we can see from Fig. 3 no clear relation $(\Delta M_b - \log k)$ could be obtained for the pairs main-sequence star with sub-

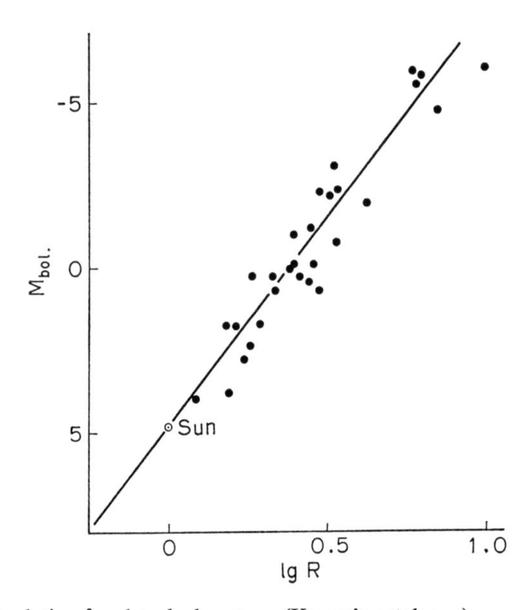

Fig. 1. $(M_{\mathrm{bol}}$-$\log R)$ relation for detached systems (KOPAL'S catalogue)

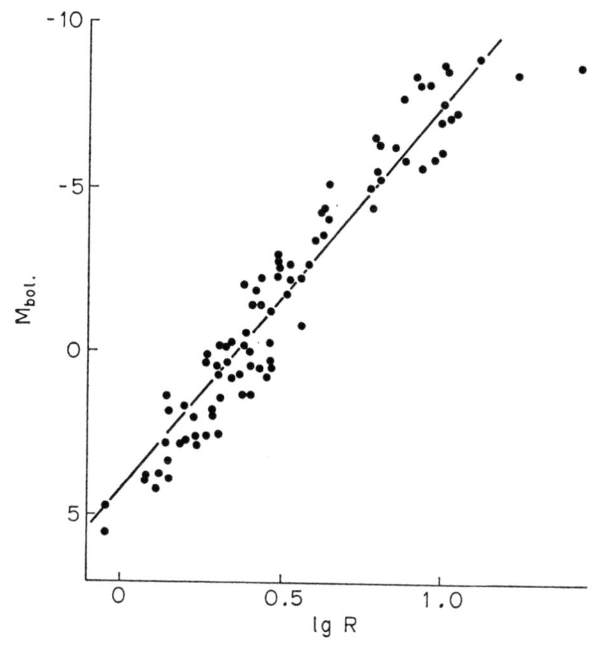

Fig. 2. (M_{bol}-log$R$) relation for detached systems (SVETCHNIKOFF's catalogue)

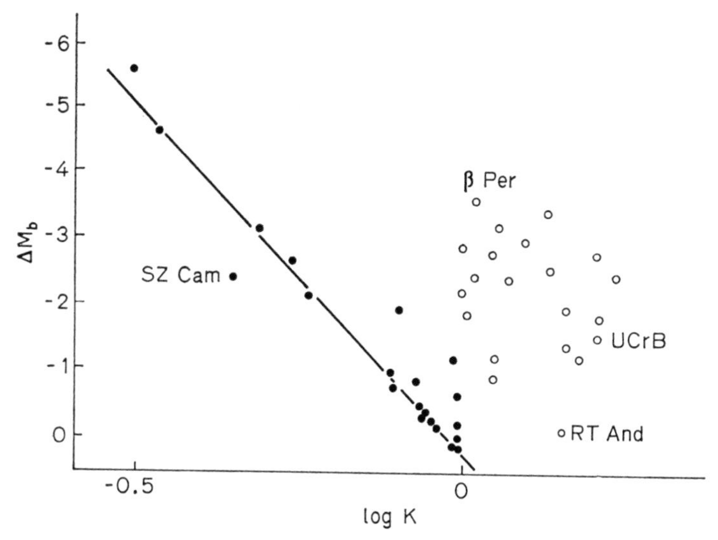

Fig. 3. (ΔM_b-log$k$) relation for detached systems (filled circles) and semi-detached systems (open circles) (KOPAL's catalogue)

giant in semi-detached systems (open circles in Fig. 3), due to different x in the ($M_b - \log R$) relations, for main-sequence and subgiant stars.

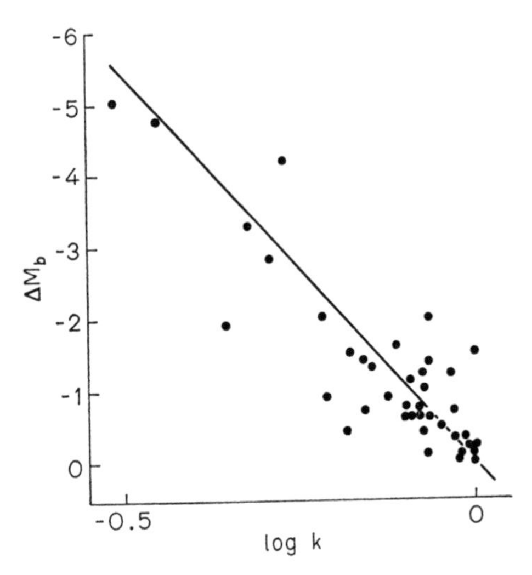

Fig. 4. (ΔM_b-log$k$) relation for detached systems (SVETCHNIKOFF's catalogue)

The subgiants could be in different evolutionary stages and their x values could be different. In this case:

$$\begin{aligned} \Delta M_{p,s} &= -x_s \log k + (x_p - x_s)\log R_p + c_p - c_s, \\ &= -x_p \log k + (x_p - x_s)\log R_s + c_p - c_s. \end{aligned} \tag{5}$$

Some uncertainty comes in (5) also from the bolometric correction used in computing ΔM_b from ΔM_v.

If we consider the semi-detached systems, we can try to obtain separately a x value for the main-sequence stars of these pairs and a mean x_p value for the subgiant

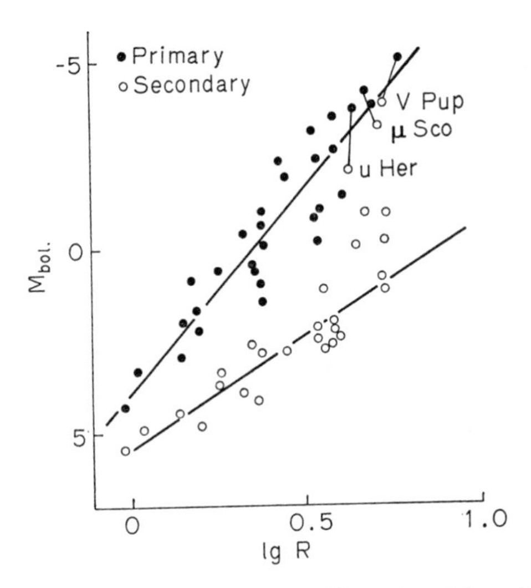

Fig. 5. (M_b-log$R$) relation for semi-detached systems (KOPAL's catalogue)

or giant component of these pairs. From 29 main-sequence stars of semi-detached systems of KOPAL and SHAPLEY's catalogue we find: $x_p = -11.12 \pm 0.70$, $c_p = +4.03 \pm 0.35$ and for the subgiant and giant secondaries $x_s = -6.09 \pm 0.32$, $c_s = 5.58 \pm 0.40$ (26 stars, Fig. 5). For the 77 main-sequence stars in SVETCHNIKOFF's catalogue $x_p = -9.91 \pm 0.15$, $c_p = +4.04 \pm 0.07$ and for the 84 subgiants and giants of the semi-detached pairs of the same catalogue, $x_s = -6.03 \pm 0.10$, $c_s = +5.51 \pm 0.08$ (Fig. 6). In Figs. 5 and 6 some discordant pairs are noted.

The decrease of module x to half its value as we pass from main-sequence stars of the pairs to subgiants, is a clear indication of the present work. We interpret this result as an evolutionary effect due to the passage of the star from the main sequence to the subgiant branch.

In Fig. 7 are given the evolutionary tracks of population I stars computed by IBEN (1964), in which we have represented the diagonal lines for the value $x = -16$ for the theoretical zero-age main-sequence stars, and for $x = -6$, which we find for subgiants secondaries. This latter line can be interpreted as an approximate isochrone for intermediate-age stars as are the subgiants in the considered semi-detached systems.

This qualitative proof due to the evolution of stars of $m > 1.5\,m_\odot$ which evolve at approximate constant L, to greater R towards the right in the HR-diagram can be verified more exactly if we compute the ratios $-2.5 \frac{\log L/L_\odot}{\log R/R_\odot}$ for stars of different masses after time intervals significant for stellar evolution. For a $15\,m_\odot$ star the module x decreases from 16.0 to 4.5 in 1.2×10^7 years; for a $1.0\,m_\odot$ star $|x|$ decreases from 16.0 to 3.2 in 1.09×10^{10} years. For a supergiant as the ζ Aurigae component, x is ≈ 2.

If we try to calculate in the same manner the $|x|$ values for the evolutionary tracks with mass exchange from the KIPPENHAHN and WEIGERT's (1967) diagram

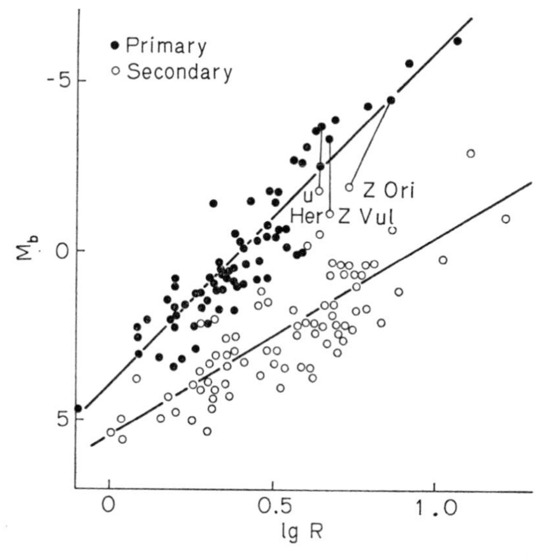

Fig. 6. (M_b-log R) relation for semi-detached systems (SVETCHNIKOFF's catalogue)

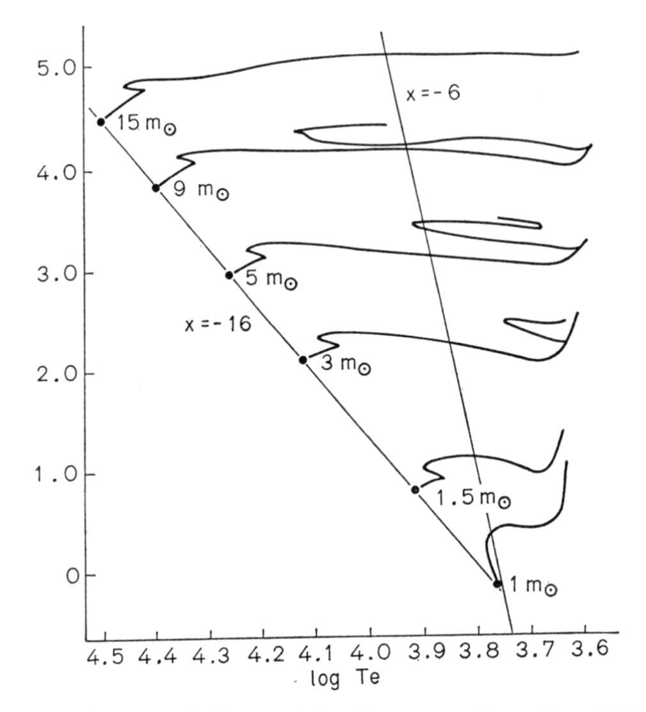

Fig. 7. Evolutionary tracks for population I stars according to IBEN (1964)

(case *A*) for the end of the track of an initial $9 m_\odot$ star and a final $3 m_\odot$ star we find $|x| = 10$. For the PACZYNSKI's (1967) diagram for an initial $16 m_\odot$ star and a final $6 m_\odot$ star we find also $|x| = 10$. These values are greater as those deduced directly from the data of the analysed catalogues mentioned above. This comes from the fact that these stars are more luminous and hotter than evolved single stars by constant mass. For the same luminosity the stars evolved with mass loss by mass exchange have a smaller radius as the subgiants in semi-detached systems ($x = -6$) we have studied.

The method given here could be utilised in a more precise quantitative way if we could have the complete lists of the computed *L*, *R* values for different initial and final masses and for different times in the evolutionary tracks with mass exchange, to permit an individual approach to each system.

References

BINNENDIJK, L.: In The Position of Variable Stars in the Hertzsprung-Russell Diagram. Third Colloquium on Variable Stars, p. 36 Bamberg 1965.

IBEN, I., JR.: Astrophys. J. **140**, 1631 (1964).

KIPPENHAHN, R., WEIGERT, A.: Z. Astrophys. **65**, 251 (1967).

KOPAL, Z.: Close Binary Systems, p. 486. London: Chapman and Hall Ltd. 1959.

KOPAL, Z.: Publ. Astron. Soc. Pacific **83**, 521 (1971).

KOPAL, Z., SHAPLEY, M. B.: Ann. Jodrell Bank **1**, fasc. 4 (1956).

PACZYNSKI, B.: Acta Astron. **17**, 356 (1967).
POPOVICI, C.: Bul. Stiint. Acad. R.P.R., Secţia pentru Matematică si Fizică, **7**, 4 (1956).
POPOVICI, C., DUMITRESCU, A.: Studii Cerc. Astron. **16**, 123 (1971).
POPPER, D. M.: Ann. Rev. Astron. Astrophys. **5**, 85 (1967).
SVETCHNIKOFF, M. A.: Catalog orbitalnii elementov, mass i svetimostii tesnii dvoinih svezd. Sverdlovsk 1969.

Discussion

RYSBERGEN:

Why did you connect the six points in pairs, on the slide showing the primary and secondary sequences?

POPOVICI, C.:

The points connected are those for some exceptional stars which depart from the general (M_b-log R) relation, but which were not excluded from the x computation.

Kinematic Properties of Selected Visual Binaries

By GUSTAV A. BAKOS
Department of Physics, University of Waterloo, Ontario, Canada

With 2 Figures

Abstract

A relation between the age of 40 visual binaries and their space-motion components has been derived. It appears that in the U, V plane the distribution of the young stars is at right angle to that of the old and very old stars.

In an earlier investigation the author (BAKOS, 1959) made photometric and spectrographic observations of a number of wide visual binaries (with a separation greater than 10 seconds of arc) and constructed a color-magnitude diagram ($C-M$ diagram). The absolute visual magnitude, M_v, of the visual binaries was determined from microphotometric tracings of the spectra of the primary components (dispersion 33 Å/mm at $H\gamma$) by measuring on the log intensity scale the ratios of selected pairs of lines for the late-type stars following the method of OKE (1957) and by measuring, on a direct intensity scale, the equivalent widths of the $H\gamma$ and $H\delta$ lines for the early-type stars using the author's own calibration curves. The M_v of the secondary component was derived by adding the photoelectrically observed magnitude difference between the components to the M_v of the primary. For the abscissa of the $C-M$ diagram the observed color, $B-V$, was used except for a number of supergiants in which case the observed color was corrected for interstellar reddening.

The spectrograms of the primary components were also measured for radial velocities by DOUCET (1971) and the galactic space-motion components, U, V and W computed for stars for which proper motions were given in the Smithsonian Star Catalogue. Finally, the age of the visual binaries was determined (making the reasonable assumption that the two stars of a binary system are of the same age) by comparing the position of the binaries in the $C-M$ diagram with the $C-M$ diagram of galactic clusters of known age. The $C-M$ diagram of visual binaries has been represented in Fig. 1 and the appropriate data collected in Table 1. The columns contain the following information:

1. The ADS number of the binary.
2. $B-V$ of the primary component.
3. M_v of the primary derived by spectrographic methods.
4. $B-V$ of the secondary component.
5. M_v of the secondary derived by adding the magnitude difference between the components to the magnitude of the primary.

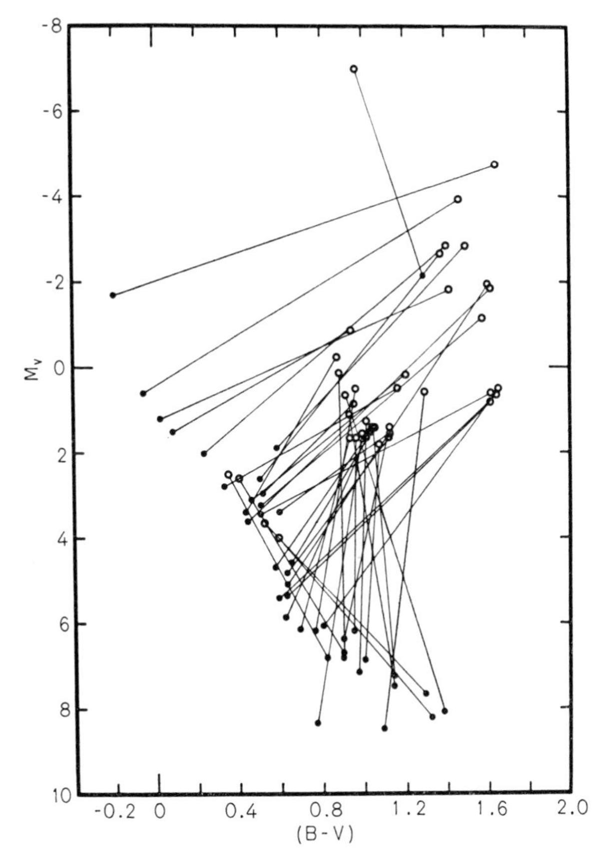

Fig. 1. The Color-Magnitude diagram of visual binaries. The open circles belong to the primary components, the filled circles to the secondaries. The pairs are connected by a line

6. The logarithm of the age of the binary.

7, 8, and 9. The U, V and W components of the space motion of the primary with respect to the local standard of rest, U being in the direction of the galactic anticenter, V in the direction of the galactic rotation and W in the direction of the north galactic pole.

In the $C-M$ diagram of Fig. 1 four groups of binaries have been recognized:

1. Binaries whose primary components are late-type stars with $M_v \lesssim 0$; their secondaries are main-sequence stars of early spectral type. These are the youngest stars with an age estimate between 10^6–10^8 years. One system, ADS 15602, if physical at all, has for its primary a very luminous supergiant and a giant secondary.

2. Binaries whose primary components are main-sequence stars of spectral type earlier than G0 and their secondary components are also main-sequence stars of later spectral type. The age of this group of binaries is intermediate, about 10^9 years.

Table 1. Colors, Magnitudes and Space Motion of Visual Binaries

ADS	Primary		Secondary		Log Age	U	V	W
	$B-V$	M_v	$B-V$	M_v				
546	1.65	0.50	0.80	6.06	10.1	10.5	−25.3	− 8.0
548	1.29	0.58	1.09	8.49	9.7	11.4	−28.2	−15.2
990	1.01	1.65	0.62	5.86	9.7	13.8	8.0	−35.8
1268	0.99	1.54	0.77	8.35	9.7	10.0	−23.7	−11.4
1534A	1.04	1.40	1.14	7.24	9.7	25.9	38.7	−27.9
1630	1.41	−1.80	0.02	1.20	7.6	10.0	−35.7	−15.3
1753	0.88	0.15	0.90	6.81	8.0	3.3	−36.2	0.2
1904	0.40	2.60	0.90	6.71	9.0	10.5	−32.3	−15.0
1964	1.40	−2.85	0.23	2.01	7.3	10.0	−30.6	−15.4
2157	1.46	−3.93	−0.06	0.60	6.8	17.8	−23.4	−24.2
8470	1.12	1.55	0.59	3.38	9.7	11.6	−24.9	−26.1
8489	1.61	0.50	0.60	3.44	10.0	9.3	−40.7	−23.8
8516	1.03	1.40	0.57	4.69	9.7	12.0	−22.7	−11.0
8992	0.58	3.65	1.29	7.67	9.7	14.6	−13.1	−24.2
9962	1.49	−2.85	0.58	1.89	7.6	58.8	34.5	−24.6
10259	0.93	1.65	0.69	6.13	9.0	12.6	−23.8	−13.7
10535	0.87	−0.23	0.43	3.40	8.7	49.8	11.2	− 9.2
10663	0.96	1.65	0.95	6.17	9.7	63.4	−31.4	−36.9
10715	1.20	0.18	0.46	3.13	9.3	66.7	14.9	−10.4
11271	0.91	0.65	1.38	8.07	9.0	6.1	−18.4	− 5.4
11773	1.05	1.40	0.65	4.55	9.7	− 2.0	−28.9	−14.3
12425	1.01	1.25	0.97	7.13	9.7	5.5	−29.7	−17.6
12445	1.61	0.80	0.63	5.34	10.1	34.7	−46.6	−18.9
12750	1.64	−4.76	−0.20	−1.70	6.8	31.9	15.6	6.9
12882	0.59	3.98	1.32	8.21	9.7	22.9	−20.4	− 7.7
12992	0.93	1.10	1.14	7.49	9.0	8.8	−19.2	− 9.9
13014	1.61	−1.85	0.44	3.60	7.7	126.9	40.3	56.7
13240	1.12	1.65	0.63	4.81	9.7	18.9	−24.4	− 9.4
14027	1.16	0.50	0.38	2.79	9.7	17.2	− 1.4	4.0
14909	1.13	1.40	0.90	6.33	9.7	72.0	−51.1	26.4
14998	1.57	−1.12	0.50	3.23	7.7	7.0	−40.2	−22.4
15431	0.95	0.85	0.51	2.98	9.0	8.7	−17.9	− 8.2
15602	0.97	−7.00	1.29	−2.14	6.0	57.5	33.8	99.8
15690	1.07	1.80	1.00	6.84	9.7	39.2	−48.4	12.6
15764	0.94	−0.87	0.08	1.50	7.8	29.4	−23.7	4.4
16140	1.60	−0.95	0.63	5.07	7.8	2.6	−17.9	−24.7
16227	0.96	0.50	0.76	6.16	9.0	9.1	−10.0	− 6.0
16681	1.64	0.64	0.59	5.41	10.1	10.5	−16.4	− 8.4
16690	1.37	−2.68	0.50	2.60	7.3	16.5	−42.5	− 4.9
16913	0.35	2.50	0.82	6.80	9.0	11.4	−47.4	−19.0

3. Binaries whose primary components are G- and K-type giants; their secondaries are main-sequence stars of late spectral type. Among these binaries two, ADS 548 and ADS 1268, appear to have subdwarf secondaries. These binaries are, in general, old systems, about 5×10^9 years old.

4. Binaries whose primary components are M-type giants and their secondaries are late-type main-sequence stars. These are believed to be the oldest systems, at least 10×10^9 years old.

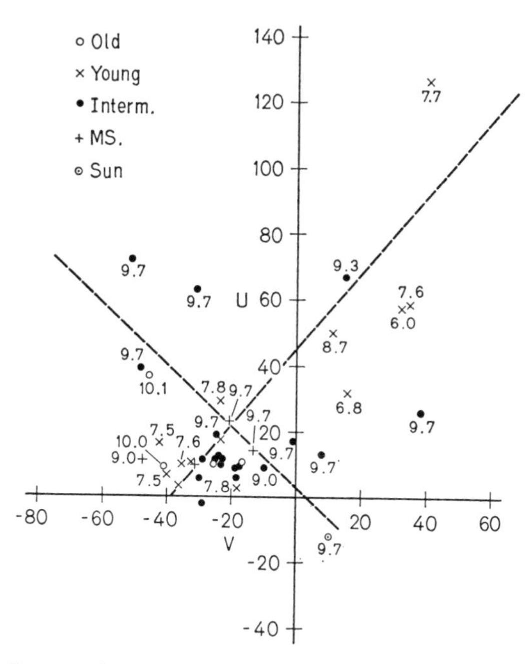

Fig. 2. The velocity diagram of visual binaries. Different symbols have been used for binaries of different age groups. For details see the text

Considering the kinematic properties of these four groups of stars their velocity components U and V have been plotted in Fig. 2 using four different symbols, according to their age category. It appears from the diagram that nearly all the binaries are confined to a small area with approximate coordinates $U=10$, $V=-25$ and containing a mixture of all four groups of stars. This is about the same area as occupied by the visual binaries of EGGEN (1963) having a color excess smaller than 0.04 mag. In his classification the stars belong to the group of the Hyades and the Pleiades which he believes to be helium-rich stars.

Thus, it would appear that kinematically there is no difference between the very young and the very old stars plotted in Fig. 2. On the other hand, it appears from the diagram that the velocity components of these giant and supergiant stars differ from the same set of velocity components of the Sun whose coordinates in the diagram are $U=-12$ and $V=10$. This, again, is in accordance with EGGEN (1963) who finds a sharp division of the velocity components between the Hyades-Pleiades stars and the Sun-Sirius stars.

Finally, attention is called to the few stars scattered in the first and second quadrant of Fig. 2. It is interesting to note that all the young stars appear to be situated along a line originating in the central area and making an angle of about 45° with the V-axis. On the other hand, the older stars, including the Sun (5×10^9 to 10×10^9 years old) are scattered along a line perpendicular to the former. Such a difference in velocities is expected to exist if it is kept in mind that the supergiants belong to the population of the spiral arms while the very old stars are a part of the local disk population of our Galaxy.

To sum up the results of this investigation it has been found that, in accordance with EGGEN's conclusion, the motions of the more evolved and metal-richer (BAKOS, 1971) giant and supergiant stars differ from those of the solar-type stars. On the other hand, on account of the small sample of stars indications are that the motions of the younger and more distant supergiants are different from those of the older and local group of stars. However, additional evidence for such a conclusion will be required.

References

BAKOS, G. A.: Thesis, University of Toronto (1959).
BAKOS, G. A.: J. Roy. Astron. Soc. Can. **65**, 222 (1971).
DOUCET, C.: Thesis, University of Waterloo (1971).
EGGEN, O. J.: Astron. J. **68**, 483 (1963).
OKE, J. B.: Astrophys. J. **126**, 509 (1957).

Discussion

MCCARTHY, M. F.:

In the very excellent data presented to us here can you give any estimate for the late-type $(K+M)$ giant stars of the absolute visual magnitude and the dispersion on absolute magnitudes? The lack of information concerning these two quantities imposes a serious limitation on statistical studies of density distribution of the late-type stars.

Your method is most promising for a reliable determination of these quantities. Can you tell us how many late-type giants have been studied in your research? How many other visual binaries could be selected further to refine determination of M_v and σ for late-type giants?

BAKOS, G. A.:

About 10% of the sample of stars are M-type giants. The spectrographic methods of luminosity determination seam to break down for these stars. However, a few lines are sensitive to luminosity. These criteria will be published elsewhere. The dispersion in absolute magnitude determination is about 0.7 mag. The problem is to find the distances of a sufficient number of stars to construct a calibration curve.

Most likely there exist more visual binaries with late-type primaries and if the statistics holds, in EGGEN's paper some 50 stars are expected to be of this type.

Gould's Belt

(Invited Lecture)

By PER OLOF LINDBLAD
Stockholm Observatory, Salts Jöbaden, Sweden

With 5 Figures

Abstract

Local properties of the Galaxy as related to the Gould Belt system of bright B stars are reviewed. In particular a first order model for the distribution of neutral hydrogen is described.

I. The Gould Belt System

In his book, presenting the results of five years of observations at the Cape of Good Hope, Sir JOHN HERSCHEL (1847) remarks upon "... the zone of large stars which is marked out by the brilliant constellation of Orion, the bright stars of Canis Major, and almost all the more conspicuous stars of Argo – the Cross – the Centaur, Lupus, and Scorpio. A great circle passing through ε Orionis and α Crucis will mark out the axis of the zone in question, whose inclination to the galactic circle is therefore about 20°, and whose appearance would lead us to suspect that our nearest neighbours in the sidereal system (if really such), form a part of a subordinate sheet or stratum deviating to that extent from parallelism to the general mass which, seen projected on the heavens, forms the Milky Way." GOULD (1874, 1879) traced this belt of bright stars all along a great circle through the brightest stars in Taurus, Perseus, Cassiopeia, Cepheus, Cygnus and Lyra, although he states that in the northern hemisphere its course is less distinctly marked. This band of stars is generally referred to as GOULD's Belt. Our intention is here to review the properties of the local system that may be revealed by the phenomenon of GOULD's Belt, and to discuss how the observations should be interpreted and reconciled with our ideas about spiral arm formation.

SHAPLEY and CANNON (1922b, 1924) showed that the phenomenon of concentration towards GOULD's Belt is very pronounced for B stars brighter than visual magnitude 5.26 (Fig. 1), while it gets less pronounced for B stars of fainter magnitudes and is practically absent for magnitudes 7.26 to 8.25, where the B stars are highly concentrated to the galactic plane. SHAPLEY and CANNON (1922a) also showed the early A stars to display a similar effect but with less inclination. We will here define the "Gould Belt system" as this system of bright B-type stars. Judged from SHAPLEY's and CANNON's results the radius of this system should be of the order of 500 pc.

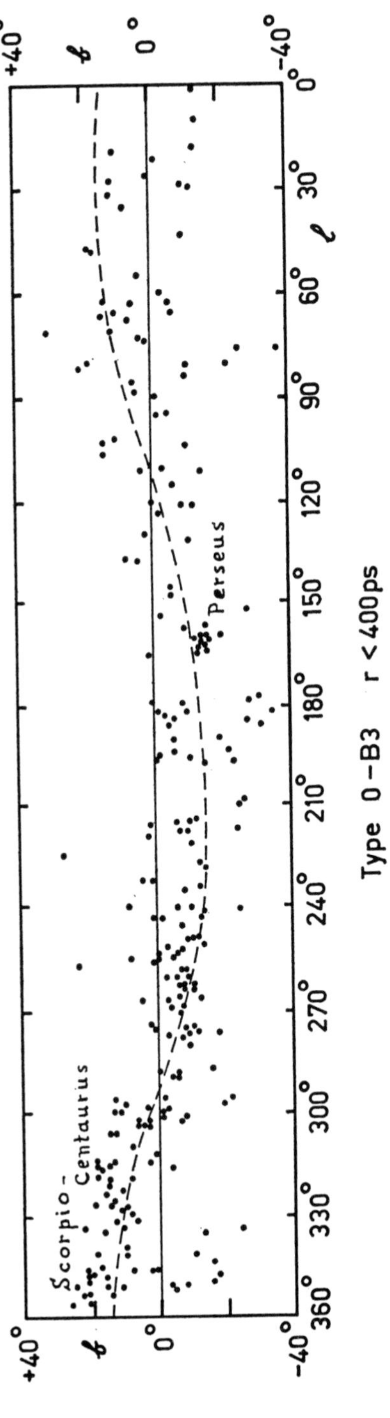

Fig. 1. Distribution on the sky of young stars closer than 400 pc. (After BLAAUW, 1965)

The motions of the B-type stars are characterized by the K-effect. Already PLASKETT (1930) showed that this effect is restricted to B stars brighter than a magnitude of about 5.5, and thus a property of the Gould Belt system as defined above.

Numerous investigations of the distribution and motions of the bright B stars have been published. Many authors have stressed the dominating role of the Scorpius-Centaurus association in giving the impression of the Gould Belt distribution and the expanding motion. The most thorough analysis is that of LESH (1968), who refined and extended analyses by BLAAUW (1956, 1965) and BONNEAU (1964). It includes a rediscussion of the observational data and is based on the equations of motion for expanding groups derived by BLAAUW (1952). Fig. 2 is taken from BLAAUW (1965) and shows how the observed radial velocity gradients as a function of galactic longitude fit the expected relation for an expanding group with an expansion age of 40×10^6 years. From the gradients of the U- and V-velocities for stars of spectral type earlier than B5 and $\delta \geqq -20°$ with well-determined distances within 600 pc LESH finds two possible solutions, one in which the entire sample consists of a mixed population with an over-all expansion age of 90×10^6 years and another in which the associations constitute a subset with an expansion age of 45×10^6 years. Recently LESH (1972) has presented complementary observational data also for the southern bright O and early B stars.

The conclusion that the Gould Belt system is a local expanding group immediately raises the question how this concentration of stars was initially formed some 50×10^6 years ago. We will return to this problem in the last section and first discuss what other types of objects may be related to the Gould Belt system.

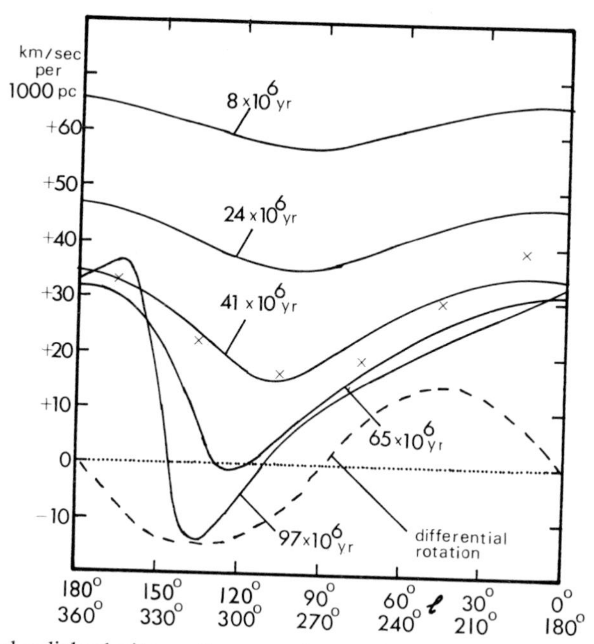

Fig. 2. Theoretical radial velocity gradients as a function of longitude for expanding systems of various expansion ages. Crosses mark observed gradients. (After BLAAUW, 1965)

II. Later-Type Stars

EGGEN (1961) has shown that B stars to spectral type B7 and brighter than visual magnitude 5.0 show the Gould Belt characteristics. EGGEN (1965, 1972) also suggests that the stars of the Pleiades group are part of the local association which we here call the Gould Belt system.

SHAPLEY and CANNON (1922a) showed that the distribution of early A stars displays to some extent a similar tilt to the galactic plane as the Gould Belt stars. According to McCUSKEY (1956) the local density distribution of B8 – A0 stars as well as that of the giant F8 – K3 stars shows a pronounced concentration towards $l = 140°$ at a distance of about 200 pc. The interstellar absorption in this direction is also quite high.

McCUSKEY'S results also show a concentration of early F stars to a local region within a few hundred pc from the Sun. RYDGREN (1970) finds this concentration to be real.

III. Interstellar Dust

HUBBLE (1922) points out that the diffuse bright nebulae are concentrated along two belts. One is the Milky Way and the other approximately the belt of bright helium stars "which defines the local cluster" and with an inclination of about 20° to the Milky Way. He finds that nebulae with continuous spectra show a decided tendency toward clustering and that most of them favour the tilted plane of bright early-type stars. The concentration of nearby bright nebulae to GOULD'S Belt is illustrated by DAVIES (1960, Fig. 8).

It is also well known that the dark absorbing clouds favour GOULD'S Belt in addition to a concentration to the galactic plane. This is clearly brought out by LYNDS' (1962) survey. The absorption in the galactic plane in the solar neighbourhood seems to be larger in the northern than in the southern sky (FERNIE, 1962). FITZGERALD (1968) has plotted the distribution of absorbing material in the galactic plane. It seems that the absorbing clouds within a distance of 1 kpc lie closer on the southern sky than on the northern.

Recent developments have brought forward the possibility to measure velocities of dust clouds through molecular lines in the radio spectrum. We will return later to some results from measurements of the formaldehyde line.

IV. Neutral Hydrogen Gas

Already LILLEY and HEESCHEN (1954) pointed out that the brightness distribution of neutral hydrogen shows two maxima, one coinciding with the galactic plane and one with GOULD'S Belt. This was confirmed by DAVIES (1960).

LINDBLAD (1967) has called attention to a local feature, called "feature A", of the 21 cm radiation defined and separated from other 21 cm features by its wide extent in galactic latitude and its small velocity dispersion, and pointed out that the velocity-longitude relation was that characteristic for a configuration

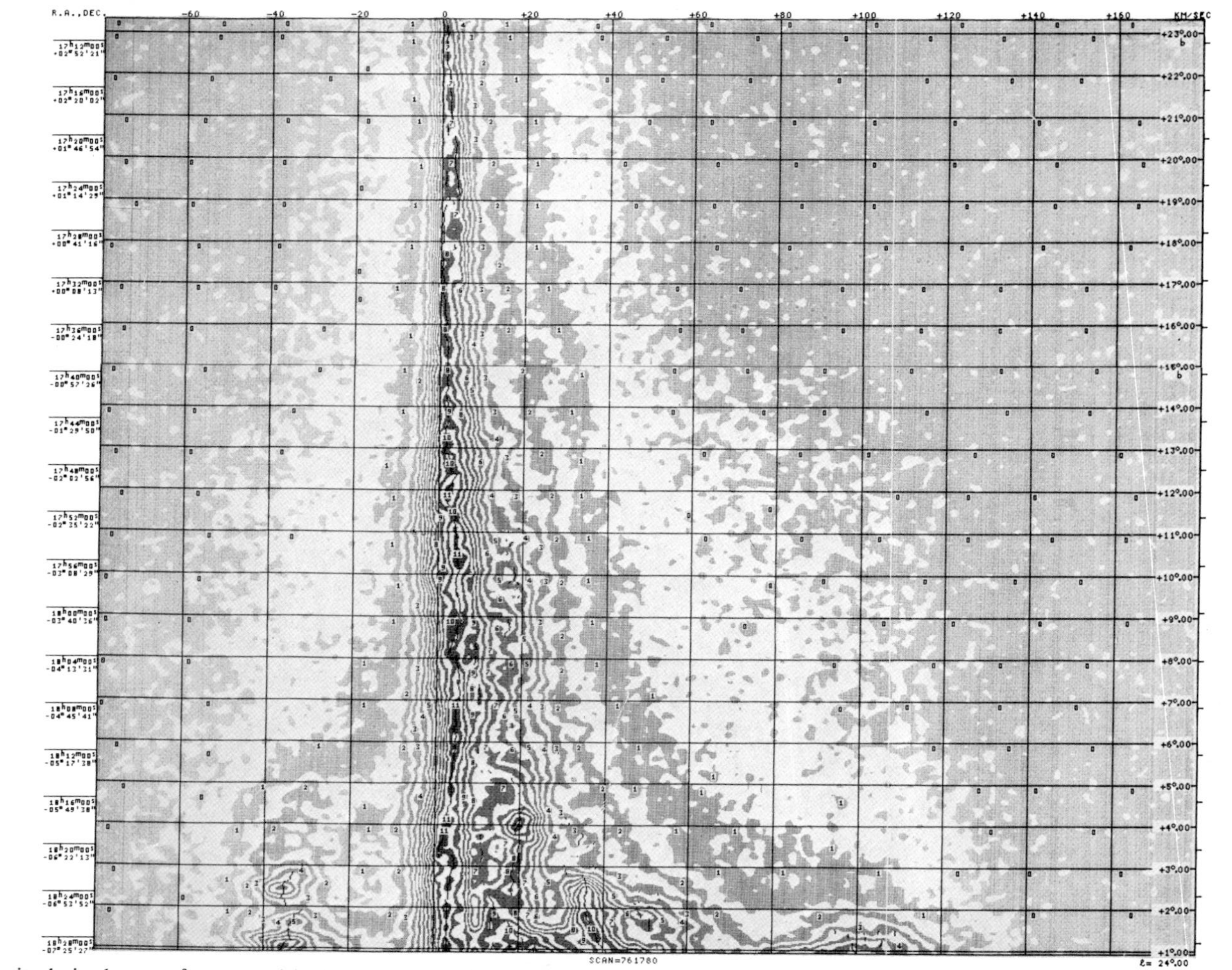

Fig. 3. Velocity-latitude map for neutral hydrogen at $l = 24°$ as derived from 21 cm observations with the 300-foot telescope of NRAO. The latitude extends from $b = +1°$ to $+23°$ and the velocity interval between vertical lines is 20 $km.s^{-1}$

expanding in the field of differential rotation. An expansion age of 65×10^6 years was tentatively derived, and it was suggested that this expansion might be related to the expansion of the local group of early-type stars studied by BLAAUW and BONNEAU. In Fig. 3 this feature stands out as the seemingly local component, with an average velocity with respect to the local standard of rest of $+4$ km.s^{-1} and a rather small velocity dispersion, extending all over the latitude region covered. At a velocity of $+18$ km.s^{-1} we note another, more irregular, local feature with less extent in latitude. It seems that what is called feature A covers a very large part of the sky. However, to separate this component from other local features may in many cases be difficult.

From a number of published observations by various authors as well as from new observations with the 300- and 140-foot telescopes at NRAO.[1] LINDBLAD *et al.* (1973) has derived the velocity-longitude diagram of feature A given in Fig. 4 (filled circles). The velocities for the "other local feature" are tentatively given by open circles. For $90° < l < 220°$ this other local feature is identical with the Orion arm. As has been pointed out by HARTEN (1971) the radial velocity of feature A generally decreases for higher values of $|b|$. The velocities for this feature given in Fig. 4 have been interpolated to $b=0$.

As a first order model that will reproduce the 21 cm data in the galactic plane, LINDBLAD *et al.* suggest a gaseous shell initially expanding with one and the same velocity, s_0, in all directions, and where no interaction with matter outside the

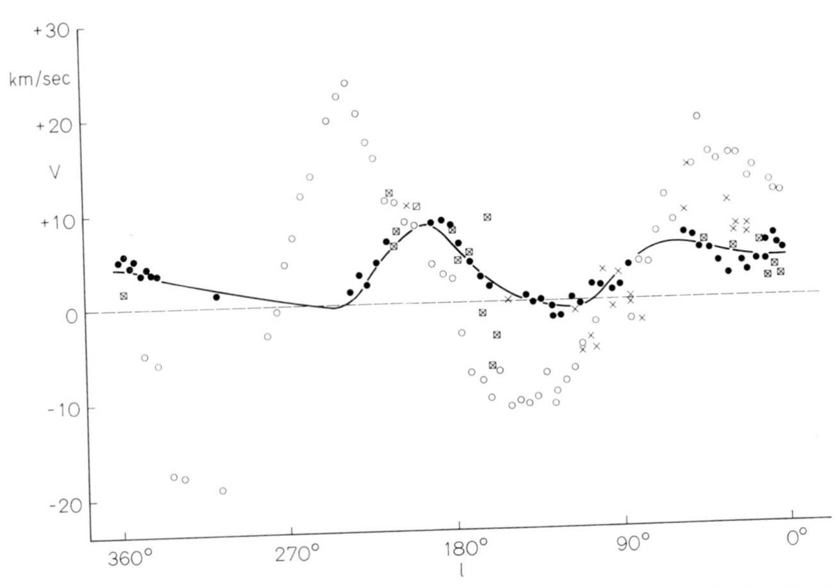

Fig. 4. Velocity-longitude diagram for "feature A" (filled circles) and for the "other local feature" (open circles). Crosses mark formaldehyde velocities, crosses within squares representing clouds falling in the Gould Belt region separated from the galactic plane. The full-drawn line shows the theoretical relation for the ring model described in the text

1 The National Radio Astronomy Observatory is operated by Associated Universities, Inc., under contract with the National Science Foundation.

shell is assumed. In the galactic plane their model is then an expanding ring, the motion of which is described by the equations given by BLAAUW (1952). By varying the parameters a best fit to the observations was found with the following values for the set of independent parameters:

expansion age $t = 60 \times 10^6$ years,
initial velocity of expansion $s_0 = 3.6$ km.s^{-1},
galactic longitude of the direction to the centre of expansion $= 150°$,
distance to centre of expansion $= 140$ pc,
velocity of centre of expansion with respect to local standard of rest $\Delta U = -0.8$ km.s^{-1}, $\Delta V = -0.5$ km.s^{-1}.

The two-dimensional model, which thus may represent a first order approximation to the structure of the expanding local system of neutral hydrogen is shown in Fig. 5. The semi-axes of the configuration are 330 pc and 160 pc. The Sun would

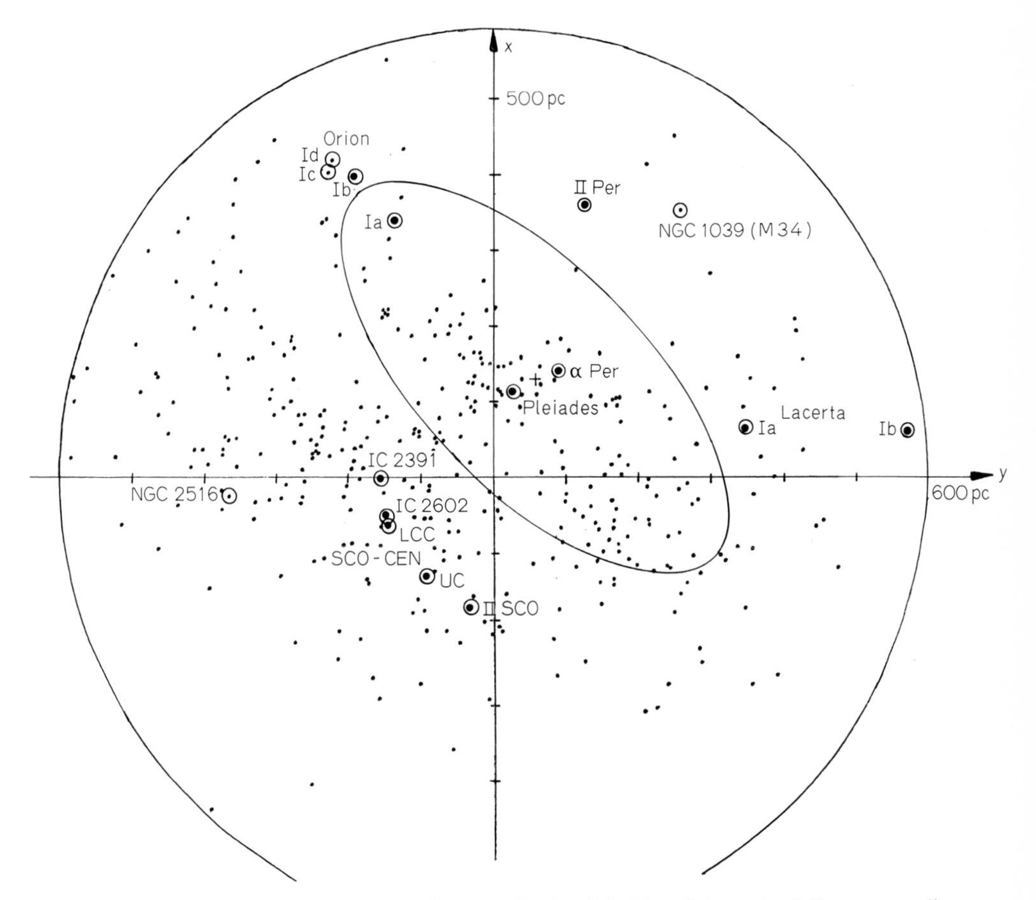

Fig. 5. Projections on the galactic plane of stars earlier than B5 with well determined distances smaller than 600 pc according to LESH (1968, 1972). The centre of the coordinate system markes the position of the Sun and the positive X-axis points towards the galactic anticentre. The ellipse with its centre marked by a cross represents the ring model for the neutral hydrogen described in the text

be situated just inside the ring, and the centre of expansion falling in the neighbourhood of the Pleiades and the α Perseus cluster would be practically at rest with respect to the local standard of rest. The young associations tend to fall just outside the ring, while MCCUSKEY's maximum for B8–A2 stars and later-type giants falls close to the centre.

The full-drawn line in Fig. 4 shows the radial velocity-longitude relation of this model as compared with the observations. HARTEN (1971) has made a similar analysis in three dimensions.

An obvious extension of the model would be to add a number of shells with different s_0 in order to get a smooth distribution of initial expanding velocities and to compare the radial velocities given by such a model with the set of line profiles of feature A.

It appears that the gas belonging to feature A and the young stars outlining the Gould Belt system occupy about the same volume of space in the galactic plane and show the same kinematic behaviour. To investigate further the relation between this gas component and the Gould Belt system SANDQVIST and LINDBLAD (LINDBLAD *et al.*, 1973), observed a number of dark clouds in the 6.2 cm formaldehyde line with the 25.6 m telescope at the Onsala Space Observatory. The clouds chosen to be of optimum size for the antenna beam were selected from LYNDS' (1962) catalogue. The radial velocities are plotted as crosses in Fig. 4. The radial velocities of the clouds concentrated to GOULD's Belt seem to show a preference for the relation for feature A.

All these circumstances give a strong indication that the local component of interstellar hydrogen defined as feature A is closely related to the Gould Belt system.

Before any conclusions can be drawn about the mass of neutral hydrogen in the Gould Belt system, its temperature must be known. At low latitudes in the general region of the galactic centre this feature is seen in absorption, and thus its temperature may be rather low.

V. Magnetic Field

Information about the galactic magnetic field can be derived from observations of polarization of starlight, polarization of galactic non-thermal radio emission, or Faraday rotation of polarized light from extragalactic radio sources. In the two first cases the transverse component of the field is observed, and in the last case the line-of-sight component.

GARDNER and DAVIES (1966) concluded from their observations of Faraday rotation of linearly polarized radio source emission that at low latitudes ($|b|<20°$) a magnetic field is directed towards $l=95°$. At intermediate latitudes ($20°<|b|<60°$) the field is directed towards $l=95°$ below the plane but reversed and directed towards $l=275°$ above the plane. DAVIES (1968) and CLUBE (1968) have suggested that this reversal is connected with the Gould Belt system.

MATHEWSON (1968) worked out a detailed model of the local magnetic field based on polarization measurements of stars mostly within 500 pc of the Sun. This model has the magnetic lines of force forming tightly wound right-handed helices of pitch angle 7° lying on the surface of tubes which have elliptical cross-sections of axial ratio 3 with semi-major axis parallel to the galactic plane. The helices have been sheared through an angle of 40° on the galactic plane in an anti-clockwise sense, looking down from the north galactic pole. The axis of the helices is in the direction $l = 90°$ and 270°. The Sun is 100 pc toward the galactic centre from the magnetic axis. In order to get a satisfactory fit with measurements of the rotation measure MATHEWSON and NICHOLLS (1968) extended this model by adding a straight-line magnetic field which is directed toward $l = 90°$. They suggest that the nearly helical field is associated with the Gould Belt condensation and that the longitudinal field is the general disk magnetic field.

It may be noted that the osculating plane of the helices in this model would be roughly parallel with the major axis of the ellipse in Fig. 5 and that the magnetic axis would pass close to the centre of expansion of the ring model.

VI. Further Observations

Many objections may be raised against the coarse ring model for the interstellar neutral hydrogen pictured in Fig. 5. One is the disputable correlation between the distribution of stars and gas, another is the non-investigated influence of surrounding interstellar matter on the expansion. A further puzzle is the fact that HOBBS (1971) already in the Pleiades stars observes interstellar line components corresponding to feature A of the 21 cm line.

A fruitful way of research to learn more about the local expanding system may be to observe the relation between radial velocity and distance at certain selected longitudes. If the latter are properly chosen, the velocity gradient should give an estimate of the expansion age, and a discontinuity of the run of velocity with distance should give a measure of the extension of the Gould Belt system in that direction. Such observations could also give a better picture of what type are the stars that share the kinematic behaviour of the Gould Belt system.

VII. The Origin of the Gould Belt System

If an expanding local system of gas, dust and early-type stars is established, the question about the mechanism of its original concentration comes up. A possible mechanism might be provided by the spiral shock model of the density-wave theory of spiral structure worked out by ROBERTS (1969, 1972). The concentration of the interstellar gas would occur in the shocks formed along the spiral arms, whereafter the so condensed systems are left to expand while leaving the arm. Accepting LIN's (1971) value for the pattern velocity an expansion age of 5×10^7 years would roughly fit an original concentration of the local system in the Carina spiral arm. However, a detailed analysis then needs to be worked out applying the dynamics of the spiral shock model.

References

BLAAUW, A.: Bull. Astron. Inst. Neth. **11**, 414 (1952).
BLAAUW, A.: Astrophys. J. **123**, 408 (1956).
BLAAUW, A.: Koninkl. Ned. Akad. Wetenschap. **74**, No. 4 (1965).
BONNEAU, M.: J. Obs. **47**, 251 (1964).
CLUBE, S. V. M.: Observatory **88**, 243 (1968).
DAVIES, R. D.: Monthly Notices Roy. Astron. Soc. **120**, 483 (1960).
DAVIES, R. D.: Nature **218**, 435 (1968).
DIXON, M. E.: Monthly Notices Roy. Astron. Soc. **151**, 87 (1970).
EGGEN, O. J.: Roy. Obs. Bull. **41**, 245 (1961).
EGGEN, O. J.: Ann. Rev. Astron. Astrophys. **3**, 235 (1965).
EGGEN, O. J.: Astrophys. J. **173**, 63 (1972).
FERNIE, J. D.: Astron. J. **67**, 224 (1962).
FITZGERALD, M. P.: Astron. J. **73**, 983 (1968).
GARDNER, F. F., DAVIES, R. D.: Australian J. Phys. **19**, 129 (1966).
GOULD, B. A.: Proc. Amer. Assoc. for Adv. Sci. **1874**, 115 (1874).
GOULD, B. A.: Uranometria Argentina. Result. Obs. Nac. Argentino Cōrdoba I, p. 355 (1879).
HARTEN, R. H.: Thesis, University of Maryland (1971).
HERSCHEL, J. F. W.: Results of Astronomical Observations Made During the Years 1834, 5, 6, 7, 8, at the Cape of Good Hope, p. 385. London: Smith, Elder and Co 1847.
HOBBS, L. M.: Astrophys. J. **166**, 333 (1971).
HUBBLE, E.: Astrophys. J. **56**, 162 (1922).
LESH, J. R.: Astrophys. J. Suppl. **17**, 371 (1968).
LESH, J. R.: Astron. Astrophys. Suppl. **5**, 129 (1972).
LILLEY, A. E., HEESCHEN, D. S.: Publ. Nat. Acad. Sci. **40**, 1095 (1954).
LIN, C. C.: In C. DE JAGER (ed.), Highlights of Astronomy, vol. 2, p. 88. Dordrecht-Holland: D. Reidel 1971.
LINDBLAD, P. O.: Bull. Astron. Inst. Neth. **19**, 34 (1967).
LINDBLAD, P. O., GRAPE, K., SANDQVIST, Aa., SCHOBER, J.: Astron. Astrophys. **24**, 309 (1973).
LYNDS, B. T.: Astrophys. J. Suppl. **7**, 1 (1962).
MATHEWSON, D. S.: Astrophys. J. **153**, L 47 (1968).
MATHEWSON, D. S., NICHOLLS, D. C.: Astrophys. J. **154**, L11 (1968).
McCUSKEY, S. W.: Astrophys. J. **123**, 458 (1956).
PLASKETT, J. S.: Monthly Notices Roy. Astron. Soc. **90**, 616 (1930).
ROBERTS, W. W.: Astrophys. J. **158**, 123 (1969).
ROBERTS, W. W.: Astrophys. J. **173**, 259 (1972).
RYDGREN, A. E.: Astron. J. **75**, 35 (1970).
SHAPLEY, H., CANNON, A. J.: Harvard Circ. 229 (1922a).
SHAPLEY, H., CANNON, A. J.: Harvard Circ. 239 (1922b).
SHAPLEY, H., CANNON, A. J.: Harvard Repr. 6 (1924).

Discussion

PIŞMIŞ, P.:

1) I take it that the age of the GOULD's Belt is determined with the help of the expansion velocity and the extent (or a representative radius) of this structure. Is this not so?

2) Another independent way to estimate the age of the GOULD's Belt would be by using the shearing effect of galactic rotation. Assuming that at the start of the expansion, the shape of the structure was circular, one could estimate the time necessary for the shearing effect to bring the GOULD's Belt to the elongated shape as of to-day.

LINDBLAD, P. O.:

1) The expansion age as far as the gaseous component is concerned is determined by fitting a model of an expanding configuration moving according to the equations given by BLAAUW, to the observed radial velocities of 21 cm. It is mainly determined by the ratio between amplitude and average level of the velocity-longitude relation.

2) We don't observe the elongated shape of the neutral hydrogen cloud or its orientation, on the contrary we get this from the expansion age derived from the observed radial velocities.

On the Distribution of Stars in the Carina – Centaurus Region

By A. SUNDMAN
Stockholm Observatory, Saltsjöbaden, Sweden

With 3 Figures

Abstract

A large survey of the southern Milky Way is beeing carried out at the Stockholm Observatory. As a first result we have obtained a list of 13300 stars of mainly spectral type O5–B9 and M, R, N, S. By means of this preliminary material it is possible to perform statistical studies of the distribution of the relevant stars with respect to various parameters. In this contribution is described the apparent density of early-type stars along the line of sight up to a distance of 1.4–1.8 kpc and the surface distribution of the registered objects.

I. Introduction

A spectral survey of the southern Milky Way is carried out at the Stockholm Observatory. It is based on observations obtained in the years 1956–1965 at the Boyden Observatory. The material and the classification principles were previously described (NORDSTRÖM and SUNDMAN, 1973), and the first volume of the catalogue will be published shortly (SUNDMAN, 1974). The major purpose of the work is to prepare finding lists and charts for early- and very late-type stars and objects with some peculiar appearence. At present more than 13000 stars in the Carina-Crux-Centaurus region with spectral types O–B9 and M, R, N, S are recorded. They appear in a catalogue which, besides the estimated spectral type, gives also coordinates and rough B magnitudes.

In spite of the preliminary condition of the material some results can be obtained directly from the catalogued stars. These are:

the surface distribution of stars of certain spectral types and apparent magnitudes,
clustering tendencies found with various methods,
the space distribution.

I will here present some results of the investigation of surface and space distributions.

II. Selection of Objects and General Distribution

The distribution of all recorded objects along the galactic equator is shown in Fig. 1. The most striking feature is the drop in star numbers from $l = 285°–290°$. The distribution of objects across the galactic equator is illustrated in Table 1.

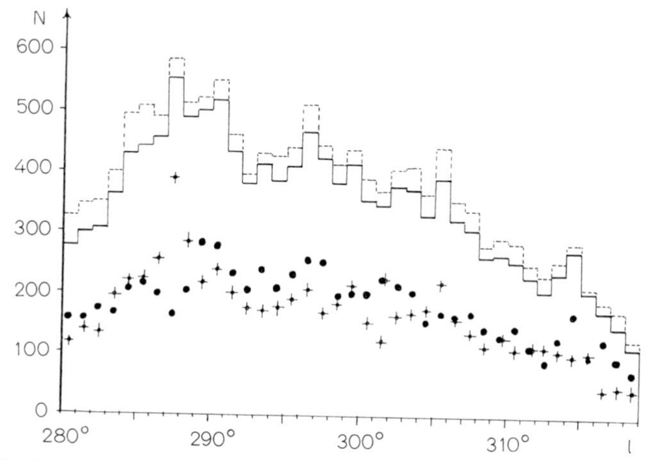

Fig. 1. Distribution along the galactic equator of all the stars recorded in the present survey. The continuous line gives the total number of stars between latitudes $-4°$ and $+4°$. The dots indicate the southern and the crosses the northern part separately. The dashed line represents a tentative extension to $\pm 5°$ where the material is somewhat incomplete and hence subject to occasional extrapolation

Table 1. Distribution of the recorded objects in galactic longitude and latitude

l \ b	$-4°$	$-3°$	$-2°$	$-1°$	$0°$	$+1°$	$+2°$	$+3°$	$+4°$
280°–294°	564	675	833	895	712	1029	698	455	
293°–307°	430	633	767	652	608	973	821	528	
306°–319°	212	348	347	439	404	453	437	370	

There is a noticeable density drop in the immediate neighborhood of the equator in the $l=280°–307°$ region but no such effect at $l=306°–319°$. However, M stars do not show any tendency to avoid the galactic plane—the density drop seems to be entirely due to the early-type stars.

The distribution according to spectral type and magnitude of the objects recorded in the Carina-Centaurus region is shown in Table 2. 85% of the early-type stars are classified as B4 or later. It has been found desirable to concentrate the investigation of the space density to the intermediate- and late-type B stars for several reasons:

These objects are of great importance for mapping the spiral structure optically. They also constitute the most observed group and there is today other observational material available better than this one suited for and already used for the same purpose.

The special advantage of the present survey is the great number of stars, dominated by intermediate- and late-type B stars, homogeneously selected and classified.

The M stars are classified in a sequence from M0 to M8 but there is no reliable luminosity classification of the M stars and no attempt has been made to compute space densities for this group.

Certain problems are connected with a material of this type. One of them is connected with the fact that the absolute magnitudes must rely on objective prism

Table 2. Distribution of the recorded objects according to spectral type and magnitude

	O–B3	B4–B7	B8	B9	A0–A2	A3–K	M	R,N,S	Me	E, OBe	?	Total
< 7.5	90	22	27	18	5	8	8	1	3	4	30	216
7.5—7.9	64	32	30	14	5	2	1	1	1	3	7	160
8.0—8.4	77	70	69	26	9	5	8	0	0	2	6	272
8.5—8.9	102	141	105	19	15	5	7	1	3	3	8	409
9.0—9.4	175	275	217	42	14	16	24	0	3	4	18	788
9.5—9.9	190	428	253	55	19	13	40	0	3	13	20	1035
10.0—10.4	187	608	365	102	36	28	48	0	0	7	29	1410
10.5—10.9	188	773	461	153	46	23	54	1	2	10	39	1750
11.0—11.4	219	1006	690	199	85	44	98	2	6	8	82	2439
11.5—11.9	123	810	498	144	80	22	125	1	6	3	89	1901
12.0—12.4	55	442	299	90	45	17	218	1	9	2	94	1272
12.5—12.9	18	123	92	47	11	7	236	1	10	3	46	594
13.0—13.4	11	46	32	8	4	2	236	3	13	2	28	385
13.5—13.9	3	17	9	3	0	0	278	4	5	2	19	340
>13.9	0	1	3	1	0	1	241	14	20	3	12	296
Total	1502	4795	3150	921	374	193	1622	30	84	69	527	13267

spectral classification only. Spectra are obtained with the ADH Baker-Schmidt telescope at Boyden and the dispersion is about 200 Å/mm at $H\gamma$. Since we are faced with star-rich areas there are inevitable variations in limiting magnitude due to overlap and disturbances from the unresolved background. Variations in plate quality also affect the limiting magnitude and the spectral classification. It is a well known fact that the B5–B8 stars are difficult to classify accurately in the MK system. However, the group is distinctly separated from the natural group dB and the later types showing *K*-lines. Obviously, there might be deviations from MK type equal

Table 3. Space distribution of the B5–A0 stars

Interval kpc	280°–288°				288°–296°			
	no giants		20% giants		no giants		20% giants	
	ex 3	ex 5	ex 3	ex 5	ex 3	ex 5	ex 3	ex 5
0.1—0.2	1.32	1.79	1.08	1.86	0.96	1.15	0.79	0.74
0.2—0.3	2.01	1.71	1.54	1.30	1.07	1.39	0.89	0.97
0.3—0.4	1.60	1.31	1.44	1.13	1.47	1.58	1.11	1.50
0.4—0.5	1.38	1.08	1.08	0.80	1.58	1.69	1.14	1.29
0.5—0.6	1.27	1.10	0.95	0.81	1.29	1.60	0.98	1.07
0.6—0.7	1.09	1.06	0.88	0.78	1.16	1.23	0.84	0.90
0.7—0.8	0.97	0.95	0.73	0.76	0.97	1.07	0.67	0.79
0.8—0.9	0.90	0.96	0.64	0.71	0.93	0.91	0.67	0.69
0.9—1.0	0.77	0.85	0.58	0.63	0.86	0.79	0.59	0.53
1.0—1.1	0.66	0.74	0.43	0.54	0.77	0.66	0.51	0.42
1.1—1.2	0.54	0.70	0.38	0.47	0.66	0.54	0.41	0.37
1.2—1.3	0.43	0.62	0.29	0.40	0.54	0.45	0.34	0.28
1.3—1.4	0.34	0.51	0.23	0.32	0.43	0.36	0.27	0.22
1.4—1.5	0.29	0.38	0.17	0.25	0.34	0.30	0.22	0.19
1.5—1.6	0.16	0.24	0.11	0.18	0.20	0.23	0.13	0.14
1.6—1.7	0.11	0.16	0.08	0.12	0.15	0.18	0.09	0.12
1.7—1.8	0.09	0.06	0.05	0.06	0.12	0.14	0.07	0.10

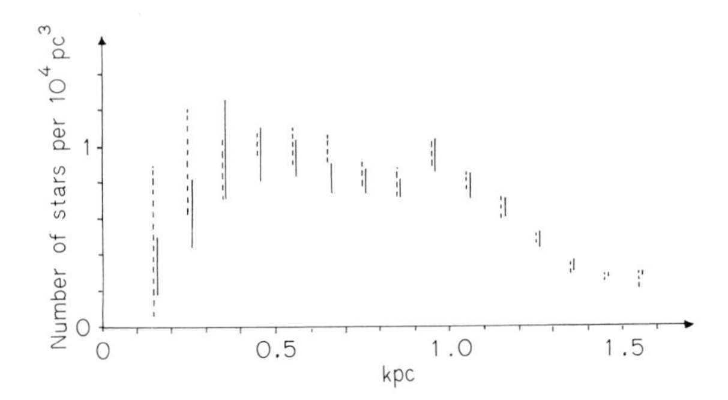

Fig. 2. Number of stars per 10^4 pc^3 in the section $l=304°-312°$ computed under the assumptions that the B7 stars have a mean absolute magnitude M_B equal to $-0\overset{m}{.}6$ (dashed lines) or equal to $-0\overset{m}{.}9$ (continuous lines). The bars give the uncertainty in space density as obtained by the standard deviation. Here the uncertainty in magnitudes but not in the correction for interstellar extinction is considered

to one or a couple of subtype units within the range B5–B9. At present this possibility cannot be properly considered since the comparison material of accurately MK classified stars is insufficient. As an illustration I have computed the space density for a part of the region under the assumption that all the B7 stars are $0\overset{m}{.}3$ fainter than what is generally assumed (Fig. 2). The deviation is not serious compared to the effects introduced by uncertainties in the magnitudes and the correction for interstellar extinction.

Some of the stars classified B5–B9 are probably not of luminosity class V as initially assumed but of luminosity classes IV or III. I chose 0.2 as a tentative maximum value of the fraction luminosity classes IV+III to luminosity class V

296°–304°				304°–312°				312°–319°			
no giants		20% giants		no giants		20% giants		no giants		20% giants	
ex 3	ex 5	ex 3	ex 5	ex 3	ex 5	ex 3	ex 5	ex 3	ex 5	ex 3	ex 5
2.41	1.75	1.34	1.80	0.33	0.52	0.37	0.54	0.36	0.52	0.49	0.23
1.44	0.46	1.16	1.10	0.63	0.61	0.69	0.53	0.65	0.50	0.38	0.30
1.19	1.30	0.92	1.07	0.98	0.71	0.58	0.48	0.69	0.71	0.47	0.49
1.05	1.25	0.73	0.91	0.95	0.67	0.67	0.48	0.79	0.83	0.51	0.60
0.91	1.23	0.70	0.90	0.93	0.73	0.69	0.54	0.88	0.82	0.52	0.53
0.87	1.20	0.60	0.81	0.82	0.79	0.52	0.57	0.70	0.84	0.46	0.59
0.80	1.01	0.56	0.73	0.80	0.90	0.52	0.60	0.71	0.90	0.49	0.53
0.78	0.95	0.46	0.61	0.76	0.93	0.51	0.57	0.89	0.77	0.57	0.44
0.89	0.82	0.59	0.53	0.93	0.92	0.59	0.57	0.71	0.64	0.41	0.36
0.81	0.72	0.54	0.43	0.77	0.83	0.48	0.52	0.60	0.52	0.32	0.30
0.69	0.57	0.40	0.35	0.65	0.74	0.37	0.43	0.40	0.40	0.20	0.21
0.57	0.47	0.34	0.26	0.47	0.60	0.27	0.32	0.27	0.27	0.15	0.15
0.44	0.38	0.27	0.22	0.33	0.44	0.20	0.26	0.17	0.19	0.10	0.10
0.42	0.29	0.24	0.17	0.28	0.28	0.17	0.20	0.10	0.10	0.06	0.08
0.46	0.21	0.29	0.12	0.28	0.15	0.20	0.12				
0.20	0.15	0.16	0.10								

and computed space densities first under the assumption that all the stars belong to the main sequence and then that 20% of the stars are giants. When 20% of the stars are assumed to be giants the value of space density goes down compared to the case when all stars belong to the main sequence, but the deviation is rather uniform along the line of sight (Table 3).

There is generally no U, B, V photometry for the stars in the present investigation and no attempt has been made to determine colours from photographic photometry. The determination of the extinction situation is based on U, B, V photometry for altogether 1 500 stars covering all parts of the region and occasionally coinciding with the programme stars. The region is divided in two ways forming 5 or 3 sections, and the relation between distance modulus and extinction is determined for each section (LODÉN and SUNDMAN, 1972). Plotting E_{B-V} as found from the intrinsic colours and the measured $B-V$ values against apparent distance moduli $m-M$ one finds a statistical increase of E_{B-V} towards the Centaurus end of the galactic belt but no obvious corresponding tendency perpendicularly to the equator. The value of R_B which has been used is 4.0. It is clear from the photometric material that local variations are very important and the choice of good mean values of the correction for an extended area becomes ambiguous. It is also obvious that individual distances obtained in this way would be of no value.

III. Methods of Computation and Results for the Space Distribution

For each section, approximately $8° \times 8°$, space distribution is determined in the following manner:

The distance to each star is computed by means of the formula

$$\log r = (m_B - M_B + A_B + 5)/5.$$

Here m_B, M_B and A_B are affected by certain errors. The standard deviation of m_B is $0\overset{m}{.}26$ and for M_B $\sigma = 0\overset{m}{.}5$ is chosen for luminosity class V and $\sigma = 1\overset{m}{.}0$ for classes IV+III. In the case of the absolute magnitude the uncertainty in spectral type enters the computation in such a way that the actual dispersion increases when the uncertainty in classification does. The error in A_B is not considered at this stage.

r-values are then grouped together in distance intervals of 100 pc.

The mean number of stars per interval is determined. The standard deviation is taken as a measure of the uncertainty in this mean value. A transformation into number of stars per 10^4 pc^3 in each distance interval is performed. The results appear in Table 3. Here the results from two different ways of correction for interstellar extinction are shown. They are based on the same photometric material but the averaging is made over 5 or 3 partial areas as mentioned above. One method cannot be expected to give better results than the other, in fact these two sets of values can be used as an illustration to the uncertainty introduced by interstellar extinction.

It is reasonable to average over the altogether 4 alternative results i.e. two different corrections for interstellar extinction and two different assumptions concerning the number of stars above the main sequence (Fig. 3). It should be

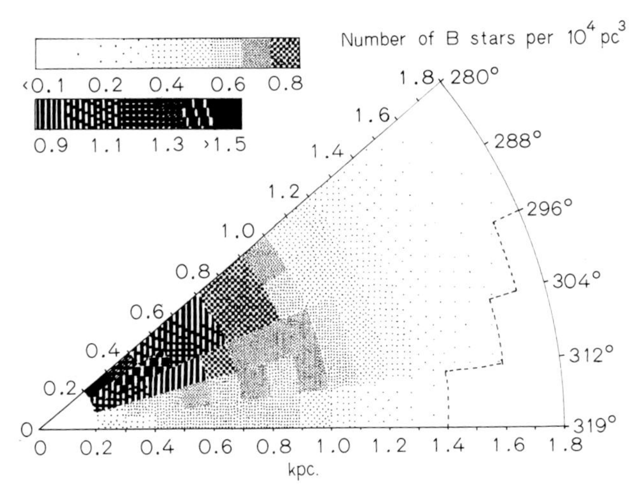

Fig. 3. Number of B stars in the whole area $l = 280° - 319°$. The dashed line indicates the limit of completeness of the data

noted that although there is a steady decrease from Carina to Centaurus of recorded stars per square degree without respect to magnitude or spectral type (Fig. 1) the space density for distances greater than 1.0 kpc shows rather small variations along the galactic equator except in the $l = 312°–319°$ section where the density seems to be lower than in any other section. The higher density around 0.5 kpc in the $l = 280°–304°$ sections is probably real but the values themselves are naturally rather uncertain at this distance.

Because of the higher extinction in the Centaurus region the reach in space is reduced. Most of the early-type stars are on the average getting too faint for completeness beyond $m_B = 11.5$. For the most opaque areas of Centaurus this means that space density values beyond 1.4 kpc are not reliable while in the Carina region 1.8 kpc is a probable limit. No corrections are attempted for the loss of stars in areas with extreme overlap of spectral images. Hence, the space density beyond 1.0 kpc is probably not going down quite as rapidly as is found here. Data will soon be availabe for the section $l = 319°–330°$ and the statistical studies will be extended to this area.

References

Lodén, L. O., Sundman, A.: Astron. Astrophys. **20**, 49 (1972).

Nordström, B., Sundman, A.: In Ch. Fehrenbach and B. E. Westerlund (Eds.), Spectral Classification and Multicolour Photometry, IAU Symp. No. 50, p. 85, Dordrecht-Holland: D. Reidel 1973.

Sundman, A.: Astron. Astrophys. 1974 (to be published).

Discussion

Williams, P.:
In your survey, did you come across many stars with abnormally strong or weak helium lines?

SUNDMAN, A.:
Yes, there are some. Not published but available on request.

MCCARTHY, M. F.:
Have you had an opportunity to analyze your star densities according to narrower spectral groups than B5–A0? One would expect that a separation of the stars in the intervals B5–B7 and again B8–A0 would permit a more detailed study. The method of natural groups as described by MORGAN and by B. LINDBLAD should find special application to such very rich material as yours is.

A second question, what values of M_V and σ have you adopted for your analysis of the B5–A0 stars studied here.

SUNDMAN, A.:
1) This has not been done. I hesitate to reduce the number of stars too much, but I agree it should be considered.

2) The dispersion of absolute magnitudes for stars which are accurately classified is assumed to be $0^m.5$. When the star is less accurately classified for some reason this is taken into account in each individual case.

Conclusions on Galactic Kinematics from Proper Motions with Respect to Galaxies

By WALTER FRICKE
Astronomisches Rechen-Institut, Heidelberg, GFR

Abstract

A comparison has been made between the results for precessional corrections and parameters of galactic rotation derived from fundamental proper motions and those measured with respect to galaxies at Lick and at Pulkovo. It is found that the results derived from the proper motions in declination are in fair agreement with each other. Since lunisolar precession and OORT's constant B follow mainly from μ_δ, it can be concluded that the fundamental system and the reference system of galaxies are identical within the range of errors of measurement. No hitherto undetected systematic motion of stars can be reliably deduced. The proper motions in right ascension measured at Lick and at Pulkovo differ from each other by more than one second of arc per century, and the fundamental μ_α lie very near to the mean of both. Although the total discrepancy cannot be explained at present, part of it appears to be due to systematic errors depending on the magnitude.

I. Introduction

During the past few years first results of measurements of proper motions of stars with respect to galaxies have become available. The measurements were carried out at the Lick Observatory and at Pulkovo on plates taken with astrographs at epoch intervals of about 20 years. At both observatories the plates covered two large regions of the northern sky separated by the zone of avoidance; the southern limit of the plates taken at Lick is declination $-23°$. The average standard deviation of an individual proper motion component is about $\pm 0''.7$ per century from the Lick plates and about $\pm 0''.9$ from the Pulkovo plates. Included in the measurements are AGK3 stars, whose proper motions are known with an accuracy of about $\pm 0''.8$ in the system of the FK4, and fainter stars with unknown proper motions. At the Lick Observatory a catalogue giving the data of about 8800 stars has been published, while no Pulkovo catalogue is yet available.

On the basis of their own material FATCHIKHIN (1970a, 1970b) and VASILEVSKIS and KLEMOLA (1971) have derived precessional corrections from the comparison of the proper motions given in the AGK3 and those measured with respect to galaxies for the same stars. Furthermore, all available proper motions relative to galaxies were used for the determination of OORT's constants A and B.

The purpose of this paper is to compare the extragalactic reference frame with the fundamental system, to describe some limitations of the available proper motions relative to galaxies and to present conclusions on galactic kinematics.

II. Comparison of the Extragalactic Reference Frame with the Fundamental System

The advantage of proper motions relative to galaxies is their independence from the assumptions underlying the construction of the conventional fundamental system. They are not affected by any kind of motion of the Earth's axis of rotation with respect to an inertial system; that is, they are free of errors of precession and of errors which may arise from the assumption that precessional corrections alone transform the fundamental system into an inertial system.

The fundamental proper motions are affected by an erroneous value of NEWCOMB'S lunisolar precession and by an erroneous, non-precessional motion of NEWCOMB'S equinox. NEWCOMB'S planetary precession is free of any appreciable error as has been shown recently by LAUBSCHER (1972). The fundamental proper motions may, in principle, also be affected by a rotation of the equatorial plane about an axis passing through the vernal equinox as suggested by AOKI (1967). The differences between proper motions of stars relative to galaxies and in the fundamental system yield the correction Δn to be applied to NEWCOMB'S value of the general precession in declination and the correction $\Delta k = \Delta m - \Delta e$, where Δm is the correction to general precession in right ascension, and Δe describes the erroneous, non-precessional motion of Newcomb's equinox. In solutions based on the proper motion differences the values Δn and Δk are not affected by galactic rotation.

Table 1 gives the corrections Δn and Δk which were derived by FATCHIKHIN (1970b) and by VASILEVSKIS and KLEMOLA (1971) from the differences $\mu_{\text{fund}} - \mu_{\text{gal}}$. For comparison, the results of the determinations by DIECKVOSS (1967) on the basis of the AGK3 and by FRICKE (1967) on the basis of distant fundamental stars are also given. DIECKVOSS divided about 160000 AGK3 stars into 42 groups according to spectral type and magnitude and carried out solutions for each group. His values given in the table were found by weighted averaging of the group solutions; this procedure explains the small standard errors which do not form a realistic description of the accuracy. FRICKE'S values are averages of two solutions carried out in the systems of the FK4 and N30 separately. The results in FK4 and N30 do not differ significantly and have the same standard deviation which is given in Table 1. In all solutions the value Δn is determined almost entirely from the proper motion components μ_δ, since μ_α does not contribute significantly; Δk results entirely from μ_α, even if the solutions are carried out from a combination of both proper motion components.

The most obvious result of an inspection of Table 1 is the fair agreement among the values of Δn and the pronounced disagreement among the values of

Table 1. Precessional corrections from proper motions with respect to galaxies and in the fundamental system. Units: Seconds of arc per century. Errors: Standard deviations

	Δn		Δk		Authors
Pulkovo	+0″.41	±0″.12	+0″.43	±0″.12	FATCHIKHIN (1970a, 1970b)
Lick	+0.31	0.07	−0.78	0.09	VASILEVSKIS, KLEMOLA (1971)
AGK 3	+0.51	0.01	−0.32	0.01	DIECKVOSS (1967)
FK 4/N 30	+0.44	0.06	−0.19	0.09	FRICKE (1967)

Table 2. OORT's constants from proper motions with respect to galaxies and in the fundamental system. Units: $km.s^{-1}.\ kpc^{-1}$. Errors: Standard deviations

	A		B		Authors
Pulkovo	+ 3.8	±4.9	−10.4	±4.1	FATCHIKHIN (1970a, 1970b)
Lick	+ 8.9	3.6	−12.5	4.1	VASILEVSKIS, KLEMOLA (1971)
AGK 3	+14.8	0.7	−11.3	0.6	DIECKVOSS (1967)
FK 4/N 30	+15.6	2.8	−10.9	2.8	FRICKE (1967)

Δk. Although the values of Δn do not agree among each other within the range of their formal errors, the deviations are not larger than can be expected from the systematic accuracy of the underlying observations. Therefore, the conclusion is permitted that the fundamental system in declination does not deviate significantly from the extragalactic reference frame. This result can be considered as most satisfactory. The situation is different with the proper motion components μ_α. In particular, the discordance between the Pulkovo and Lick results for Δk indicates systematic errors of measurements in right ascension with respect to galaxies which will be commented upon later.

Table 2 gives the values of OORT's constants A and B as derived from the proper motions with respect to galaxies and, for comparison, the values of the constants which followed from the proper motions in AGK3 and in FK4/N30. DIECKVOSS' procedure is the same as mentioned previously and explains again the small formal error of his results. The most remarkable feature of the list of values is the close agreement among the results for OORT's constant B. Both determinations of B from proper motions with respect to galaxies are not only in agreement with each other, but also in complete agreement with the determinations based on fundamental proper motions. Since B is mainly determined from μ_δ, even if the solutions are carried out from a combination of μ_α and μ_δ, the agreement confirms that the fundamental system in declination does not deviate significantly from the extragalactic reference frame. Furthermore, since the value of B is mainly determined from the rotational component about the axis passing through the vernal equinox, the agreement implies that the fundamental system cannot be affected noticeably by a rotation as suggested by AOKI (1967).

The determinations of the constant A from the Pulkovo and Lick material which differ from the results based on the fundamental system, suffer at least partly from insufficiencies in the proper motion components μ_α measured with respect to galaxies, when A was determined by both components. From μ_δ alone, VASILEVSKIS and KLEMOLA (1971) found $A = +12.3 \pm 5.0\ km.s^{-1}.\ kpc^{-1}$.

III. Limitations of the Proper Motions with Respect to Galaxies and Conclusions

The emphasis VASILEVSKIS and KLEMOLA have given to the preliminary nature of their results is supported by the fact that the Lick and Pulkovo proper motions in right ascension differ at the equator by more than one second of arc as manifested by the values of Δk

$$+0''.43\ (\text{Pu}) \qquad -0''.19\ (\text{FK4/N30}) \qquad -0''.78\ (\text{Lick}).$$

The results obtained at Pulkovo and Lick differ from the fundamental proper motions in opposite directions by about the same amount. The large difference must be attributed either to systematic errors of measurements with respect to galaxies or to unsatisfactory procedure in the determinations. The AGK3 motions of the stars cannot be responsible for the deviation, although different sets of AGK3 stars may yield values of Δk which may differ slightly from each other. In searching for a possible explanation, the residuals of the Lick proper motions in VASILEVSKIS and KLEMOLA'S solution have indicated a direction towards a reconciliation of the values of Δk from μ_{gal} and from the proper motions in AGK3. The authors have divided the stars in magnitude groups around the average blue magnitudes 9.0, 11.7, and 16.0. The residuals in μ_α of the brightest group reveal a systematic component $\delta(\Delta k)$ with respect to the residuals in the faintest group of about $\delta(\Delta k) = +0''.5$. Since the AGK3 stars are among the brighter stars of the program, one may conclude that their proper motions with respect to galaxies are affected by a systematic error of this order of size. Then the application of $\delta(\Delta k) = +0''.5$ to $\Delta k = -0''.78$ gives a complete reconciliation with the fundamental proper motions. The alternative interpretation of $\delta(\Delta k)$ as a real rotation of the stars about the axis of rotation of the Earth may be excluded. Again, it is noteworthy that no such effect depending on the magnitude is significantly present in μ_δ and, therefore, not in Δn. There remains the problem of explaining the deviation of the Pulkovo μ_{gal} from the fundamental system. Magnitude equations in the opposite direction are unlikely. Then, if one may offer a provocative suggestion, it is possible that a sign error in Δk, arising from the sign of the differences between proper motions with respect to galaxies and fundamental proper motions, exists.

There are other limitations, besides possible magnitude equations in right ascension, which support VASILEVSKIS and KLEMOLA'S suggestion that the available results should not be overestimated. For the determination of precessional corrections the lack of data from the southern hemisphere is a serious handicap which, of course, also affects solutions based on the AGK3. The determination of OORT'S constants is, moreover, greatly weakened by the lack of sufficient data in the galactic belt. Under these circumstances, the results for the galactic constants may be considered in surprisingly good agreement with the commonly adopted values.

If one doubts the validity of the model of galactic rotation used in many determinations, then, as a basis of investigation, the available proper motions relative to galaxies would not be suited for finding deviations from the model. Several authors have tried to find deviations from the rotation about the galactic center and also anomalies in the shear described by OORT'S constant A. No reliable effects have yet been found. Such investigations require the knowledge of the distances, proper motions, and radial velocities of stars over the whole sky, or at least over the whole galactic belt. The area of the sky covered by the available measurements relative to galaxies is too small for this purpose.

Finally, one may ask which contribution the new proper motions may provide to the determination of peculiar stellar motions. The accuracy of three quarters of a second of arc per century in one proper motion component is satisfactory for motions of stars nearer than about 200 pc. At a distance of 1000 pc, however, this accuracy corresponds to a standard deviation of about ± 35 km.s^{-1} in one tangential component and is, therefore, insufficient for conclusions about peculiar stellar motions. Nevertheless, one may conclude that for faint stars of known distance up to a few hundred parsec the proper motions relative to galaxies will become of great importance, as soon as the deficiencies in μ_α are eliminated.

If reliable tangential motions of stars at distances of the order of 1000 pc or larger are required, the accuracy of measurement has to be considerably increased. The aim should be an accuracy of about $\pm 0''.20$ per century, which is the accuracy of fundamental proper motions at present. It appears doubtful that this aim can be reached by the repetition of plates at later epochs with the instruments and the methods used at Lick and Pulkovo. It is more likely that in future more powerful telescopes with much fainter limiting magnitude will permit the determination of accurate proper motions of faint stars relative to extragalactic quasi stellar objects. Nevertheless, the continuation of the enormous efforts at the Lick Observatory and at Pulkovo and the completion of the analogous plan of the Lick program in the southern sky will certainly contribute effectively to further progress.

References

AOKI, S.: Publ. Astron. Soc. Japan **19**, 585 (1967).
DIECKVOSS, W.: Astron. Nachr. **290**, 141 (1967).
FATCHIKHIN, N. V.: Izv. Gl. Astron. Obs. Pulkovo No. 185, 93 (1970a).
FATCHIKHIN, N. V.: Astron. Z. **47**, 619 (1970b). Engl. transl. in Soviet Astron. **14**, 495.
FRICKE, W.: Astron. J. **72**, 1368 (1967).
LAUBSCHER, R. E.: Astron. Astrophys. **20**, 407 (1972).
VASILEVSKIS, S., KLEMOLA, A. R.: Astron. J. **76**, 508 (1971).

Molecules as Probes of the Interstellar Matter

(Invited Lecture)

By Peter G. Mezger
Max-Planck-Institut für Radioastronomie, Bonn, GFR

With 10 Figures

Abstract

26 interstellar molecules have now been observed. Some of the observational problems pertaining to detection of molecules by radio spectroscopy are discussed in section II. Our present knowledge concerning the interstellar matter, especially with regard to its cloud structure, is reviewed in section III. The observational results pertaining to the formation of stars from the interstellar gas are discussed in Section IV. Sections V and VI present some examples of how molecules can be used as probes of the physical state of dense and cool interstellar clouds, and, possibly, of the UV radiation within these clouds. Use of molecules to study the large-scale structure of the Galaxy, especially the inner nuclear disk, is discussed in Section VII.

I. The Importance of Radio Spectroscopy

Most cosmic radio emission is continuum radiation from either thermal plasma (free-free or Bremsstrahlung) or a relativistic plasma (synchrotron or Magneto-Bremsstrahlung). Such continuum radiation contains little information about the physical state of the emitting plasma or the distance of the emitting radio source.

The discovery of the 21 cm hyperfine-structure line of atomic hydrogen in 1951 was a major breakthrough in radio astronomy. Observations of this line allowed measurements of radial velocities and, consequently, estimates of kinematic distances and hence of the distribution of the interstellar gas. Also, and equally important, observations of the 21 cm line allowed radio astronomers to investigate the physical state of neutral interstellar gas, which is a major constituent of our Galaxy and of galactic systems in general.

The first systematic 21 cm surveys, made in the Netherlands and in Australia, led radio astronomers to the conclusion that the interstellar gas forms a thin layer along the galactic plane, that the kinetic gas temperature was rather uniform at about 125 K, and that the density of the interstellar gas varied over a factor of perhaps ten, with an average value of about 0.7 H cm^{-3}. It was very difficult to understand how, under these conditions, stars, or small stellar clusters could form out of the interstellar gas.

In 1965, it became gradually clear that the distribution and temperature of the interstellar gas was much more inhomogeneous than previously thought, and that there were dense and cool clouds which obviously could not be investigated by means of the 21 cm line, either because the clouds are opaque, or because most of the atomic hydrogen contained in these clouds is tied up in molecules, especially the H_2-molecule.

The H_2-molecule, probably the most abundant molecule in interstellar space, has no observable transitions in the visual or radio range. However, in 1963, the first radio spectral lines from an interstellar molecule, the OH radical were identified at 18 cm (WEINREB *et al.*, 1963). Previously, interstellar CN and CH radicals were observed in absorption in the visual spectra of a number of stars. Detection of the OH 18 cm lines was followed, about five years later, by the detection of the NH_3 1.3 cm line (CHEUNG *et al.*, 1968). The year of 1968 marks the real beginning of a new era of radio astronomy, viz. the era of molecular radio spectroscopy. The number of interstellar molecules identified by means of their radio spectral lines increased nearly exponentially; to date, some 26 interstellar molecules are known and some five radio spectral lines still await identification. Up to fifteen different radio spectral lines of an individual molecule have been observed.

A variety of review papers on interstellar molecules have been published or will soon appear (see, e.g. RANK *et al.*, 1971; SNYDER, 1971, 1972, 1973; WINNEWISSER *et al.*, 1973). In this review paper, I want to stress the use of molecules as probes of the interstellar matter: firstly, as indicators of the physical state and radiation field in dense, cool clouds which are probably the first stages in the evolution of stars and stellar clusters; secondly, as tracers of the large-scale structure of the interstellar gas.

II. Observing Instruments and Observations

Although a number of very interesting observations, especially of the OH 18 cm and the H_2CO 6 cm lines, have been made with interferometers and aperture synthesis radio telescopes, the field of radio spectroscopy is, at present, dominated by single-dish parabolic reflector type radio telescopes. In fact, most of the recent work in molecular spectroscopy has been performed with the NRAO 140-ft telescope in Green Bank ($\lambda \geqq 1$ cm), the NRAO 36-ft telescope in Kitt Peak/Tucson ($\lambda \geqq 2$ mm) and the CSIRO 210-ft telescope in Parkes ($\lambda \geqq 3$ cm). To date, between 60 and 70% of the total observing time of these telescopes is devoted to radio spectroscopy. The angular resolution of these telescopes at the wavelengths of typical molecular lines ranges from 1 to 20 minutes of arc.

In theory, the sensitivity of a spectrometer is limited by its system noise temperature and by its long-term gain stability. System noise temperatures in the cm-wavelength range increase from some 30 K at low frequency to about 200 K at the frequency of the water vapor line. In the mm-wavelength range, the best Schottky-barrier mixers have noise temperatures of some 1000 K. Multi-channel spectrometers are gradually being superceeded by autocorrelation (AC) spectrometers which measure the power spectrum of the line profile. The profile proper is subsequently obtained by a Fourier transformation, performed by an on-line computer. The advantage of AC spectrometers is the great flexibility of their analysing bandwidth, frequency resolution and their long-term stability, which allows integration times of several tens of hours.

The actual limits of the measuring accuracy of spectral lines in the radio range, however, arise from differences in the instrumental baseline between "on-source" and "off-source" comparison observations. These baseline problems are the results of single and multiple reflections between feed, main reflector and (in two-reflector telescopes) subreflector, which affect feed match and antenna gain in a way which is not yet fully understood.

The molecules which have been identified in interstellar space to date are compiled in Table 1. The first column gives the chemical symbol and name of the molecule, the second column indicates in which wavelength range spectral lines of the molecule have been detected. In the third column, E stands for emission, ME for maser emission and A for absorption. The fourth column gives the ap-

Table 1. Observed interstellar molecules. (According to SNYDER, 1973)

Molecule	Wavelength range	Emission/ absorption	Number of reported sources	Column density molecules cm^{-2}
	A. Inorganic Instellar Molecules			
H_2-molecular hydrogen	UV	A	1 (ξ Per)	$\sim 10^{20}$
OH-hydroxyl radical	radio (cm)	ME/A	>200	$10^{12}-10^{16}$
SiO-silicon monoxide	radio (mm)	E	1 (Sgr B2)	$\sim 10^{13}$
H_2O-water	radio (cm)	ME	$>$ 50	?
H_2S-hydrogen sulfide	radio (mm)	E	$\sim$ 2	not known
NH_3-ammonia	radio (cm)	E	$<$ 10	$\sim 10^{16}$
	B. Organic Interstellar Molecules			
CH^+ } CH-radical	visible	A	$\sim$ 60	$\sim 10^{13}$
CH } CH-radical	visible	A	$\sim$ 40	$\sim 10^{13}$
CN-cyanogen radical	radio (mm)	E	3	$\sim 10^{15}$
	visible	A	14	$\sim 10^{12}$
CO-carbon monoxide	radio (mm)	E	many	10^{17}–10^{19}
	UV	A	1 (ξ Oph)	$\sim 10^{15}$
CS-carbon monosulfide	radio (mm)	E	4	10^{13}–10^{14}
OCS-carbonyl sulfide	radio (mm)	E	1 (Sgr B2)	$\sim 10^{16}$
HCN-hydrogen cyanide	radio (mm)	E	$\sim$ 10	10^{14}–10^{15}
H_2CO-formaldehyde	radio (cm)	A	>100	10^{12}–10^{16}
HNCO-isocyanic acid	radio (cm, mm)	E	1 (Sgr B2)	$\sim 10^{14}$
H_2CS-thioformaldehyde	radio (cm)	A	1 (Sgr B2)	?
HCOOH-formic acid	radio (cm)	E	1 (Sgr B2)	$<10^{13}$
HC_3N-cyanoacetylene	radio (cm, mm)	E	$\geqq$ 2	$\sim 10^{16}$
H_2CNH-methylene imine (formaldimine)	radio (cm)	E	1 (Sgr B2)	?
CH_3OH-methyl alcohol	radio (cm, mm)	E	$\geqq$ 3	
CH_3CN-methyl cyanide	radio (mm)	E	2	$\sim 10^{14}$
$HCONH_2$-formamide	radio (cm)	E	$\geqq$ 2	$>10^{11}$
$HCOCH_3$-acetaldehyde	radio (cm)	E	1 (Sgr B2)	$\sim 10^{14}$
CH_3C_2H-methylacetylene	radio (mm)	E	2	?
	C. Lines without Laboratory Identification [a)]			
X-ogen (C_2H?)	89,190 MHz	E	8	
"HNC" (hydrogen isocyanide)	90,665 MHz	E	$\geqq$ 5	

[a)] More than three unidentified lines not yet reported by various groups.

proximate number of sources in which the specific molecule or line has been observed. The last column gives the typical column density of molecules (see Section V). This table has been prepared by L. SNYDER (1973). Not included are a number of isotopic molecules.

H_2 is believed to be the most abundant molecule. It has observable transitions only in the far UV and IR and has been observed in the UV in the spectrum of one star (CARRUTHERS, 1970)*. Otherwise, the existence of this molecule in interstellar space is only inferred by indirect methods, e.g. absence of 21 cm line emission of atomic hydrogen from dark clouds etc.

Of all the other molecules, systematic surveys over large areas around the galactic plane have, to date, only been made for the OH radical (18 cm) and the H_2CO molecule (6 cm). Because both lines are seen predominantly in absorption against galactic continuum sources (HII regions and supernova remnants), the results may be biased.

Both the OH 18 cm and the H_2O 1.35 cm lines in non-thermal (maser) emission have been systematically surveyed in a large number of galactic sources, especially HII regions, SN remnants and IR stars. The line emission comes from sources of extremely small angular diameters.

Perhaps the most widely distributed molecule is CO. Because its 2.6 mm line is seen in emission, observational results should not be biased towards galactic sources. However, to date, there exist extensive maps only in the galactic center region and around some giant HII regions.

Column four of Table 1 shows that many molecules, especially the more complex ones, have been detected only in one or a few sources. These sources, which are sometimes referred to as "molecular clouds", are very dense and massive clouds of interstellar gas, located in spiral arms and in the very central region of our Galaxy. Because of the serious observational selection effects, one has to be very careful in converting "typical column densities" of molecules in the last column of Table 1 into relative molecular abundances.

Of special value as probes of the interstellar gas are molecules, such as CH_3OH, NH_3 or H_2CO, for which a number of transitions can be observed.

From the second column of Table 1, one can see that more than 50% of all interstellar molecules have been identified by means of their spectral lines in the mm-wavelength range. This is all the more surprising, since the effective antenna area of the 36-ft telescope is only 7% of the effective area of the 140-ft telescope and the system noise temperature of radiometers in the mm-wavelength range is, typically, by one or two orders of magnitude higher than in the cm-wavelength range. These facts suggest that, with better instrumentation becoming available, the mm-wavelength range will become more and more important for radio spectroscopy. It is, therefore, worthwhile to consider which types of molecular spectra are usually observed in the cm- and mm-wavelength range.

In the cm-wavelength range, one observes mainly K-type doublets which reflect the asymmetry of a molecule (e.g. H_2CO, NH_2CHO, CH_3CHO), hindred rotations (e.g. the inversion spectrum of NH_3), fine structure transitions like Λ-type doublets (e.g. OH), and R-type ($\Delta J = +1$) rotational transitions of heavier

* Note added in proof: The new OAO-C results show H_2 in substantial column densities in the spectra of any star with a color excess $E_{B-V} \geqq 0.03$ ($A_V \geqq 0.09$).

molecules (molecules with molecular weight $\geqq 40$; e.g. HNCO). In the mm-wavelength range, one observes mainly pure rotational transitions of R-type with low J-values of light molecules (e.g. HCN, H_2O, CO) and with high J-values of heavier molecules (e.g. OCS, CH_3CN, CH_3CCH).

Intensity considerations clearly favour the mm-wavelength range. For an optically thin line emitted in LTE (=Local Thermodynamical Equilibrium), the integrated line profile is determined by the absorption coefficient α' (corrected for stimulated emission) and by the excitation temperature T_{ex} (see Eq. 8a, b)

$$T_L = T_{ex}\,\tau_L = \int_0^L d\,l \int_0^\infty d\,\nu\,\alpha' \propto \frac{g_u}{g_l}\,\nu_{lu}^2\,|\mu_{lu}^2|\,e^{-\frac{E_l}{kT}}\,N \quad \text{for } \tau_L \ll 1 \tag{1}$$

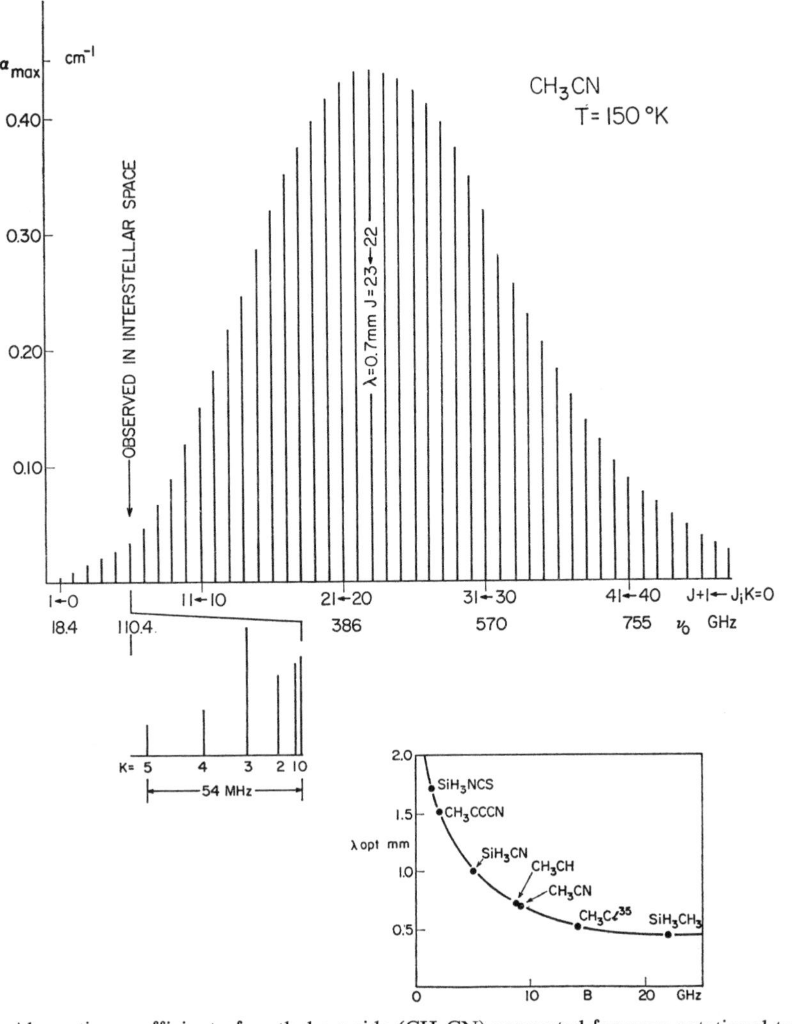

Fig. 1. Absorption coefficient of methyl cyanide (CH_3CN) computed for pure rotational transitions with $K=0$ (G. WINNEWISSER, private communication). Note that in the insert it should read CH_3C_2H rather than CH_3CH

with u and l referring to the upper and lower state of the transition, g the statistical weigth, $|\mu_{lu}^2|$ the square of the dipole moment matrix element, E_l the energy of the lower level and $N=\int n dl$ the column density of molecules along the line of sight. Evaluation of Eq. (1) shows that, for a given type of transition with lines both in the cm- and mm-wavelength range and for excitation temperatures prevailing in interstellar space, the peak line intensity occurs at mm-wavelengths.

The behaviour of the absorption coefficient α (i.e. not corrected for stimulated emission) for the pure rotational transitions (with $K=0$) of CH_3CN, computed for $T_{ex}=150$ K by G. WINNEWISSER (private communication), is shown in Fig. 1. The K-structure of the only transition observed to date from interstellar molecules, $J=6\rightarrow 5$, is also indicated. For this molecule the absorption peak occurs at $\lambda_{opt}=0.7$ mm. This optimum wavelength depends on both the excitation temperature and the molecular weight. For a given molecule, the optimum wavelength increases with decreasing temperature. For a given excitation temperature, the optimum wavelength decreases with increasing molecular weight, as shown in Fig. 1. Here, the molecular weight is reflected in the rotational constant B.

III. Interstellar Matter and UV Radiation Field

Most of our present knowledge of the distribution and physical state of the interstellar gas in our Galaxy comes from observations of the 21 cm hyperfine structure line of atomic hydrogen. The hydrogen gas extends out to distances of about 20 kpc from the galactic center. The thickness of the gas layer increases from about 100 pc in the galactic center region to about 200 pc between galactic radii 4 and 10 kpc and reaches values of about 600 pc in the outer regions of the Galaxy. The interstellar gas may account for about 10% of the total mass of our Galaxy. The main spiral structure, as defined by a high formation rate of massive stars, occurs between the 3 kpc arm and a distance of about 13 kpc from the galactic center.

Abundances and first and second ionization potentials of the most abundant elements are given in Table 2. The element abundances, with the exception of the abundance of He which is derived from our own radio observations (as quoted by SEATON, 1971), are those compiled by UNSÖLD (1972, private communication) and probably represent the best abundance values currently available. A comparison of Tables 1 and 2 shows that the interstellar molecules detected at present are composed of the most abundant elements, i.e. (in the sequence of their abundance) H, O, C, N, Si and S. Mg and Fe, whose abundances fall between Si and S have not been found in interstellar molecules, but olivine, which contains Mg and Fe, has been identified in interstellar dust grains by means of its spectral line in the far IR ($\lambda \simeq 10$ μm).

However, rough estimates of molecular abundances, derived from observed column densities (see Table 1), show clearly that there must be other factors besides the element abundances that govern the molecular abundances. As will be discussed in more detail in Section VI, the abundance of a given molecule probably reflects an equilibrium between formation and destruction rates; both of which depend on the UV radiation field and the interstellar dust.

Table 2. Abundances and ionization potential of the most abundant elements

Element	Abundance relative to hydrogen	First ionization eV	potential Å	Second ionization eV	potential Å
H	1.0	13.6	911.3		
He	1.0×10^{-1}	24.6	504.2	54.4	227.9
C	2.3×10^{-4}	11.3	1101.1	24.4	508.1
N	1.3×10^{-4}	14.5	853.0	29.6	418.8
O	6.0×10^{-4}	13.6	910.3	35.1	353.2
Ne	2.0×10^{-4}	21.6	574.9	41.1	301.6
Na	4.4×10^{-6}	5.1	2412.1	47.3	262.1
Mg	4.4×10^{-5}	7.6	1621.3	15.0	826.5
Al	2.2×10^{-6}	6.0	2071.1	18.8	659.5
Si	4.8×10^{-5}	8.1	1520.9	16.3	760.6
S	2.1×10^{-5}	10.4	1196.6	23.4	529.8
Ar	7.1×10^{-6}	15.8	786.6	27.6	449.2
Ca	2.5×10^{-6}	6.1	2028.1	11.9	1041.8
Fe	3.2×10^{-5}	7.9	1574.8	16.2	765.3

Element abundances—with exception of He—are average values from UNSÖLD (1972, private communication).
Ionization potential as given in ALLEN (1963).

The dependence of the interstellar extinction due to dust has been studied from the far IR to the far UV (i.e. to the Lyman continuum limit). Fig. 2 shows a typical extinction curve; it can be explained by a mixture of grains that includes at least two different grain sizes. The grains responsible for extinction in the optical range appear to be dielectric grains of typical grain size $a_1 = 0.15$ μm. To explain the extinction in the far UV, one needs typical grain sizes between 0.02 μm and 0.05 μm depending on the refractive index of the grain material. However, no material so far suggested for interstellar grains can explain the extremely high albedo which was observed in the far UV by WITT and LILLIE (1973) and which is shown in Fig. 3. The number ratio of small to large particles has to be of order 1000 in order to match the observed extinction curve. The total surface of the small grain particles could thus be by a factor of 10 larger than that of the large particles. This fact may be important if chemical surface reactions are considered.

It has been found (SAVAGE and JENKINS, 1972) that there is a proportionality relationship between the strength of the $L\alpha$-absorption line in the UV spectra of bright stars and the color excess of the stars; this confirms that gas and dust are, on the whole, well mixed. Assuming the ratio of total to selective absorption of $R = 3.0$, one obtains a relation between column density of hydrogen and visual extinction A_v

$$\frac{N_H}{A_v} = 1.7 \times 10^{21} \text{ [atoms cm}^{-2} \cdot \text{mag}^{-1}]. \tag{2}$$

The interstellar UV radiation field is made up of the dilute radiation of all the stars, attenuated by absorption by the interstellar dust. It may be calculated if the energy distribution of the stars, the extinction properties of the dust and the

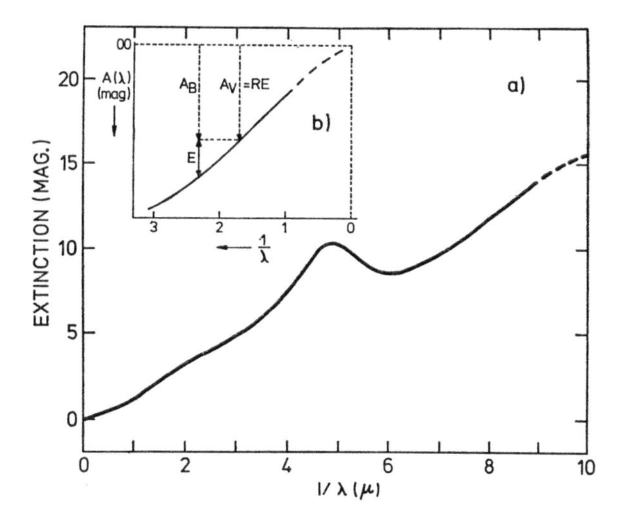

Fig. 2. Typical extinction curve of interstellar dust. Observations are now available from the far IR close to the Lyman continuum limit

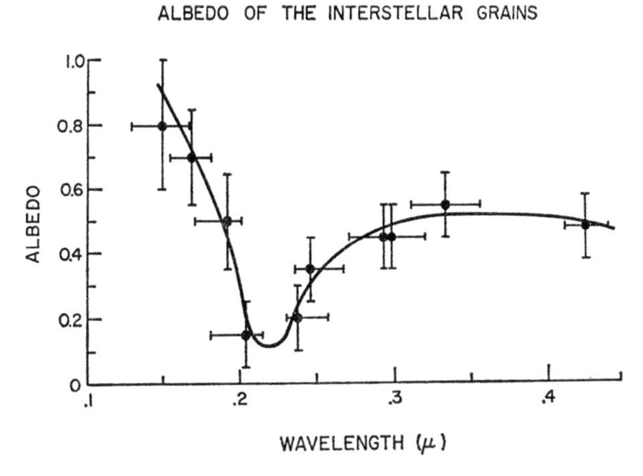

Fig. 3. Albedo of dust grains in the UV, as derived by WITT and LILLIE (1973) from OAO-2 measurements of the diffuse galactic light

relative distribution of stars and dust are known (or assumed). All estimates are made for the vicinity of the Sun, where at least the stellar distribution is well determined. The most recent estimate of the interstellar radiation density is that by WITT and JOHNSON (1973) (Fig. 4); it yields, in the wavelength range between 1000 and 2000 Å, values which exceed earlier estimates by HABING (1968) by a factor 2.

Inside a young association, or in the galactic center region, where many O-stars are near-by, the radiation density may be considerably higher than that in the solar neighborhood, especially in the far UV. Likewise, in dust clouds the radiation

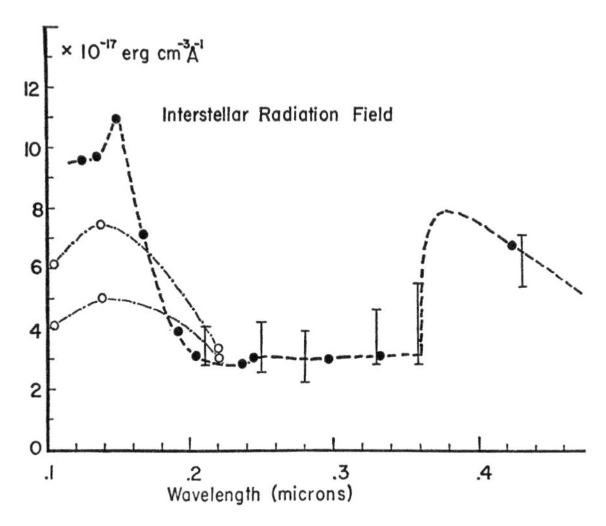

Fig. 4. Interstellar UV radiation field in the vicinity of the Sun, estimated by WITT and JOHNSON (1973) on the basis of the albedo of dust grains shown in Fig. 3. Open circles refer to earlier estimates by HABING (1968)

density may be less. The picture is complicated, however, by the fact that dense interstellar clouds, in which molecules are abundant, are also regions, in which stars are born. Thus, we do not know what the radiation field in these clouds is likely to be (see Section VI).

Since molecular abundances are determined by formation and destruction processes, molecular lines are not very useful in determining interstellar *element abundances*. However, the chemical properties of isotopic molecules are identical; hence, the abundance ratios of isotopic molecules do indicate the *isotopic element abundances*, which in turn will give us information about the nuclear processes by which elements were synthesized. At the Charlottesville Symposium on interstellar molecules, ZUCKERMAN (1971) has summarized interstellar abundance determinations of the elements C, O, N and S. Most of the isotopic element abundances determined from interstellar molecular lines (which still have large uncertainties) are compatible with the much more accurately determined terrestrial values. An important exception may be the C^{12}/C^{13}-ratio determined in the center region of our Galaxy, which appears to be considerably smaller than the terrestrial value of 90.

IV. Cloud Structure and the Formation of Stars out of the Interstellar Gas

About ten years ago, from a careful analysis of 21 cm line absorption and emission profiles, it was concluded that the distribution of interstellar gas is highly inhomogeneous, both in its density and its kinetic temperature (CLARK, 1965). Firstly, there are the large-scale structures such as the spiral arms (both density wave and material arms) and the nuclear disk (a rapidly rotating disk of gas, that extends from the galactic center out to a distance of about 800 pc). These large-scale structures are governed by gravitational forces. Superimposed on the large-scale structure is a "cloud structure" which is governed by the thermal properties of the interstellar gas.

For the interstellar gas in an equilibrium state, the kinetic temperature is determined by the condition that the energy input, which is primarily via ionization, is equal to the energy loss, which is primarily due to inelastic collisions, i.e. to collisional excitation of atoms or ions, resulting in radiation which escapes. The most important collisionally excited lines are Lyman α, at high temperatures, and CII $\lambda 156$ µm and CI $\lambda 610$ µm, at low temperatures. Since the interstellar gas in the interarm region has temperatures above 1000 K, the heating of the gas must be due to relatively energetic particles or photons such as subcosmic rays or soft X-rays. Once the ionization rate ξ is fixed, the temperature of the gas depends only on the density. Because the heating of the gas is proportional to the gas density, while the cooling by collisional excitation is proportional to the square of the density, the temperature of the gas decreases with increasing gas density. Detailed calculations predict (see, e.g. FIELD, 1970) that a hot and tenuous gas ($T \simeq 10^4$ K $n \simeq 10^{-1}$ cm^{-3}) can be in pressure equilibrium with cool and dense condensations ($T < 10^2$ K, $n > 10$ cm^{-3}). This model explains qualitatively the co-existence of dense and cool gas clouds in the hot interarm gas.

Observations confirm the existence of such "cloudlets" in the interstellar gas. The cloudlets have average masses of some 10 $M_\odot$ and an average extinction of $A_v \simeq 0\overset{m}{.}25$. They are observed by their 21 cm line absorption against continuum sources. In the denser and more massive cloudlets, one also observes the OH 18 cm and the H_2CO 6 cm lines in absorption. It is important to realize that this two-component structure of the interstellar gas is not correlated with the large-scale structure; it is common to both the spiral arm and the interarm gas.

However, there are other clouds that are much more dense and massive than the cloudlets; these appear to be predominantly located in spiral arms and in the nuclear disk. Densities and total masses of these clouds are so high that subcosmic particles and soft X-rays will be absorbed in the outer layers. Thus, the thermal balance of these clouds is not the same as that in the cloudlets. Fig. 5 shows a qualitative picture of the change of kinetic gas temperature in a gas cloud as a function of distance to the surface of the cloud. I have assumed a linear increase of the density towards the center of the cloud but have neglected attenuation of the subcosmic radiation. Thus, the decrease in temperature is primarily due to the increase in density. In the outer layers, the kinetic gas temperature is still determined by the heat input provided by ionizing subcosmic particles. At point A, however, this heat input becomes negligible compared to heat input by photons in the wavelength range between 912 and 1100 Å which can ionize carbon. This mechanism provides an equilibrium temperature of about 12 K, independent of the density. At point B, all photons that can ionize carbon are absorbed and the CII region ends. The temperature then drops to a value limited by the 3 K black body background radiation and controlled by the temperature of the grains and the ionization by high energy cosmic rays on the one hand, saturation of the cooling transitions on the other hand.

The temperature of dust grains is an equilibrium between heating by absorption of stellar light and re-emission in the far IR. In free interstellar space, dielectric

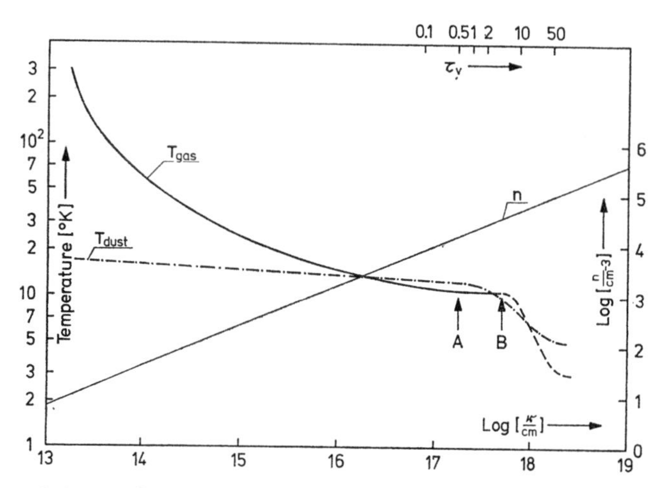

Fig. 5. Estimated change of kinetic gas temperature and equilibrium temperature of dust grains as a function of the distance r to the surface of the cloud. The calculation assumes a linear increase of density $n \propto r$, a constant ionization rate by subcosmic particles and penetration of UV photons $\lambda > 912$ Å into the cloud only from the outside. Absorption of subcosmic particles is neglected The upper scale indicates the optical depth of the dust grains in the visual

grains attain a temperature of about 17 K. With increasing optical depth of the gas layer, the heating of the dust grains and consequently their temperature, decreases, but this decrease is very slow. Only at optical depth of dust grains $\tau_v \simeq 50$ (in the visual) will the grain temperature drop to values of the order of 5 K (WERNER and SALPETER, 1970). Beyond point B, it is not quite clear if dust grains will provide the dominant heating or cooling mechanism of the gas.

As long as one does not know more about the interstellar particle and photon radiation fields which provide the heating of the interstellar gas, one should not put too much faith in estimated equilibrium temperatures of dense gas clouds such as shown in Fig. 5. However, at least qualitatively the result appears to be correct, that dense condensations in the interstellar gas have low kinetic temperatures. In fact gas kinetic temperatures as low as 5 K have actually been observationally determined in the center of dark clouds (see Section V).

Whenever the gravitational forces in a condensation become large enough to overcome the internal pressure of the gas, the condensation will begin to contract; the superiority of the gravitational forces is thereby increased and the contraction must eventually lead to the formation of stars. Gravitational contraction will start if the mass of a gaseous condensation of hydrogen density n_H and kinetic temperature T_k exceeds a critical value, known as Jeans mass

$$\frac{M_{\text{jeans}}}{M_\odot} = 32 \, \frac{T_k^{3/2}}{n_H^{1/2}}. \qquad (3)$$

In interstellar clouds of high density and low temperatures, stellar masses can according to Eq. (3), become gravitationally unstable and contract. Obviously, these clouds must represent the first stages in the evolutionary path which leads

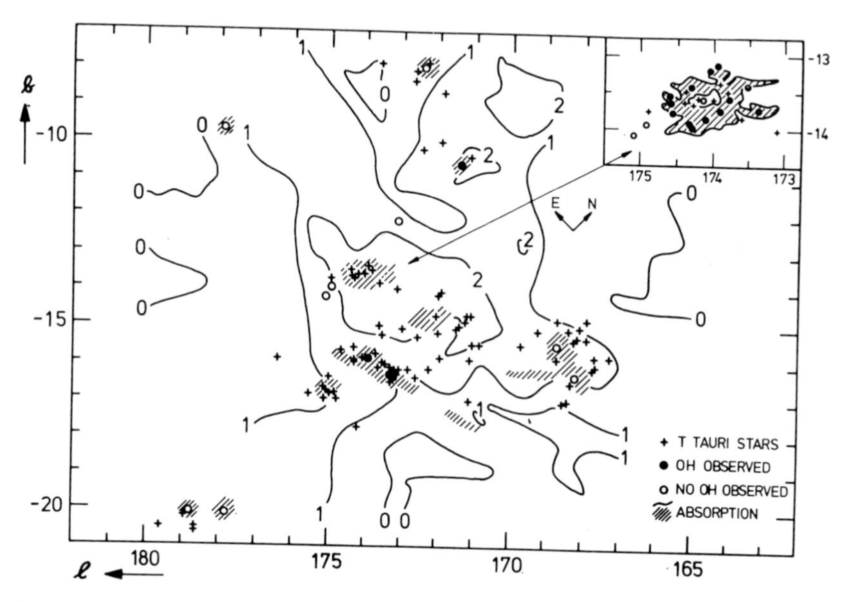

Fig. 6. Contour map representation of the Taurus cloud (MEZGER, 1971). Contours are lines of constant visual extinction (in mag.). Hatched areas indicate areas of increased extinction ($A_v \geq$ 5 mag), referred to as "dark clouds". Dots refer to OH observations. Crosses indicate positions of T-Tauri stars. Note correction of the galactic coordinates of the isophotes; 1971 diagram contained a zero point error

from condensations in the interstellar gas to main sequence stars. Computations show that, in denser interstellar clouds, most of the atomic hydrogen is tied up in H_2 molecules (HOLLENBACH *et al.*, 1971). It is, therefore, impossible to investigate the physical state of such a cloud by means of 21 cm line observations. Molecular lines, on the other hand, are excellent probes for the very cool and dense clouds. To me, it is one of the most fascinating aspects of molecular spectroscopy that it provides, at last, an observational method for the investigation of objects which lie along the evolutionary path which leads from a cloud of interstellar gas to a star cluster.

Within 1 kpc distance from the Sun, dense clouds can be observed optically by means of their extinction of the light of stars located behind the cloud. Fig. 6 shows, as an example, a contour map representation of the Taurus cloud, located in the direction of the anticenter and at a distance from the Sun of about 100 pc (MEZGER, 1971). The contour lines are lines of constant extinction in visual magnitudes. We see that the Taurus cloud extends over several tens of degrees, with an average extinction between 1^m and 2^m. There are smaller regions in this cloud (indicated in the diagram as hatched areas) where the extinction attains values of 5^m to 8^m. These are the "dark clouds" proper referred to in the reports of many radio astronomical investigations. In these dark clouds the two central OH 18 cm lines are observed in quasi-thermal emission, the CO 2.6 mm line is seen in LTE emission and the H_2CO 6 cm line is seen in anomalous absorption against the

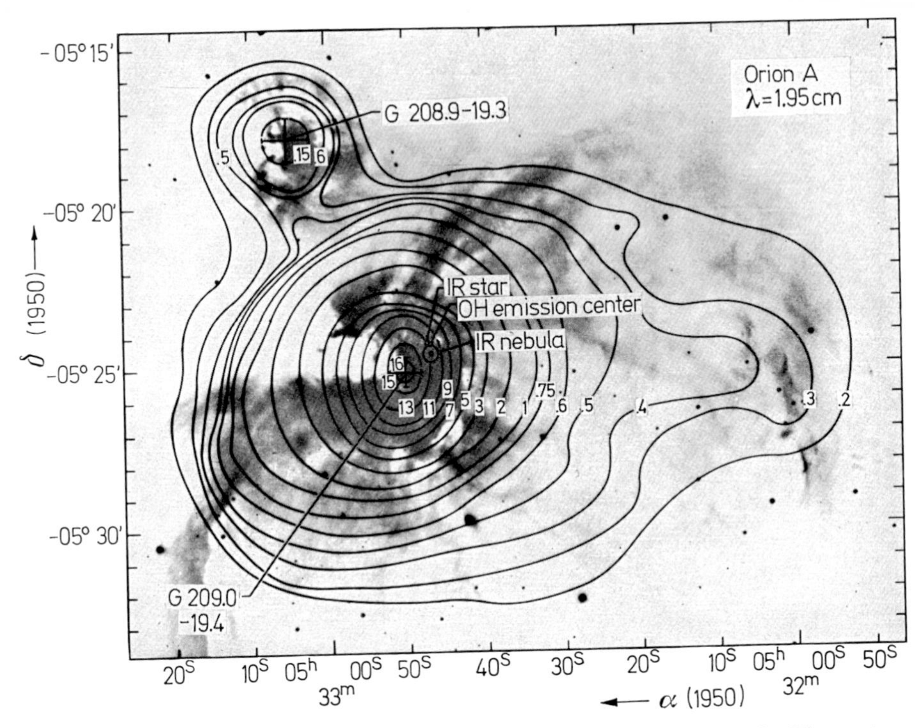

Fig. 7. Overlay of radio contours on an $H\alpha$-photograph of the Orion Nebula. The contours pertain to the free-free emission of the plasma. Indicated are the positions of OH/H_2O emission and of the Becklin-Neugebauer IR star, which are at the center of the Orion molecular cloud. This molecular cloud has an apparent diameter of $\simeq 50$ minutes of arc and appears to be located behind the Orion Nebula

3 K background radiation. Interpretation of the emission lines indicates that densities in dark clouds are of order 10^4 particles cm^{-3} and kinetic gas temperatures are as low as 5 K (MEZGER, 1971; HEILES, 1971). Eq. (3) tells us that under those conditions single stellar masses can become gravitationally unstable and contract. At least in some cases, the contraction will lead to the formation of stars of lower mass, as wittnessed by the presence of many T-Tauri stars, whose positions in the Taurus cloud are indicated by crosses in Fig. 6.

Why have only low-mass stars formed in clouds of the Taurus type? Frankly, we do not know. From the width of molecular lines, $\simeq 1$ km.s^{-1}, we can only conclude that turbulence in the dark clouds is relatively high as compared to the kinetic gas temperature. If we want to learn more about the formation of massive stars, we have to investigate molecular clouds in the vicinity of compact HII regions; these are believed to be the ionized shells of recently formed O-stars.

Fig. 7 shows the Orion nebula. The contour lines pertain to the free-free radio emission. About 1 minute of arc northwest of the Trapezium, there is a strong source of IR emission and several sources of maser emission of the OH 18 cm and the H_2O 1.3 cm lines. By maser emission, I mean a quasi-coherent emission of these lines from sources of apparent diameter of 10^{-3} seconds of arc and less. The equivalent excitation temperatures of these lines range from 10^{11} to 10^{13} K. The Trapezium star cluster is the youngest of four recognizable sugroups in the Orion OB star association. The Trapezium stars probably reached the main sequence only some 10^4 years ago. The Orion nebula itself is the ionized remnant of the protocluster out of which the O-stars were formed.

The existence of the maser sources puzzled us for quite a while. These objects have about stellar mass and the gas densities within them are high, somewhere between 10^9 and 10^{12} particles cm^{-3}, but the gas kinetic temperature can not be higher than some 100 K. How could such a neutral condensation survive if it were embedded in an HII region? Subsequent observations showed that the Maser and IR sources are not inside the HII region, but form the center of a dense molecular cloud, which apparently is located behind the Orion nebula. This molecular cloud shows emission lines of a large number of molecules, including some of the most complex molecules found to date in interstellar space.

The proximity of the center of this molecular cloud to the center of the Orion Nebula appears to be a projection effect. I have suggested that this dense molecular cloud, with molecular maser and with IR sources at its center, is the fifth and youngest stellar subgroup in the Orion association, and that we see it in a stage where its O-stars are in their pre-main sequence contraction (MEZGER, 1971).

In genuine density-wave spiral arms or in the center of the nuclear disk cloud, the formation of O-stars occurs at a much higher rate than in the Orion O-star association. Giant HII regions in spiral arms, such as W49, W51 or G0.7–0.1 in the nuclear disk, are composed of a number of compact HII regions each of which needs one or a few O-stars to account for its ionization. Whereas the formation of O-stars in the subgroups of the Orion association occurs at random and at quite different times, the O-stars in giant HII regions all form at approximately the same time.

What are the type of clouds out of which giant HII regions and their associated O-stars form? This question has been investigated by SCOVILLE (1972a) by means of the CO 2.6 mm line. He finds that the column density of hydrogen inferred from the CO 2.6 mm line observations ranges from 10^{22} to 3×10^{23} H cm^{-3}. Such clouds would have a visual extinction between 6^m and 180^m and the term "black clouds" used by SCOVILLE to discriminate these objects from the "dark clouds" in regions like the Taurus cloud is certainly justified. The line widths observed in black clouds are 4 to 40 times wider than line widths of about 1 $km.s^{-1}$ typical of dark clouds. These large line widths appear to be the result of large-scale motions; in fact, for one of these black clouds in the nuclear disk, observations made with high angular resolution revealed a quasi-rigid-body rotation. Kinetic gas temperatures in black clouds appear to be higher than in dark clouds, a fact which implies that they contain some sources of heating. The average density in black clouds is of order 10^3 particles cm^{-3} but, in some condensations, may reach values as high as 10^5 to 10^7 particles cm^{-3}. Radii of black clouds range from 10 to 30 pc and masses range from 10^4 to 10^6 $M_{\odot}$.

In some cases, these black clouds are associated with giant HII regions. Obviously, this means that part of such a cloud has collapsed and O-stars have formed. However, the total amount of ionized hydrogen is usually small compared to the total amount of neutral gas contained in the black cloud. In these black clouds, the problem is not so much how massive stars form but rather what prevents the rest of the cloud from gravitational contraction also.

V. Molecular Lines as Probes of the Physical State of Cool and Dense Interstellar Clouds

How can the physical state of these black clouds be inferred from observations of molecular lines? The brightness temperature of an emission line depends on both its excitation temperature and its optical depth

$$T_L = \{T_{ex} - T_{bb}\}(1 - e^{-\tau}) \tag{4}$$

with $T_{bb} = 2.7$ K the brightness temperature of the microwave background radiation. The excitation temperature is (u=upper, l=lower)

$$T_{ex} = \begin{cases} T_{kin} & \text{for } C_{ul}/A_{ul} \gg 1 \\ T_{bb} & \text{for } C_{ul}/A_{ul} \ll 1 \end{cases} \tag{5}$$

with A_{ul} the Einstein transition probability for spontaneous emission and C_{ul} the probability of a collisionally induced transition. If radiative transitions dominate $T_{ex} \simeq T_{bb}$ and the line becomes undetectable. If collisions dominate (in dark or black clouds the collisions will be predominantly with H_2 molecules), the excitation temperature becomes equal to the kinetic gas temperature of the colliding particles. The condition $C_{ul}/A_{ul} \gg 1$ implies that the density n_{H_2} of the colliding particles must be

$$n_{H_2} \gg \frac{A_{ul}}{\langle \sigma v_{th} \rangle}. \tag{6}$$

Here, σ is the cross section for a collisionally induced transition, v_{th} is the thermal velocity of the colliding particles and $\langle \sigma v_{th} \rangle$ is averaged over a Maxwellian velocity distribution. These cross sections are not very well known; I have adopted a value $\sigma \simeq 10^{-16}$ cm^2 and estimated in Table 3 the H_2-densities which are necessary to give $T_{ex} = 0.9\ T_{kin}$ for some of the observed interstellar molecular lines (for the detailed computations see e.g. MEZGER, 1973). For a detection of a molecular line, $T_{ex} \simeq 0.5\ T_{kin}$ may be a sufficient condition and the required densities are decreased by about a factor of ten. In these estimates, I have assumed that excitation by collisions with free electrons is negligible. If, however, all carbon is ionized, electron collisions will dominate the excitation and this decreases the required densities again by about an order of magnitude. It is important to realize that some lines, such as the OH 18 cm line or the CO 2.6 mm line, are thermalized at relatively low densities and consequently are ideal "thermometers" of interstellar gas of medium density $n_{H_2} \simeq 10^3 - 10^4$ cm^{-3}; mm-lines of H_2CO, NH_3 and HCN, on the other hand, require much higher densities and thus are ideal probes of very dense clouds ($n_{H_2} \gtrsim 10^5$ cm^{-3}).

For a line to be seen in absorption, the background temperature T_{bb} plus the brightness temperature of a continuum source T_c located behind the cloud, must exceed the excitation temperature.

$$T_L = \{T_{ex} - (T_{bb} + T_c)\}(1 - e^{-\tau}). \tag{7}$$

Table 3. H_2 densities required to raise the excitation temperature of a given molecule to 0.9 T_k, the kinetic gas temperature. Densities are computed for $\sigma = 10^{-16}\,\text{cm}^2$ and $T_k = 20$ K. (From MEZGER 1973)

Emitting molecule	Line frequency in GHz	$\left[\frac{H_2}{cm^3}\right]$
H_2CO	4.83	5.1×10^6
	141	3.1×10^7
CO	115	4.3×10^5
HCN	88	1.86×10^6
OH	1.67	3.8×10^4
NH_3	23.7	5.8×10^6

In the case of 21 cm line observations, one combines absorption and emission profiles observed in nearly the same direction to determine both optical depth and excitation temperature. In the case of molecular lines, this procedure does not work and other methods have to be used.

If the intensity of two lines whose ratio of strength is theoretically known, e.g. the central OH 18 cm lines or the $C^{12}O$ and $C^{13}O$ 2.6 mm lines are measured, one obtaines two equations of type (4) which can be solved for T_{ex} and τ. In special cases there are also other methods: If $\tau \gg 1$ then $T_L = T_{ex}$; or if T_c, the brightness temperature of the continuum source, varies across the absorbing cloud, one can determine for which value of T_c the brightness temperature T_L goes to zero and thus determine $T_{ex} = T_{bb} + T_c$. All these methods have been more or less successfully used to estimate excitation temperatures of molecular lines. If a line is seen in absorption and $T_c \gg T_{ex}$, or if $T_{ex} - T_{bb} \simeq 0$, then Eq. (7) yields $T_L = -T_c(1 - e^{-\tau})$, and with T_c known τ and consequently the column density $N_m = \int_0^L n_m\, dl$ of the corresponding molecule can be estimated, since

$$\frac{N_m}{T_{ex}} = 6.205 \times 10^{-10} \frac{\nu_L}{A_{lu}} \frac{Q}{g_u e^{-E_l/kT}} \tau \Delta\nu_L. \tag{8a}$$

Substitution of the well known relation between Einstein transition probability and the square of the dipole matrix element

$$A_{ul} = \frac{64\pi^4 |\mu_{ul}|^2 \nu_3}{3hc^3} \tag{8b}$$

in Eq. (8a) yields Eq. (1). In Eq. (8a) ν_L and $\Delta\nu_L$ are the center frequency and the half power width of the line in Hz, Q is the partition function and E_l the energy of the lower level to which the transition goes. Relevant parameters for some molecular lines are given in Table 4.

Eq. (8a) assumes LTE for the energy levels involved and, to evaluate N_m, the excitation temperature must be known. Since deviations from LTE are very common for radio molecular lines, and since excitation temperatures for most molecular lines are only rough estimates, the column densities given in Table 1 are highly uncertain.

Table 4. Some data pertaining to molecular radio spectral lines. (From MEZGER, 1973)

Molecule	Transition	$\left[\frac{\nu_L}{\text{MHz}}\right]$	$\left[\frac{A_{lu}}{\text{s}^{-1}}\right]$	$Q/g_u \exp\{-E_l/kT_e\}$
$O^{16}H$	$J=3/2, \quad K=1, \quad F=1-2$	1,612.231	1.29×10^{-11}	16/3
	$F=1-1$	1,665.401	7.11×10^{-11}	16/3
	$F=2-2$	1,667.358	7.71×10^{-11}	16/5
	$F=2-1$	1,720.533	0.94×10^{-11}	16/5
$H_2C^{12}O^{16}$	$JK_aK_c=1_{10}-1_{11}, \quad \nu_0(F=2-2)$	4,829.660	3×10^{-8}	$(1/0.11)T_e$
	$2_{11}-2_{12}$	14,488.65	3×10^{-7}	
	$2_{12}-1_{11}$	140,839.53	5.3×10^{-5}	$(1/1.75)T_e$
H_2O^{16}	$JK_aK_c=6_{16}-5_{23}, \quad \nu_0(F=7-6)$	22,235.08	2×10^{-9}	
$C^{12}O^{16}$	$J=1-0$	115,271.2	6×10^{-8}	$\simeq\frac{T_e}{3}$
$C^{12}N^{14}$	$J=1-0, \quad F={}^5/_2-{}^3/_2$	113,492.0	1.31×10^{-5}	$(1/2.72)T_e$
$HC^{12}N^{14}$	$J=1-0, \quad F=1-1$	88,630.416	2×10^{-7}	$\sim(1/2.12)T_e$

In the preceeding section, I have summarized our present knowledge of the cloud structure of the interstellar gas. Now I will summarize in more detail the observational results pertaining to some of the denser clouds.

a) The protostellar Cloud in the Orion OB-Association

I have mentioned this cloud in the preceeding section as an example of a protostellar subgroup. WILSON *et al.* (1970) first discovered that the OH/H_2O maser sources and the IR star and IR nebula (Fig. 7) are located at the center of a molecular cloud whose CO 2.6 mm emission extends out to at least 25 minutes of arc. The color temperature of the IR star is 700 K; it is surrounded by an IR nebula of angular extent $\simeq$15 seconds of arc, whose IR emission is probably thermal radiation from dust grains. The dust grains are believed to be heated by absorption of radiation emitted by the IR star in the center of the nebula. The grain temperature is about 70 K. The gas kinetic temperature of the molecular cloud can be estimated from the CO 2.6 mm line which, at least in the central part, is optically thick, and consequently $T_L \simeq T_{ex}$ according to Eq. (4). Because of its small dipole moment, the 2.6 mm line of the CO molecule is thermalized at relatively modest densities (see Table 3). The observations by WILSON *et al.* show, that the kinetic temperature in the central part of the cloud, close to the IR nebula, attains values as high as 60 K, but that it decreases with increasing distance from the center.

The schematic representation (Fig. 5) of the physical state of a dense cloud shows a steady decrease of the temperature of both gas and dust with increasing distance from the surface of the gas and dust layer. This decrease of temperature is a consequence of the assumption that the ionization rate is constant while the gas density increases towards the center of the cloud. In the case of the Orion protostellar cluster, we are obviously wittnessing, in its center, the pre-main sequence evolution of a very massive object, probably similar to that out of which the Trapezium stars in the Orion nebula have been formed (MEZGER, 1971). It appears to be this object which provides the additional heat input to the gas in the molecular cloud. Alternatively, saturation of the cooling lines may account for the increasing gas temperature.

Because of its opaqueness and thermalization at relatively low densities, the CO 2.6 mm line is not the appropriate probe for an investigation of the dense core of this cloud. Recently, THADDEUS *et al.* (1971) have detected the 2 mm lines of ortho- and para-formaldehyde in an area 3 × 5 minutes of arc around the center of the molecular cloud. On the assumption of collisional excitation of these lines, a space density $n_{H_2} \simeq 2 \times 10^5$ cm^{-3} and a column density of $N_{H_2} \simeq 10^{23}$ cm^{-2} for the central core of the cloud is estimated. Its total mass would be of order 200 $M_\odot$ and its extinction in the visual $A_v \simeq 100^m$. These results clearly demonstrate how powerful a tool molecular lines are for an investigation of the physical state of very dense protostellar clouds. Once the emission mechanisms of the OH/H_2O masers are better understood, we may actually be able to investigate individual massive protostars in their pre-main sequence contraction.

b) Shells Surrounding Compact HII Regions

Compact HII regions represent evolutionary stages of O-star clusters in which the O-stars have recently reached the main sequence and now emit sufficient UV Lyman continuum radiation to ionize the remnants of their protostellar clouds. This leads to the formation of compact shells of ionized gas with electron densities as high as 10^5 cm^{-3} and kinetic temperatures of about 10^4 K. The hot plasma subsequently expands and, with decreasing density, more gas can be ionized. The expanding ionization front is preceded by a shockwave which compresses and heats the neutral gas. The theoretical investigation of the expansion of an HII region into a neutral gas predicts the existence of dense shells of neutral gas surrounding HII regions. Likewise, supernova remnants are supposed to lead to the formation of shells of neutral gas. Efforts have been made to observe these shells by means of 21 cm line emission and absorption, but observational results regarding the existence of such shells were rather inconclusive.

Today, we know that these attempts had to fail, since most of the hydrogen gas surrounding compact HII regions is tied up in molecules, and molecular lines, therefore, are more adequate for an investigation of these shells. DICKEL (1973) mapped the gas surrounding the supernova remnant W44 and the compact HII region W3 by means of its OH 18 cm and H_2CO 6 cm absorption. She finds dense clouds surrounding both the supernova remnant and the HII region; however, the thickness of the shell is not resolved by the single dish telescopes used. In the case of W3, a comparison of radial velocities of radio recombination lines emitted by the HII region and the H_2CO 6 cm line absorbed by the surrounding gas confirms the spatial association and also indicates similar large-scale motion (e.g. rotation) of neutral and ionized gas. It thus appears that, in the case of W3, the inner core of a massive cloud collapsed and formed stars. The O-stars, in turn, ionized the surrounding gas and in this way halted or even reversed the contraction of the outer parts of the cloud.

Other observations indicate a similar association between the giant HII regions W49 and W51 and dense molecular clouds whose total masses are an order of magnitude larger than those of the corresponding HII regions (SCOVILLE, 1972a).

This shows, once more, that apparently only a relatively small fraction of a dense cloud collapses and forms stars.

c) Molecular Clouds in the Galactic Center Region

Some of the densest and most massive clouds have been observed in the center region of our Galaxy (see Section VII). Here I summarize detailed observations of two of these clouds: The $+40$ km.s^{-1} cloud in front of Sgr A ($l \simeq 0°$) and the -60 km.s^{-1} cloud associated with the giant HII region G0.7–0.1 (also known as Sgr B2) at $l \simeq 0.7°$ (see Fig. 10). These clouds were detected by means of their OH 18 cm (ROBINSON and McGEE, 1970) and H_2CO 6 cm absorption (SCOVILLE *et al.*, 1972). The absence of 21 cm line absorption by atomic hydrogen indicates that most of the atomic hydrogen in these clouds is tied up in molecules. The details of the spatial distribution of the molecules in these clouds was obtained by means of lunar occultations (KERR and SANDQVIST, as summarized by SANDQVIST, 1974) and by aperture synthesis techniques (FOMALONT and WELIACHEW, 1973).

The strengths of these absorption lines indicate average densities of $n_{H_2} \simeq 10^4$ cm^{-3} and total masses of some 10^5 $M_\odot$. However, these may be underestimates, since both the H_2CO 6 cm line and the CO 2.6 mm line appear to be optically thick at least in the center region of this cloud. To investigate the physical state of the inner core of the Sgr B2 cloud, ZUCKERMAN *et al.* (1971) used NH_3 1.3 cm lines; these involve nonmetastable rotational levels with lifetimes for spontaneous emission of only $\simeq 50$ s. Collisional excitation of these lines would, therefore, require densities of $n_{H_2} \geqq 10^9$ cm^{-3}; however, photon trapping could reduce the required densities considerably. ZUCKERMAN *et al.* estimate for the core of the Sgr B2 cloud densities $n_{H_2} \geqq 10^7$ cm^{-3} and kinetic temperature $T_k \simeq 50$ K; the size of this core is $\geqq 1.5$ pc and, hence, the total mass is $\geqq 10^6$ $M_\odot$. The association of this cloud with the giant HII region G0.7–0.1 tells us that part of this cloud has already collapsed and O-stars have reached the main sequence.

VI. Molecules as Probes of the Interstellar Radiation Field

Let us now consider how molecules form and under which conditions they can exist in interstellar space. I mentioned that molecular abundances in interstellar space do not necessarily reflect the element abundances of their constituents. This is an indication that molecule formation is not primarily determined by collision rates.

The dissociation energy of most molecules is well below that of the Lyman continuum limit. Fig. 8 shows the lifetime of molecules in interstellar space computed by STIEF *et al.* (1972) on the basis of extensive laboratory measurements. In unprotected space, any molecule except CO would be destroyed in less than 100 years. If, however, a molecule is protected from stellar UV radiation by a dust layer whose thickness in this diagram is described by the corresponding visual extinction, its lifetime increases rapidly. This explains, qualitatively, the association of molecules with dark clouds.

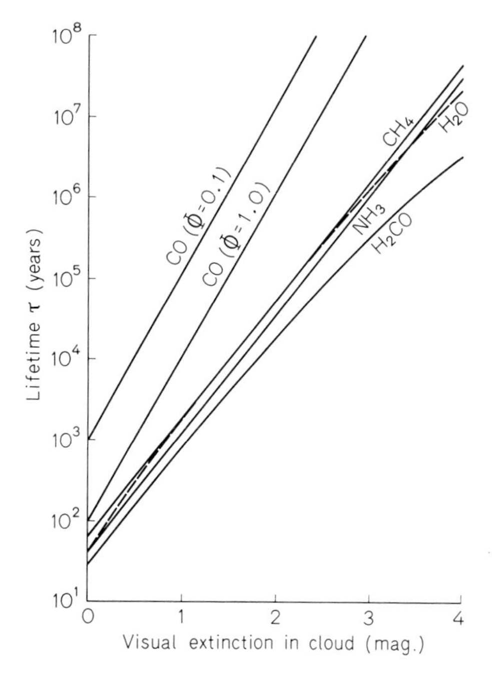

Fig. 8. Estimated lifetimes of a number of molecules observed in interstellar space. These lifetimes are estimated on the basis of laboratory measurements of dissociation rates, as a function of the dust layer (expressed in mag of visual absorption) which protects the molecules from the interstellar UV radiation field (STIEF *et al.*, 1972)

In my Saas Fee lectures (MEZGER, 1973, Section IV) I have summarized our present knowledge about the formation of interstellar molecules. Another excellent summary is given in a series of papers (KLEMPERER, 1971; SALPETER, 1971; BREUER and MOESTA, 1971; WICKRAMASINGHE, 1971; WINNEWISSER *et al.*, 1973). For the formation of molecules reactions in the gas phase are relatively unimportant compared to reactions on the surface of grains. Catalytic surface reactions which lead to the formation of molecules on dust grains have been investigated theoretically by SALPETER and associates (HOLLENBACH and SALPETER, 1971; HOLLENBACH *et al.*, 1971; WATSON and SALPETER, 1972a, 1972b). However, the molecules, once formed on grains, can only be desorbed by absorption of UV radiation shortward of 3000 Å. BREUER and MOESTA (1971) have investigated, experimentally, photochemical reactions on grain surfaces which lead to both the formation and desorption of molecules.

Thus, UV radiation is needed for the formation and desorption of interstellar molecules, but UV radiation will also destroy molecules. Therefore, interstellar molecules must represent an equilibrium between formation and destruction. If the UV radiation density is too high, as is the case in unprotected interstellar space, then destruction will be the dominant process. However, if the normal stellar UV radiation is completely shielded then we expect that the molecules will completely condense on grain surfaces.

The theory by WATSON and SALPETER (1972b) predicts complete condensation if the parameter (not to be confused with the ionization rate)

$$\xi = \left[\frac{n}{100\,\mathrm{cm}^{-3}}\right] e^{2.5\,\tau_v} > 10^4. \tag{9}$$

Here, n is the gas density in units of 100 cm^{-3}, τ_v is the optical depth in the visual from points inside a cloud to the nearest point of its surface, and $e^{2.5\tau_v}$ is the attenuation of UV radiation by the dust layer.

By substitution of appropriate space and surface densities, we find that, in practically all dark and black clouds, the condition $\xi > 10^4$ is fullfilled and, therefore, all molecules should have condensed on grains. The fact, that we actually see molecules in these clouds, with a higher abundance than anywhere else in interstellar space, is another indication that the interior of these clouds is not simply black and cool, but that many newly formed low-mass stars or massive pre-main sequence stars maintain a relatively high density of UV radiation inside these clouds.

The equilibrium condition between formation and destruction, and, therefore, the abundances of different molecules, appears to be mainly determined by the characteristics of dust grains and by the UV radiation. Once the processes which lead to the formation and desorption of molecules on grain surfaces are quantitatively better known, we may be able to use molecules not only as probes of the physical state of the interstellar gas but also as probes of the radiation field that prevails in dark and black clouds. I feel that this knowledge is highly important if we want to understand the physical processes which lead to the formation of stars.

VII. Molecules as Tracers of the Galactic Large-Scale Structure: The Gas in the Galactic Center

Let us consider the use of molecules as tracers of the large-scale structure of the interstellar medium.

Molecular clouds often represent early phases in the evolution of stars and stellar clusters. The spiral arms are regions of very active star formation. Molecular clouds, together with giant HII regions, may turn out to be the best tracers of the large-scale spiral structure of the Galaxy. The CO molecule, because of its long lifetime in interstellar space (see Fig. 8) and because the 2.6 mm line is collisionally excited at medium densities, should play a particularly important role in unraveling the galactic large-scale structure. Likewise the OH 18 cm and the H_2CO 6 cm lines, seen in absorption against galactic HII regions, will help to resolve the ambiguity of kinematic distance determinations for giant HII regions located within the solar circle.

The central region of the Galaxy also contains molecular clouds and regions of active star formation. In this section, I wish to show how, by means of molecular lines, radio astronomers begin to unravel the structure and kinematics of the gas in the center region. Inside the 3 kpc arm the surface density of atomic hydrogen decreases; then, at a distance of about 1 kpc, from the galactic center, both surface and volume densities of the interstellar gas begin to rise again. ROUGOOR and OORT (1960) found that the gas within 800 pc of the galactic center forms a thin and rapidly rotating disk, which is generally referred to as "the nuclear disk". These authors also suggested the existence of a dense star cluster in the center of the Galaxy. The existence of this cluster was later confirmed by IR observations at 2.2 μm (BECKLIN and NEUGEBAUER, 1968). The total mass of the star cluster is

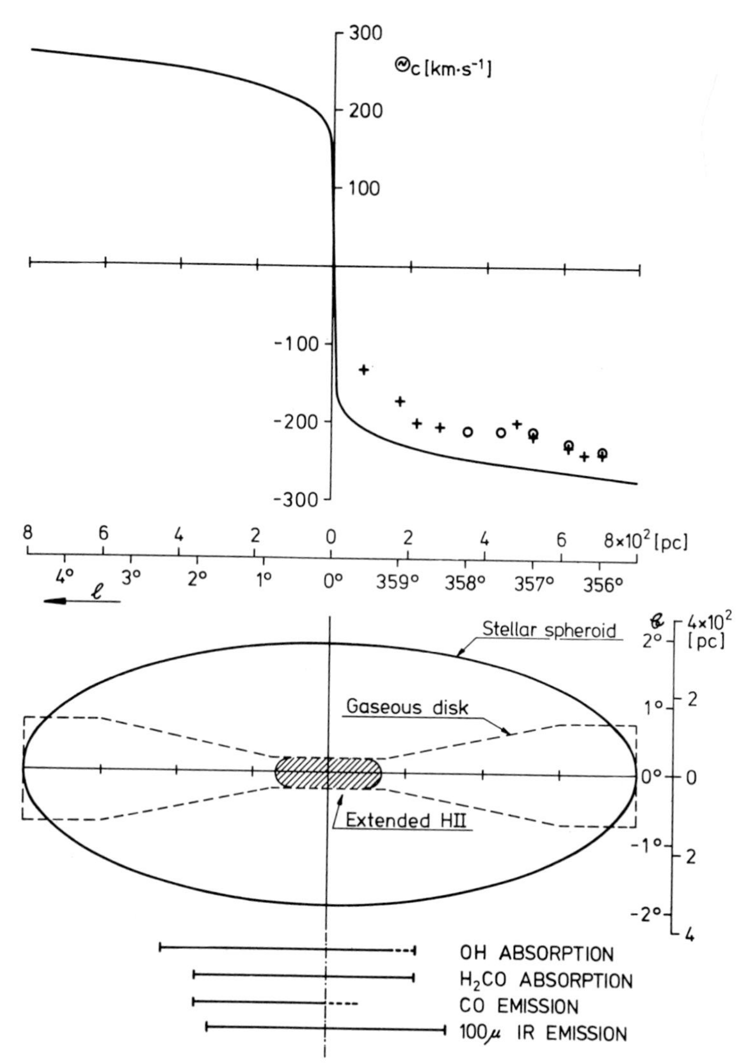

Fig. 9. Schematic representation of star cluster and gaseous "nuclear" disk in the center region of the Galaxy. Approximate extent of star cluster and nuclear disk are shown in the lower part. The extent of molecular lines in galactic longitude, as outlined by different molecules, is indicated below, together with the extent of λ 100 μm IR emission. The hatched area indicates the extent of diffuse thermal radio emission. The equilibrium rotation curve computed by SANDERS and LOWINGER (1972) is shown in the upper part, together with observed radial velocities of the gas in the nuclear disk

estimated to be about 10^{10} $M_\odot$ (SANDERS and LOWINGER, 1972). Since the total mass of the observed gas amounts to only 10^7 $M_\odot$, the gravitational field of the central cluster dominates the dynamics of the central region.

Fig. 9 is a schematic representation of the spheroidal star cluster and the gaseous disk; the latter has a width of only about 80 pc in the central part. The rotation curve, shown in the upper part of Fig. 9, is the equilibrium curve for circular rotation, computed by SANDERS and LOWINGER and extrapolated on the basis of a mass distribution derived from the observed IR 2.2 μm brightness distribution. Crosses and points are observed radial velocities of the gaseous disk derived from 21 cm line observations (ROUGOOR, 1964). The difference between observed and computed rotational velocities of the gas, if real, may be due to the fact that the gas has a significant turbulent velocity. A rather high turbulent velocity is also indicated by the finite width of the gas layer (80 to 250 pc) despite the presence of the strong gravitational field of the star cluster perpendicular to the galactic plane.

The gas in the inner part of the Galaxy appears to be composed mostly of molecules or ions. Thus, 21 cm is not the most useful tool for investigating the properties of this region. Systematic surveys of molecular lines have been made for the OH 18 cm (ROBINSON and MCGEE, 1970) and H_2CO 6 cm (SCOVILLE *et al.*, 1972) lines in absorption and for the CO 2.6 mm line in emission (at positive velocities) (SCOVILLE, 1972a). It was found that the molecules concentrate to dense, massive clouds. Their distribution as indicated by bars in the lower part of Fig. 9 is very asymmetric with respect to the dynamical center of the Galaxy—most of the molecules are found at positive galactic longitudes. The distribution of ionized gas is indicated by the hatched area. This paper is concerned with the molecules; I report on the physical state and dynamics of the ionized gas elsewhere in this meeting (MEZGER *et al.*, 1974).

Fig. 10 shows contours of equal optical depth determined by SCOVILLE (1972b), for the H_2CO 6 cm line seen in absorption against the diffuse galactic background radiation. The ordinates of the figure are galactic longitude and velocity with respect to the local standard of rest. Features at +50 km and near zero velocity are partly due to the expanding 3 kpc arm and partly due to clouds in local spiral arms. The remaining clouds have an approximately elliptical distribution; SCOVILLE gives the two curves labelled I and II as limiting approximations to the actual cloud distribution, which is highly asymmetric with respect to the dynamical center $l=0°$, $v_{LSR}=0$ km.s^{-1}. An ellipse in this diagram corresponds to a rotating and expanding (or contracting) ring; on observational grounds SCOVILLE believes that most of the observed clouds are moving away from the center. The parameters for the two fitted ellipses are given in Table 5; Case II refers to the ellipse with the higher excentricity.

Table 5. Parameters of expanding rings in the galactic center according to SCOVILLE (1972b)

	$[r/\text{pc}]$	$[\theta_c/\text{km.s}^{-1}]$	$[v/\text{km.s}^{-1}]$
Case I	218	50	(±) 145
Case II	305	50	(±) 70

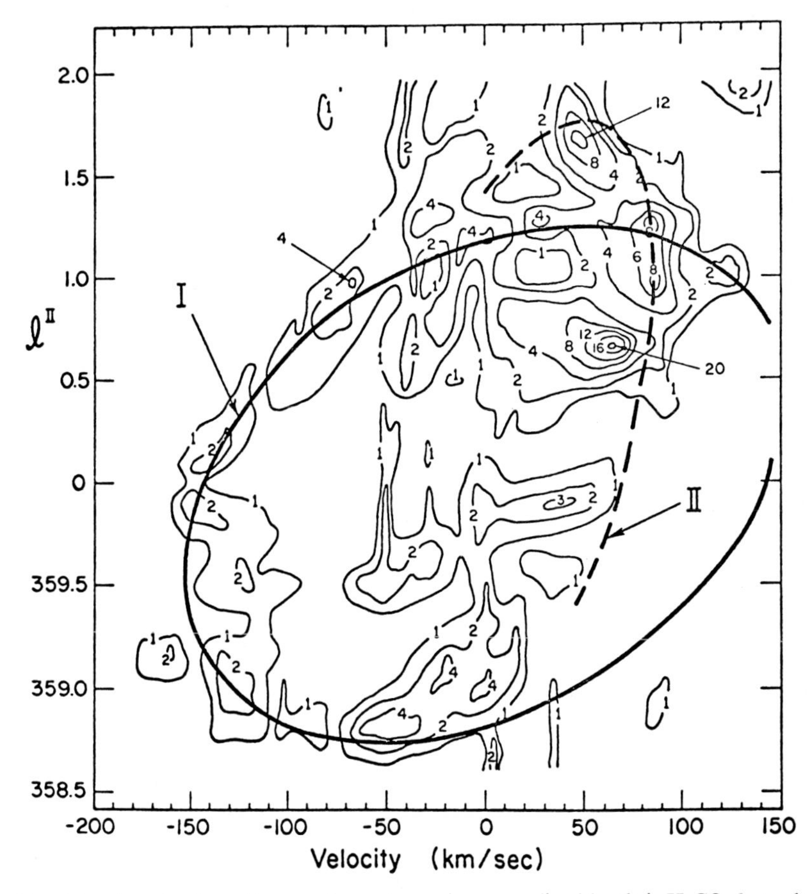

Fig. 10. Molecular clouds in the galactic center region as outlined by their H_2CO 6 cm absorption lines. Contours are lines of constant optical depth. Curve I and II represent limiting ellipses fitted by SCOVILLE (1972b) to the observed distribution of clouds

SCOVILLE estimates a total mass of gas of $3 \times 10^6\ M_\odot$. This may, in fact, be underestimated if many of these clouds contain dense and massive cores (see section V).

PAULS and PISHMISH (private communication) have investigated the dynamics of SCOVILLE's rings using the following simple model:

1. The gas contained in the rings was originally in dynamical equilibrium at a distance r_e from the dynamical center. Conservation of angular momentum requires

$$r\theta_c = r_e \theta'_c. \tag{10}$$

2. By some explosive event, the gas was accelerated to the velocity v_e and is now moving radially outward in the gravitational field of a point mass of $M_p = 10^9\ M_\odot$ [1] into essentially empty space. The radial velocity v observed to day,

1 According to the mass distribution adopted by SANDERS and LOWINGER (1972) $10^9\ M_\odot$ of stars are contained in the spheroid with $R \lesssim 120$ pc.

at the distance r, is then related to v_e and r_e by

$$v^2 = v_e^2 - 2GM_p\left(\frac{1}{r_e} - \frac{1}{r}\right) + (\theta_c'^2 - \theta_c^2). \tag{11}$$

These two relations, together with the equilibrium rotation curve shown in the upper part of Fig. 9, can be used to solve for v_e and r_e. If the values of r and v from Table 5 are substituted, one finds, that the initial radial velocities of the clouds were well below the escape velocity, and the gas contained in the ring-like structure will, after attaining a maximum distance r_{max}, fall back into the center, overshoot the equilibrium radius r_e and eventually come to rest at r_{min}. These parameters, together with the corresponding time scales, are given in Table 6.

Obviously, this simple model of circular gaseous rings oscillating in the gravitational field is, in some ways, unrealistic. The asymmetry of the observed gas distribution suggests that the gas was not uniformly ejected. The expansion will be braked by quiescent rotating gas in the nuclear disk. A certain fraction of the gas will condense and form stars. It thus appears unlikely that clouds would survive more than one oscillation between r_{min} and r_{max}. The simple calculation does, however, suggest that the gas which is now contained in the expanding rings was originally at rest at a distance of about 60 pc from the galactic center. Then, about 10^6 years ago, some explosive event occured which resulted in the observed expansion motions; SCOVILLE (1972b) estimates the energy involved in the expansion, which must, therefore, have come from the "explosive event", to be greater than 4×10^{54} ergs. The initial temperature of the expanding gas must have been of the order of several 10^6 K. However, since the density was certainly greater than 10 cm^{-3}, the plasma would have cooled and recombined in about 10^5 years. The observed thermal IR radiation (HOFFMANN *et al.*, 1971) and high molecular abundances imply that dust grains and molecules also formed on a time scale of the order of some 10^5 years.

Table 6. Parameters of a simple model describing the dynamics of the gas rings in the galactic center (PAULS and PISHMISH, private communication)

	Case I	Case II
$[r/\text{pc}]$	218	305
$[r_e/\text{pc}]$	55	74
$[r_{max}/\text{pc}]$	523	383
$[r_{min}/\text{pc}]$	14	29
$[\theta_c'/\text{km.s}^{-1}]$	198	206
$[v_e/\text{km.s}^{-1}]$	319	231
$[t(r_e \rightarrow r)/10^6 \text{ yrs}]$	0.8	1.8
$[t(r_e \rightarrow r_{max})/10^6 \text{ yrs}]$	6.3	4.2
$[t(r_{min} \rightarrow r_e)/10^6 \text{ yrs}]$	0.12	0.22

References

ALLEN, C. W.: Astrophysical Quantities. Second edition. London: The Athlone Press 1963.
BECKLIN, E. E., NEUGEBAUER, G.: Astrophys. J. **151**, 145 (1968).

BREUER, H. D., MOESTA, H.: In C. DE JAGER (ed.), Highlights of Astronomy, vol. 2, p. 432. Dordrecht-Holland: D. Reidel 1971.
CARRUTHERS, G. R.: Astrophys. J. **161**, L81 (1970).
CHEUNG, A. C., RANK, D. M., TOWNES, C. H., THORNTON, D. C., WELCH, W. J.: Phys. Rev. Letters **21**, 1701 (1968).
CLARK, B. G.: Astrophys. J. **142**, 1398 (1965).
DICKEL, H. R.: In J. M. GREENBERG and H. C. VAN DE HULST (eds.), Interstellar Dust and Related Topics. IAU Symp. No. 52, p. 277. Dordrecht-Holland: D. Reidel (1973).
FIELD, G. B.: In HABING (ed.), Interstellar Gas Dynamics. IAU Symp. No. 39, p. 51. Dordrecht-Holland: D. Reidel 1970.
FOMALONT, E. B., WELIACHEW, L.: Astrophys. J. **181**, 781 (1973).
HABING, H. J.: Bull. Astron. Inst. Neth. **19**, 421 (1968).
HEILES, C.: Ann. Rev. Astron. Astrophys. **9**, 293 (1971).
HOFFMANN, W. F., FREDERICK, C. L., EMERY, R. J.: Astrophys. J. **164**, L23 (1971).
HOLLENBACH, D., SALPETER, E. E.: Astrophys. J. **163**, 155 (1971).
HOLLENBACH, D. J., WERNER, M. W., SALPETER, E. E.: Astrophys. J. **163**, 165 (1971).
KLEMPERER, W.: In C. DE JAGER (ed.), Highlights of Astronomy, Vol. 2, p. 421. Dordrecht-Holland: D. Reidel 1971.
MEZGER, P. G.: In C. DE JAGER (ed.), Highlights of Astronomy, Vol. 2, p. 366. Dordrecht-Holland: D. Reidel 1971.
MEZGER, P. G.: In Proc. Second Adv. Cource in Astron. and Astrophys. on "Interstellar Matter" Geneva-Switzerland: Geneva Observatory 1973.
MEZGER, P. G., CHURCHWELL, E. B., PAULS, T. A.: In L. N. MAVRIDIS (ed.), Stars and the Milky Way System, p. 40. Berlin-Heidelberg-New York: Springer 1974.
RANK, D. M., TOWNES, C. H., WELCH, W. J.: Science **174**, 1083 (1971).
ROBINSON, B. J., MCGEE, R. X.: Australian J. Phys. **23**, 405 (1970).
ROUGOOR, G. W.: Bull. Astron. Inst. Neth. **17**, 381 (1964).
ROUGOOR, G. W., OORT, J. H.: Proc. of the Nat. Acad. Sci. **46**, No. 1,1 (1960).
SALPETER, E. E.: In C. DE JAGER (ed.), Highlights of Astronomy, vol. 2, p. 429. Dordrecht-Holland: D. Reidel 1971.
SANDERS, R. H., LOWINGER, Th.: Astron. J. **77**, 292 (1972).
SANDQUIST, A.: In L. N. MAVRIDIS (ed.), Stars and the Milky Way System, p. 157. Berlin-Heidelberg-New York: Springer 1974.
SAVAGE, B. D., JENKINS, E. B.: Astrophys. J. **172**, 491 (1972).
SCOVILLE, N. Z.: PhD. Thesis, Columbia University (1972a) (unpublished).
SCOVILLE, N. Z.: Astrophys. J. **175**, L 127 (1972b).
SCOVILLE, N. Z., SOLOMON, P. M., THADDEUS, P.: Astrophys. J. **172**, 335 (1972).
SEATON, M. J.: In C. DE JAGER (ed.), Highlights of Astronomy, vol. 2, p. 288. Dordrecht-Holland: D. Reidel 1971.
SNYDER, L. (ed.): Molecules in Space. Charlottesville Symp. 1971 (in press).
SNYDER, L.: MTP Interntl. Rev. of Sci. Biennial Rev. of Chemistry (1972) (in press).
SNYDER, L. E. Jr.: In J. M. GREENBERG and H. C. VAN DE HULST (eds.), Interstellar Dust and Related Topics. IAU Symp. No. 52, p. 351. Dordrecht-Holland: D. Reidel 1973.
STIEF, L. J., DONN, B., GLICKER, S., GENTIEV, E. P., MENTALL, J. E.: Astrophys. J. **171**, 21 (1972).
THADDEUS, P., WILSON, R. W., KUTNER, M., PENZIAS, A. A., JEFFERTS, K. B.: Astrophys. J. **168**, L 59 (1971).
WATSON, W. D., SALPETER, E. E.: Astrophys. J. **174**, 321 (1972a).
WATSON, W. D., SALPETER, E. E.: Astrophys. J. **175**, 659 (1972b).
WEINREB, S., BARRETT, A. H., MEEKS, M. L., HENRY, J. C.: Nature **200**, 829 (1963).
WERNER, M. W., SALPETER, E. E.: In Évolution Stellaire avant la sequence principale. Liège Symp. 1969, p. 113 (1970).
WICKRAMASINGHE, N. C.: In C. DE JAGER (ed.), Highlights of Astronomy, vol. 2, p. 438. Dordrecht-Holland: D. Reidel 1971.
WILSON, R. W., JEFFERTS, K. B., PENZIAS, A. A.: Astrophys. J. **161**, L 43 (1970).
WINNEWISSER, G., MEZGER, P. G., BREUER, H. D.: In Topics in Current Chemistry, Vol. Cosmochemistry. Berlin-Heidelberg-New York: Springer 1973 (in press).
WITT, A. N., JOHNSON, M. W.: Astrophys. J. **181**, 363 (1973).

WITT, A. N., LILLIE, C. F.: Astron. Astrophys. **25**, 397 (1973).
ZUCKERMAN, B., MORRIS, M., TURNER, B. E., PALMER, P.: Astrophys. J. **169**, L 105 (1971).
ZUCKERMAN, B.: In L. E. SNYDER JR. (Ed.), Molecules in Space. Charlottesville Symp. 1971 (in press).

Discussion

TURLO, Z.:
There exists already quite a number of data concerning molecular lines observed in the galactic center. What do you think about using molecules as probes of physical processes in the nucleus of our Galaxy. Do we understand well enough the somewhat puzzling observations of the galactic center as to draw more general conclusions?

MEZGER, P. G.:
I don't think so. To my knowledge the very central region (i.e. galactic radii $\leqq 50$ pc) is practically empty of molecular clouds.

STEPIEN, K.:
I would like to ask whether the turbulence inside a cloud will not influence the lifetimes of molecules.

MEZGER, P. G.:
The angular resolution of radio observations is usually much too coarse to discriminate between large scale and micro turbulence. However, in some cases where high resolution radio observations have been made the molecular clouds showed systematic motions such as e.g. rigid body rotation for the $+40$ km.s^{-1} cloud in front of Sgr A. I believe that most of line broadening being interpreted as "turbulence" in fact is due to large-scale motions which will affect the kinetic gas temperature only in case of collisions between clouds.

Molecules in the Widely Distributed Interstellar Clouds

R. D. Davies, R. J. Cohen and A. J. Wilson
University of Manchester, Nuffield Radio Astronomy Laboratories, Jodrell Bank, United Kingdom

With 2 Figures

Abstract

Observations showing the occurrence of neutral hydrogen, OH and H_2CO in a variety of widely distributed interstellar clouds are described. These clouds include those seen in absorption against strong radio sources and the dense dust clouds which produce strong optical obscuration. Such observations provide a useful guide to the formation processes of interstellar molecules.

I. Introduction

Molecules are found in a wide range of situations within the interstellar medium. Perhaps the most spectacular discovery in molecular radio astronomy has been the OH and H_2O masers. These are condensed clouds with diameters of about 10^{16} cm and gas densities somewhere in the range 10^6 to 10^{10} cm^{-3}. A variety of pumping mechanisms are at work in different types of maser sources. Only a small fraction of the OH in the Galaxy is in such objects.

The majority of the more abundant molecules such as OH, H_2CO and CO are more broadly distributed in the interstellar medium. OH and H_2CO are found in absorption in front of the stronger radio sources. These molecules have excitation temperatures less than the brightness temperature of the background radio sources. CO on the other hand has an excitation temperature greater than the brightness temperature of the background continuum sources at 2.6 mm, the wavelength of the CO line, and is consequently seen in emission. The clouds in which the OH and H_2CO are found are also seen in absorption in the 21 cm neutral hydrogen line. Typical densities are 10–100 neutral hydrogen atoms cm^{-3}. The dense dust clouds such as listed in Lynds' (1962) catalogue are also a rich source of these molecules in the Galaxy. Hydrogen densities here are probably 10^3 cm^{-3} or even more.

We report here recent work at Jodrell Bank using the Mark IA (250 ft diameter) and the Mark II (125 × 85 ft diameter) telescopes. Observations have been made of the H, OH and H_2CO in widely distributed clouds.

II. Molecular Clouds in Absorption in front of Strong Sources

The present work continues the investigation made by Davies and Matthews (1972) of the H, OH and H_2CO absorption spectra of the bright continuum

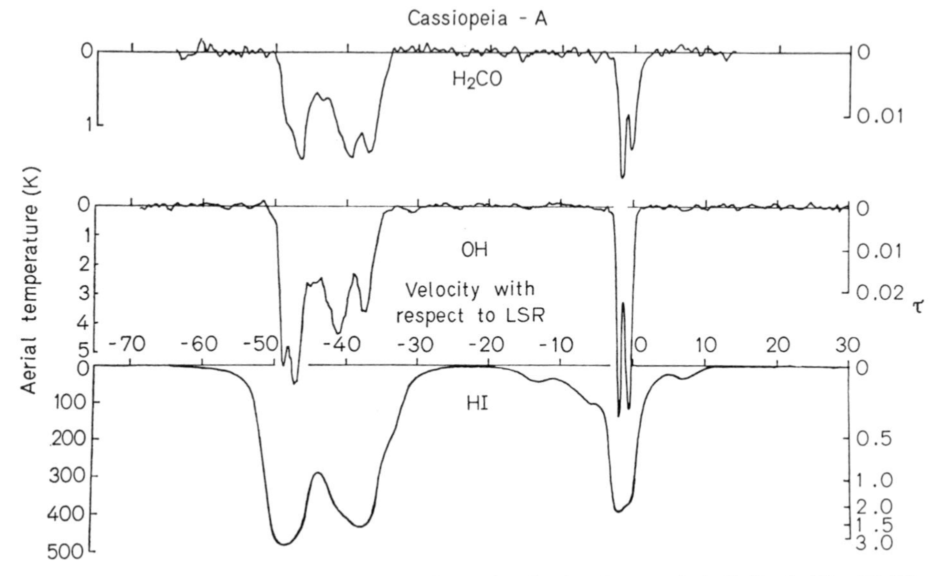

Fig. 1. The H_2O, OH and HI absorption of Cassiopeia A taken at Jodrell Bank. The H_2CO spectrum is the 4830 MHz $1_{11} \rightarrow 1_{10}$ transition and the OH spectrum is the 1667 MHz line

radio sources. They found that all the strong HI absorbing clouds lying in front of the sources Cas A, Cyg A and Tau A showed similar absorption features in the OH and H_2CO spectra. This indicated that these molecules had similar relative densities in interstellar clouds in different parts of the interstellar medium and suggested that a common formation process must be at work throughout the interstellar medium.

New high sensitivity observations have been made of the 6 cm $1_{11} \rightarrow 1_{10}$ line of H_2CO. The H_2CO spectrum of Cas A is shown in Fig. 1 along with the H and OH absorption spectra obtained previously. They allow a more detailed comparison to be made of the H_2CO and H densities in a variety of interstellar absorbing clouds. This experiment had enough sensitivity to detect the weak H_2CO absorption in the spectra of Cyg A and Tau A (with optical depths of $\tau = 0.002$ and 0.0012 respectively). The data suggest a power law relationship between the formaldehyde and neutral hydrogen line integrals of the form

$$N_{H_2CO} \propto N_H^x$$

where $x = 2$ or more. In the case of OH, DAVIES and MATTHEWS found $x = 2.1 \pm 0.2$. Such a square (or higher power) law dependence of the formaldehyde density on the neutral hydrogen density would be expected for any molecular formation process which involved collisions. The most likely formation process in these clouds which are widely distributed throughout the interstellar medium is gas collisions with dust grains that have an absorbed layer of gas. WATSON and SALPETER (1972) have considered the processes of molecule formation on dust grains and obtain predicted OH/H density ratios in substantial agreement with the observations although the predicted H_2CO/H ratio is much less than observed by us in the distributed clouds.

III. Molecules in Dense Dust Clouds

We have made H_2CO and HI observations in the direction of two dense dust clouds using similar angular resolutions ($\sim$ 10 minutes of arc) at the two line frequencies. One was the cloud L1534 of LYNDS' catalogue (=HEILES cloud 2) and the other is a cloud lying behind the Orion nebula. This latter dust cloud is clearly seen on the Mt. Palomar Sky Survey prints; formaldehyde absorption in it was first reported by KUTNER and THADDEUS (1971). The reasons why we believe this dust cloud lies behind the Orion nebula are (a) it does not produce H_2CO absorption in the radio source Orion A which lies at the centre of the nebula and (b) the optical obscuration lying in front of the Orion cluster is several magnitudes less than produced by the dust cloud itself.

The dust cloud L1534 was studied most intensively; HI and H_2CO spectra were obtained at the positions where HEILES (1970) had already obtained OH spectra and in addition HI observations were made at half beamwidth intervals in a strip crossing the centre of the object. Fig. 2 shows the HI, OH and H_2CO spectra at (1950) RA=4^h 36^m 11^s, Dec=25°45′. The formaldehyde and neutral hydrogen are seen in absorption while the OH is in emission. H_2CO is in absorption against the 2.7 K isotropic background while the HI is absorbed against the general spiral arm radiation behind the dust cloud.

The main conclusions of the study of the dense dust clouds are as follows.

(a) The OH and H_2CO spectra are very similar in shape, except that the former is in emission and the latter is in absorption. Small systematic velocity change of $\sim$1 km.s^{-1} occur across the nebula in both molecular lines.

(b) The neutral hydrogen absorption features show the presence of more than one velocity component at most positions. The mean velocity is similar to the OH and H_2CO velocity but is not identical with it in all cases. The line widths correspond to a thermal broadening in the range 10 to 20 K. If account is taken of the mass motions in the line of sight as shown by the OH and H_2CO spectra, then the kinetic temperatures could be considerably lower than these values. However, it is not easy to be sure of the precise amount of turbulent broadening affecting the HI profile since the OH and H_2CO may not necessarily be coincident in space with the atomic hydrogen concentrations as it is likely that most of the hydrogen will be in the molecular form. Nevertheless, it is possible to conclude that the kinetic temperature in these clouds is likely to be 10 K or less.

(c) The molecular densities in the dense dust clouds are greater than in the absorbing clouds discussed in section II by a factor of at least 10 and possibly as much as 100, depending on the assumed depth of the dust cloud. The hydrogen (mostly molecular) density is a factor of $\sim$10 greater than those relevant to section II (HEILES, 1970). These higher molecular densities are predicted by the calculations of WATSON and SALPETER (1972).

(d) A major difference between the two types of cloud is that ionizing and dissociating ultraviolet radiation can penetrate the clouds described in section II but not the dense dust clouds. This means that the molecules in the former type of cloud will have a lifetime of $\sim 10^2$ to 10^3 years (DAVIES and MATTHEWS, 1972) and that the molecules will have to be continuously regenerated.

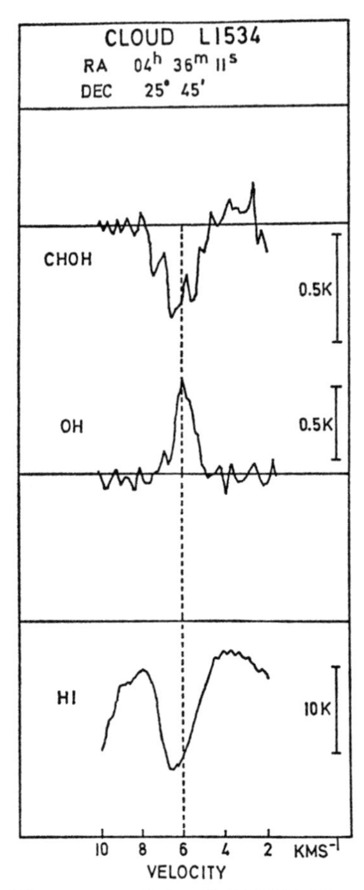

Fig. 2. The H_2CO, OH and HI spectra at RA=$4^h 36^m 11^s$, Dec=25°45′ (1950 coordinates) in the dense dust cloud L1 534. The line intensities are given as aerial temperatures. Observing beam-widths are 9′, 18′ and 11′ arc respectively. (The OH spectrum is from HEILES, 1970)

A more detailed description of these observations is being prepared for publication.

References

DAVIES, R. D., MATTHEWS, H. E.: Monthly Notices Roy. Astron. Soc. **156**, 253 (1972).
HEILES, C.: Astrophys. J. **160**, 51 (1970).
KUTNER, M., THADDEUS, P.: Astrophys. J. **168**, L 67 (1971).
LYNDS, B. T.: Astrophys. J. Suppl. **7**, 1 (1962).
WATSON, W. D., SALPETER, E. E.: Astrophys. J. **175**, 659 (1972).

Discussion

SANDQVIST, A.:
On the topic of the interstellar clouds *not* observed against continuum sources, was there any evidence of differences in velocities for a specific cloud, obtained from

the different spectral lines? I am thinking specifically of a result reported by SANCINI that for one such cloud there were some small differences.

DAINTREE, E. J.:
No difference in velocities was observed in the one case for which I have information.

SANDQVIST, A.:
How many such clouds were in fact observed?

DAINTREE, E. J.:
Ten clouds containing molecules were observed in absorption.

On the Homogeneity and the Kinetic Temperature of the Intercloud HI-Gas

By O. Hachenberg and U. Mebold
Max-Planck-Institut für Radioastronomie, Bonn, FRG.

With 2 Figures

Abstract

The brightness temperature fluctuations which can be observed between the 21 cm intercloud gas emission from adjacent positions are interpreted to come from a random variation of the number of turbulence cells in the antenna beam. We find about 200 cells in our $36' \times 36'$ beam. The velocity dispersion σ within one cell can be separated from the velocity dispersion τ of the centre velocities of the cells. We find $\tau/\sigma \simeq 1.1$. From τ/σ and from the observed velocity dispersion of the emission lines we find $\sigma = 6.5\,\mathrm{km} \cdot \mathrm{s}^{-1}$. This corresponds to a kinetic temperature $T_\mathrm{k} \simeq 5000$ K for the intercloud HI-gas.

According to the two component model for the interstellar neutral hydrogen gas, this gas exists in predominantly two states:

In cold and dense clouds and in a diffusely distributed hot and thin intercloud medium. In intermediate galactic latitudes both states can be distinguished quite well in the HI-emission spectra. The clouds appear as narrow and intense lines while the intercloud gas emits comparatively broad and faint lines (c.f. e.g. Mebold, 1972).

In order to get new information on the physical parameters of the intercloud HI-gas the authors in 1971 observed the 21 cm line emission of the intercloud HI-gas in 10 fields of the sky where it is not blended by the emission of HI-clouds. Preliminary results for 11 positions of one of these fields shall be discussed in the present paper.

The field considered here is $2.^\circ5 \times 2.^\circ5$ wide and centered at the galactic position $l = 72^\circ$, $b = 21^\circ$. The profiles were observed with the Bonn 25 m telescope. It was equipped with a cooled parametric amplifier frontend, and a spectrometer with 25 channels of 8 kHz bandwidth. The observed r.m.s. noise of the profiles was about 0.2 K; baseline errors were smaller than 0.1 K and the temperature scale was stable within about 1%. Further details on the receiver and the reduction procedure will be published elsewhere.

Typical line profiles are shown in Fig. 1. The faint emission with high negative velocities will not be considered here. If we talk about the "profiles" henceforward, only the main peak emission is concerned. The main peak has a maximum brightness temperature T_0 of about 10.0 K and a velocity dispersion b of about $9.6\,\mathrm{km} \cdot \mathrm{s}^{-1}$. The average profile of the field may quite well be approximated by a single Gaussian while the individual profiles show significant deviations from this. The deviations of the individual profiles from the average profile of the

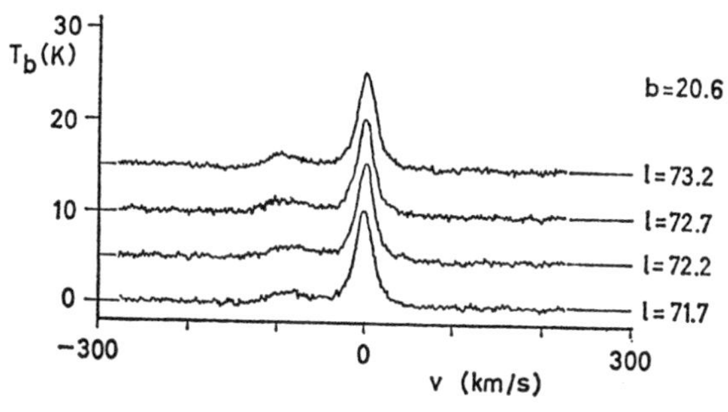

Fig. 1. The brightness temperature T_b of the 21 cm lines versus radial velocity v at the galactic positions given by l and b

field will be used now to separate thermal and turbulent gas motions (cf. e.g. BAKER, 1972).

If we assume that the main peak emission is produced by a relatively large number N of emitting elements, this number can be calculated from the r.m.s. scatter ΔF of the observed profile areas F around the area $\bar{F}$ of the average profile of the field. Observations give

$$\bar{F} = 240 \pm 5\,\mathrm{K \cdot km \cdot s^{-1}},$$

and

$$\Delta F = 17 \pm 0.8\,\mathrm{K \cdot km \cdot s^{-1}}.$$

From

$$\bar{F}/\Delta F = N^{1/2},$$

we find

$$N = 200 \pm 20$$

elements in the antenna beam. With suitable values for the scale height of the gas perpendicular to the galactic plane we find an average diameter of 6 pc and an average density of $0.12\,\mathrm{cm^{-3}}$ for the HI-elements, if a filling factor of 1 is assumed.

Similarly as in the case of N, we can determine the number N_0 of elements which contribute noticeably to the maximum brightness temperature at the velocity v_0. If $\bar{T}_0$ is the maximum brightness temperature of the average profile of the field and ΔT_0 is the r.m.s. scatter of the corresponding individual brightness temperatures around $\bar{T}_0$, we have

$$\bar{T}_0/\Delta T_0 = N_0^{1/2}. \tag{1}$$

The observations give

$$N_0 = 166 \pm 10.$$

$N > N_0$ means that there are elements which contribute to the total area of the main peak but not to the maximum brightness temperature T_0. This necessarily implies that the standard deviation $\tau\,[\mathrm{km \cdot s^{-1}}]$ of the mean radial velocities of the HI-elements is not small compared to the thermal velocity dispersion $\sigma\,[\mathrm{km \cdot s^{-1}}]$ of the HI-elements. An analytic description of this situation shows that the ratio N/N_0 is a unique function of the Mach number $\alpha = \tau/\sigma$:

As the average profile of the field may quite well be approximated by a single Gaussian, it is reasonable to write the observed line shape $T_b(v)$ as the convolution

$$T_b(v) = \int_{-\infty}^{\infty} n(u)\, t(v-u)\, du, \tag{2}$$

of the distribution

$$n(u) = \frac{N}{\tau\sqrt{2\pi}}\, e^{-\frac{u^2}{2\tau^2}},$$

of the mean radial velocities u [km · s^{-1}] of the HI-elements and the line shape

$$t(v-u) = t_0\, e^{-\frac{(v-u)^2}{2\sigma^2}},$$

of an HI-element. Here t_0, which is assumed to be a constant for all elements, can be normalized by

$$t_0 = \bar{F}/(\sqrt{2\pi}\,\sigma N) = \bar{T}_0 \sqrt{1+\alpha^2}/N,$$

as Eq. (2) implies that the relations $\bar{F} = \sqrt{2\pi}\,\bar{T}_0\, b$ for the area of the main peak and

$$b^2 = \sigma^2 (1+\alpha^2), \tag{3}$$

for the observed velocity dispersion b are valid. If we further combine Eqs. (1) and (2) noting that (1) is valid only, if elements of equal heights are counted by N_0, we use the integral

$$(\Delta T(v))^2 = \int_{-\infty}^{\infty} n(u)\, t^2(v-u)\, du,$$

and find

$$(\Delta T(v_0))^2 = \bar{T}_0^2/N_0 = \frac{\bar{T}_0^2}{N}\, \frac{1+\alpha^2}{\sqrt{1+2\alpha^2}}.$$

The final function

$$N/N_0 = (1+\alpha^2)/\sqrt{1+2\alpha^2}, \tag{4}$$

is plotted in Fig. 2.

From the observed ratio $N/N_0 = 1.20 \pm 0.15$ and Eq. (4) we find

$$\alpha = 1.1 \pm 0.3.$$

Solving Eq. (3) for σ, with $\alpha = 1.1 \pm 0.3$ and $b = 9.6$ km · s^{-1}, we find

$$\sigma = 6.45 \pm 0.85 \text{ km} \cdot \text{s}^{-1},$$

corresponding to a kinetic temperature

$$T_k = 5000 \pm 1200 \text{ K}.$$

The error in T_k is mainly due to baseline uncertainties. Changes in the temperature scale strongly tend to be cancelled out, because in $N/N_0 = (\bar{F}^2/T_0^2)(\Delta T_0^2/\Delta F^2)$ scale changes are equally involved in ΔT_0 and in ΔF. It should be noted further that N and N_0 are biased towards higher values, because the F- and T_0-values from adjacent positions are not statistically independent but are correlated due to a certain

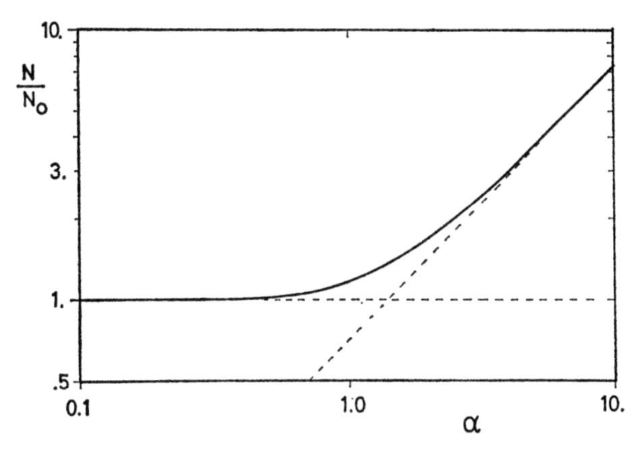

Fig. 2. The ratio N/N_0 of the number N of HI-elements producing the main peak of the profiles in Fig. 1 to that N_0 of HI-elements contributing noticeably to the maximum brightness temperature plotted against $\alpha = \tau/\sigma$, the ratio of the turbulent τ to the thermal σ velocity dispersions of the HI-gas

angular extension of the HI-elements. These systematic errors, however, strongly tend to be cancelled out, too.

For the calculation of ΔF and ΔT_0 one should use the number i_{eff} of statistically independent positions instead of i, the number of observed positions. But as i as well as i_{eff} are about the same numbers for both ΔF and ΔT_0 they are cancelled out in performing $\Delta T_0^2/\Delta F^2$.

Hence the calculated values for α and T_k are probably free of systematic errors within the quoted limits.

T_k fits into the range of temperatures set by the assumption of purely thermal line broadening ($T_k \leqq 9000$ K) (Mebold, 1972) and set by the observed emission and absorption of the intercloud HI-gas ($T_k \geqq 700$ K) (Radhakrishnan *et al.*, 1971). It is, however, smaller than the lower limit of about 7000 K set by the cosmic- or X-ray heating theories (cf. Field *et al.*, 1969; Vogel, 1971). Whether or not this difference is significant can possibly be decided, when the results for the other 9 fields of our survey become known.

References

Baker, P. L.: A Statistical Investigation of Neutral Hydrogen Line Profiles. Preprint (1972).

Field, G. B., Goldsmith, D. W., Habing, H. J.: Astrophys. J. **155**, L 149 (1969).

Mebold, U.: Astron. Astrophys. **19**, 13 (1972).

Radhakrishnan, V., Murray, J. D., Lockhardt, P., Whittle, R. P. J.: Astrophys. J. Suppl. **24**, 15 (1971).

Vogel, U.: PhD Thesis, Universität Bonn (1971).

The Outer Spiral Structure of the Galaxy and the High-Velocity Clouds

By R. D. Davies
University of Manchester, Nuffield Radio Astronomy Laboratories, Jodrell Bank, United Kingdom

With 2 Figures

Abstract

Outer spiral arms have been mapped using high sensitivity observations. Their velocity and location suggest a close association with the high-velocity clouds (HVCs).

I. Observations

A series of high sensitivity measurements have been made of the outer spiral structure of the Milky Way. The survey employed a 32′ arc angular resolution in a series of latitude cuts across the galactic plane at 10° or 20° intervals in longitude. The r.m.s. noise on each spectrum was 0.04 K thus making it possible to detect signals of 0.1 K or greater. The velocity resolution was 7.4 km$.$s^{-1}. These data provide a high sensitivity, high angular and frequency resolution survey of the tenuous outer spiral structure. They are presented as contour maps of brightness temperature in the latitude-velocity plane at fixed longitudes.

II. The Spiral Structure

Spiral structure can be recognized in the outer regions of the Galaxy in the same way as in the regions of denser spiral arms. An arm or arm-like feature is seen as (1) a component on the spectra (2) confined in latitude to a region near the galactic plane (3) extended in longitude and (4) it shows a variation of velocity with longitude compatible with galactic rotation.

On the longitude range 20° to 140° the outer spiral structure is displaced above the galactic plane in conformity with the well-known tilt of the galactic structure. The upper boundary of features can be traced to between 15° and 20° above the galactic equator in this sector of the plane.

The velocity range encountered in the outer spiral arms extends to -150 km$.$s^{-1} at longitudes between 20° and 160°. Outer edges of some velocity components can be traced to -180 km$.$s^{-1}. At the southern extremity of the present survey ($l=260°$) velocities as high as $+150$ km$.$s^{-1} were found.

Distances were estimated for the outer spiral structure by assuming that the gas was in circular motion about the galactic centre with the velocity given by the Schmidt (1965) rotation model. A sketch of the derived positions of the outer

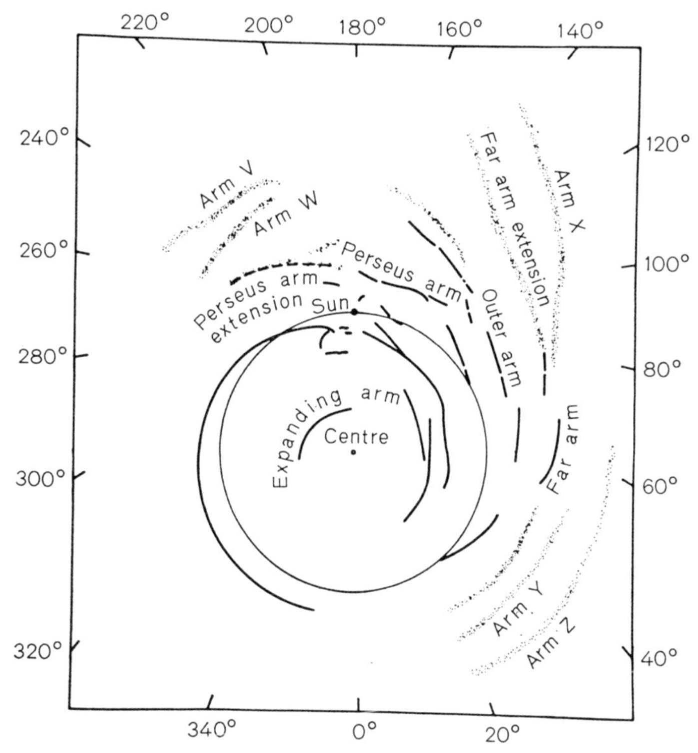

Fig. 1. A schematic diagram of the outer spiral structure of the Galaxy. Distances were assigned on the basis of the 1965 Schmidt rotation model. The inner spiral structure (heavy lines) is taken from WEAVER (1970)

arms in shown in Fig. 1 superposed on the model of the inner spiral structure published by WEAVER (1970). No great accuracy is claimed for the detailed locations of the outer spiral structure; the spiral arms must lie in the relative positions shown.

The high velocity material connected to the galactic plane described here has also been seen in some limited and rather less sensitive surveys such as those of HABING (1966) and KEPNER (1970). Mrs. DIETER's (1972) extensive survey shows most of the features plotted here.

III. The High-Velocity Clouds (HVCs)

Most of the brighter HVCs found in surveys of the northern sky are at negative (approach) velocities and are found in the longitude range 60° to 180° and at latitudes between +10° and +60°. Their velocities are mostly in the range −70 to −200 km.s^{-1} (OORT, 1966). Until recently very few positive velocity clouds were known; they were mostly restricted to $l > 230°$. However, more sensitive surveys (WANNIER *et al.*, 1972) show extensive regions in this longitude range with similar positive velocities.

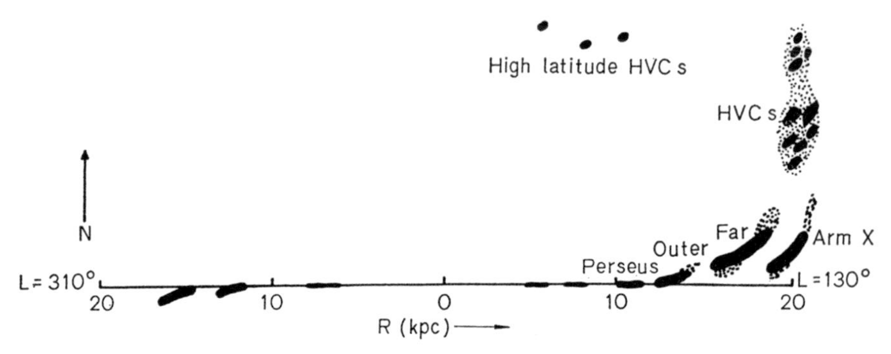

Fig. 2. The location of the neutral hydrogen arms and HVCs in a section through the galactic centre along the line of galactocentric longitude $L=130°$, $310°$. The HVCs have been placed at the distance of the associated spiral arms

The above properties of HVCs show a more than chance similarity to those of the outer spiral arms described in the previous section .They exhibit mostly negative velocities in the range $l=20°$ to $160°$ and mostly positive velocities in the range $l>220°$. The velocity range in the outer arms and the HVCs is similar. Also the upwards tilt at longitudes of 20° to 140° is shown by both phenomena.

These facts and the apparent connections of the bands of HVCs to the galactic plane lead me to postulate that the two phenomena are intricately connected. The main body of negative velocity HVCs would appear to be an extension of the upward tilt of the outer spiral arms. Fig. 2 shows the relationship between the spiral arms and HVCs along a cut through the galactic centre along galactocentric longitudes $L=130°$, $310°$. The HVCs have been placed at a distance equal to that of the outer spiral arms.

IV. The Origin of the HVCs

On this picture the event which produced the tilt of the outer parts of the Galaxy will also have been responsible for the HVCs. The most favoured explanation of the galactic tilt is the gravitational effect of the Large Magellanic Cloud during its most recent perigalactic passage (Hunter and Toomre, 1969; Toomre, 1970). The form of the tilt shown in Fig. 2 is entirely expected on the tidal theory.

The first report of this work has been published in Nature (Davies, 1972) and a more extensive presentation of the observational material and discussion is in course of publication in the Monthly Notices of the Royal Astronomical Society, volume 160.

References

Davies, R. D.: Nature **237**, 88 (1972).
Dieter, N. H.: Astron. Astrophys. Suppl. **5**, 21 (1972).
Habing, H. J.: Bull. Astron. Inst. Neth. **18**, 323 (1966).

HUNTER, C., TOOMRE, A.: Astrophys. J. **155**, 747 (1969).
KEPNER, M.: Astron. Astrophys. **5**, 444 (1970).
OORT, J. H.: Bull. Astron. Inst. Neth. **18**, 421 (1966).
SCHMIDT, M.: In A. BLAAUW and M. SCHMIDT (eds.), Galactic Structure, p. 513. Chicago: Chicago Univ. Press 1965.
TOOMRE, A.: In W. BECKER and G. CONTOPOULOS (eds.), The Spiral Structure of Our Galaxy. IAU Symp. No. 38, p. 334. Dordrecht-Holland: D. Reidel 1970.
WANNIER, P., WRIXON, G. T., WILSON, R. W.: Astron. Astrophys., **18**, 224 (1972).
WEAVER, H. F.: In W. BECKER and G. CONTOPOULOS (eds.), The Spiral Structure of Our Galaxy. IAU Symp. No. 38, p. 126. Dordrecht-Holland: D. Reidel 1970.

Discussion

OORT, J.:
I think the interpretation which Dr. DAVIES has given of the high negative velocities observed in longitudes between about 120° and 200° as being due to spiral arms at distances of several tens of kpc from the galactic centre is open to considerable doubt. From observations around the anticentre it is known that high radial components (directed toward the centre) occur over large regions near the outer edges of the Galaxy. As a consequence one cannot derive distances in these regions on the assumption that the measured radial velocities reflect circular motions. This takes away the basis for the determination of distances such as surmised by DAVIES.

The conclusion that the high-velocity clouds would generally be parts of such arms extending to very high latitudes is contradicted by the observations of some high-velocity clouds near the galactic poles as well as by observations made by MUNCH and ZIRIN which showed some high-latitude clouds in absorption in the spectra of OB stars at high z, proving that the distances of these clouds do not exceed 1 kpc.

DAVIES, R. D.:
The point of my paper is that the high velocity emission seen at $l=120°$ is not an isolated phenomenon but is part of a distribution extending from $l=20°$ to at least $l=250°$. The average velocities vary over this entire longitude range in the manner expected for the outer regions of the Galaxy. Also the observed fall-off in the brightness of the emission with distance from the galactic centre is a continuation of the trend shown by the well-known arms. I agree with Professor OORT that there will be some uncertainty in deriving an accurate location for this material; I believe, however, that the distribution of the outer arms shown in Fig. 1 is a reasonable interpretation of the data at present available.

A small amount of the high velocity gas does not fit into this picture. It includes (a) the negative and *positive* velocity material in the anticentre and (b) the *high latitude* HVCs. I prefer to think of this residual gas as arising from the same cause as that which produced the twist in the plane shown by the bulk of the high velocity gas. The favourite explanation at present seems to be tidal interaction by the Large Magellanic Cloud.

As regards the clouds investigated by MUNCH and ZIRIN at distances up to 1 kpc above the plane, I believe that they are high z clouds associated with the well-known spiral arms. My reasons are (1) Their velocities are comparable with those found in these arms and moreover they change sign with galactic longitude in the sense expected for galactic rotation. (2) The intermediate-velocity clouds (IVCs) seen in neutral hydrogen surveys have similar velocities and are found at similar values of z. They are known to be associated with spiral arms.

Very Long Baseline Interferometer Measurements of H_2O Masers in Orion A, W49N, W3OH and VY CMa

By G. D. PAPADOPOULOS
Research Laboratory of Electronics, Massachusetts Institute of Technology, Cambridge, Mass., U.S.A.

Abstract

A review is presented of the results of very long baseline interferometer measurements of the H_2O emission regions in Orion A, W49N, W3OH and VY CMa. The observations were made in June 1970, February 1971, and March 1971 using the 120-foot antenna of the Haystack Observatory, the 140 and 36-foot antennas of the NRAO and the 85-foot antenna of the Naval Research Laboratory.

Orion A exhibited strong variations in this time interval. The main spectral features of the June observations at 9.5 and 3.7 km $\cdot$ s^{-1} had decreased in intensity by factors of 4 and 2, respectively in February 1971. The 9.5 km $\cdot$ s^{-1} feature was unresolved in June but partially resolved in February; this could imply an increase in size from 0.″0008 to 0.″003 in this period, if we assume a uniformly bright disk model for the emission region. Prominent among the new features that became observable in the February and March experiments were the 11.0 km $\cdot$ s^{-1} and -0.6 km $\cdot$ s^{-1} lines. The former, which was the most intense amongst all Orion A features, was completely resolved while the latter was partially resolved; their sizes were estimated to be 0.″01 and 0.″0036 respectively. The H_2O emission regions in Orion A fall within a square having a side of approximately 25″; the OH complex coincides with the inner H_2O regions located around the BN IR point source.

W49N is composed of many features, all of which are less than 0.″0005 in size and lie within a circle having angular diameter of approximately 1.″5 (JOHNSTON *et al.*, 1971). The intensity, size and location of the features did not change in the period considered. The profile of the main feature at -1.9 km $\cdot$ s^{-1} is asymmetrical. This profile could be decomposed into two features with the same position and having a separation of 33 KHz suggesting that they are two hyperfine rotational components. The position of the H_2O complex is coincident with that of the OH sources to within $\pm 2''$.

The emission spectrum of W3OH obtained from the March 1971 measurements exhibited five features in the frequency range from -46.7 to -51.5 km $\cdot$ s^{-1}. All features appear to be unresolved with diameters less than 0.″002 assuming a uniformly bright circular disk model. The map indicates that the emission regions are distributed along a line of length 2″ parallel to the galactic plane.

The spectrum of VY CMa has two emission features at 18 and 14.5 km $\cdot$ s^{-1}. These two emission regions have an angular size smaller than 0.″002 and are separated by 0.″15 with a position angle of 210°.

References

JOHNSTON, K. J., KNOWLES, S. H., SULLIVAN, W. T. III, BURKE, B. F., LO, K. Y., PAPA, D. C., PAPADOPOULOS, G. D., SCHWARTZ, P. R., KNIGHT, C. A., SHAPIRO, I. I., WELCH, W. J.: Astrophys. J. **166**, L 21 (1971).

Discussion

MEZGER, P. G.:
Could you summarize your results from comparing OH and H_2O maser sources in terms of: i) positional correlation, ii) velocity correlation ?

PAPADOPOULOS, G. D.:
The absolute positions of the H_2O sources coincide with those of the OH to within 2'' for W49, 5'' for Orion A and 6'' for VY CMa. Our absolute position estimates for W3OH are not conclusive because of the large experimental errors.

Correlation in velocity on a feature-to-feature basis does not exist. The velocity ranges, however, over which the molecular emissions for OH and H_2O spread, correlate quite well. For instance, in Orion A the H_2O emission features range from -2 to $+17\ \mathrm{km \cdot s^{-1}}$ while the OH from $+4$ to $+20\ \mathrm{km \cdot s^{-1}}$. In W49 the H_2O features range from -9 to $+12\ \mathrm{km \cdot s^{-1}}$ while the OH from $+10$ to $+22\ \mathrm{km \cdot s^{-1}}$. In VY CMa the velocities of OH and H_2O fit adequately in the model of an expanding disk of gas and dust, proposed by HERBIG.

High Resolution Search for Interstellar Lithium

By H. E. Utiger
Middle East Technical University, Ankara, Turkey

With 4 Figures

Abstract

New upper limits are given for the interstellar lithium abundance in the direction of HD 37128 (ε Ori) and HD 197345 (α Cyg). The upper limit on the equivalent width in the direction of HD 197345 is 2.54 mÅ and in the direction of HD 37128 is 4.06 mÅ. The possibility exists that lithium was detected in the clouds of HD 37128 but the statistics of the data preclude a positive claim of the detection of the lithium resonance absorption lines at 6708 Å.

I. Introduction

Modern spectroscopic techniques developed in the last few years have reached the stage at which much more precise measurements can be made on the abundances of naturally occuring elements and the environment in which they reside. In particular as Hobbs (1965, 1969a, b, c) has shown, the PEPSIOS instrument lends itself well to the study of interstellar lines when used with modern detectors and recording techniques, because of its high photometric efficiency at high resolution and high spectral purity (or at least the possibility of measuring or calculating the parasitic light).

Several attempts have been made to observe interstellar lithium absorption lines, but all of these attempts have failed. Spitzer (1949) observed several background stars in a search for interstellar lithium. In particular, he studied both HD 197345 (α Cyg) and HD 37128 (ε Ori) which were the stars used in the present observations. Spitzer was unable to identify any lithium absorption lines on his plates and could only establish an upper limit on the equivalent width of a lithium line which could be detected.

In addition to the observation of stellar lithium in T Tauri stars, Bonsack and Greenstein (1960) searched for a lithium absorption line formed in the circumstellar material of the star T Tau, from which this star presumably condensed. Here too, no interstellar absorption line was detected. This is rather surprising considering the large equivalent width of the lithium feature observed in the stellar spectrum.

Herbig (1968) made extensive measurements on the foreground material of ξ Oph, but again there was no evidence of the presence of any lithium. Herbig's limit of detection is still greater by a factor of 9 than the lithium content found in chondritic meteorites.

In all attempts to observe interstellar lithium, except that of BONSACK and GREENSTEIN, it is not surprising that the experiments yielded negative results. Unless there had been an abnormally high lithium to hydrogen abundance ratio, these experiments probably could not have detected the lithium in any of the interstellar clouds observed.

II. Observations

The observations were made on the foreground material of α Cyg and ε Ori during five consecutive nights (Oct. 4–9, 1968) at the Coudé focus of the Kitt Peak National Observatory 84-inch telescope. Data were taken using a PEPSIOS spectrometer operating in a photon counting mode. The basic spectrometer has been extensively discussed elsewhere (MACK *et al.*, 1963; ROESLER and MACK, 1967) and only minor modifications were made to the reference system to allow for photon counting.

A relatively low resolution (for a PEPSIOS) was chosen for the instrument, because both stellar and interstellar observations were to be taken. The Fabry-Pérot spacers were in the MCNUTT (1965) ratios of 1.000:0.8831:0.7244 with the longest spacer having a length of 3.000 mm. These spacers were optically contacted to the Fabry-Pérot plates and were extremely stable. The plate reflectance was chosen as the limiting factor of the resolution. A theoretical resolution of 2×10^5 was given with coatings of 89% reflectance. The actual resolution was of the order of 1.75×10^5 and well within the design limits of the instrument. The best estimate of the FWHM width of the instrumental passband was taken from scans of a liquid nitrogen cooled hollow cathode. This width is approximately 38 mÅ.

The instrument and associated optical equipment as was used on the telescope is shown in Fig. 1. Light coming down the polar axis of the telescope was intersected by a dichroic mirror DM. Light with a wavelength shorter than 6400 Å was transmitted through the mirror while the light of longer wavelength was reflected back through the PEPSIOS coupling lens L, which formed the stellar image on the PEPSIOS entrance aperture. The light transmitted through the dichroic mirror formed an image on the entrance slits of the normal Coudé spectrograph. It was hoped that by making the crosshairs on the Coudé spectrograph observations telescope conjugate to the entrance aperture of the PEPSIOS the telescope could be guided using only the unwanted blue light. This method was not altogether satisfactory due to diffraction in the atmosphere. The interference filter F used in this experiment had a halfwidth of 4.4 Å. The photomultipliers in both the signal channel PM_1 and the reference channel PM_2 were ITT FW130 having S-20 photo response cathodes.

Fig. 2 shows a block diagram of the system used for photon counting in this experiment .The electronics in the signal and reference channels are identical. The pulse counting equipment is a commercial unit manufactured by ORTEC, consisting of a preamplifier, active filter-amplifier, scaler and print-out control unit. The discriminator has both an upper and lower bound which allow only pulses with a height falling within these bounds to be passed on to the scaler. This allows only those pulses which originate at the photo-cathode to be counted. These

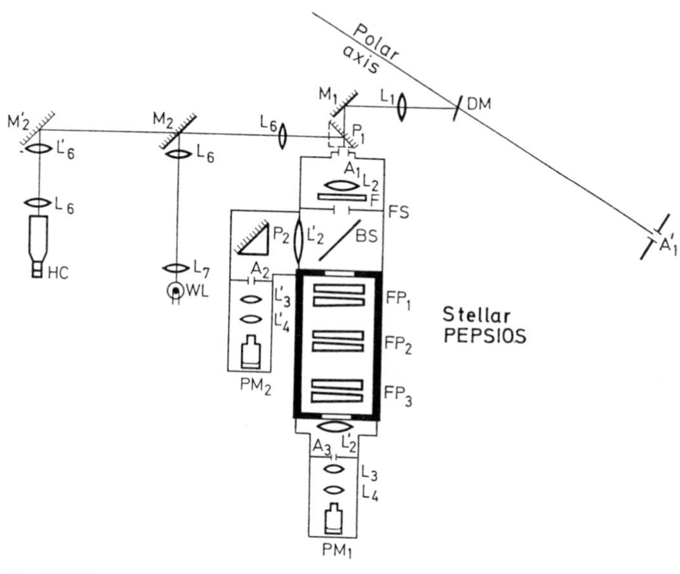

Fig. 1. The stellar PEPSIOS as it was coupled to the 84" Kitt Peak telescope at the Coudé focus

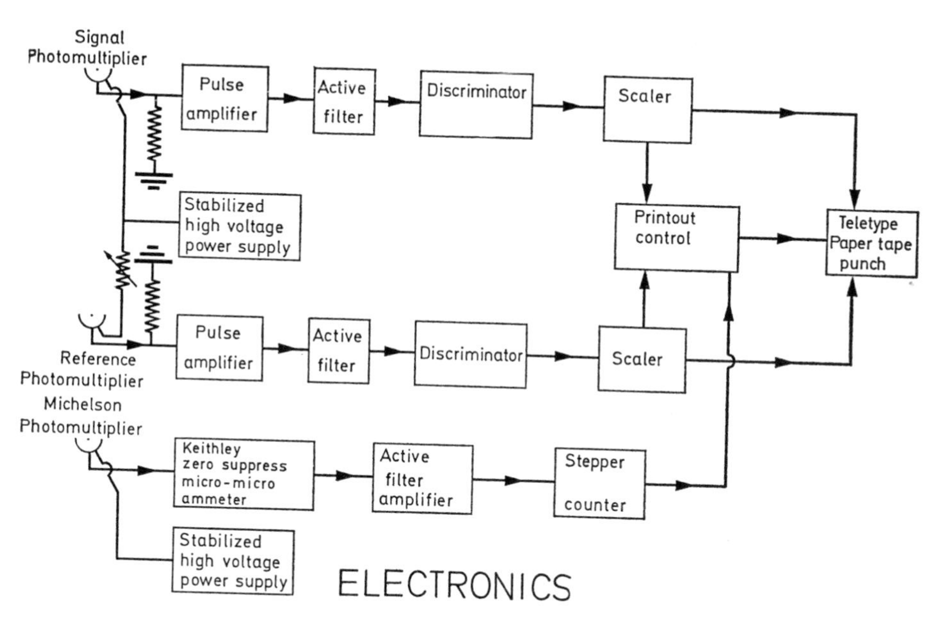

Fig. 2. A block diagram of the photon counting electronics used with the PEPSIOS spectrometer

pulses are counted and stored in the scalers until a command from the print-out control causes the data in the counters to be transferred out of each counter into a buffer storage, after which the counters are reset to zero and counting begins again. The data from the buffers is then transferred serially to a teletype where it is written out on the printer and punched on paper tape. The total dead time of the counters in this operation is 50 µs.

The print-out control commands can be initiated by either an internal timer or an external signal. In this experiment the print-out was triggered externally by pulses originating from the fringes of a Michelson interferometer. This interferometer and associated electronics gave a pulse to the print-out control every 9 mÅ in the wavelength region of the lithium resonance line.

III. Data

Each individual channel on usable scans is summed together for both signal and reference channels. This reduces both the photon statistics and the seeing scintillation noise. At this point the summed average photomultiplier dark count and parasitic light are subtracted from the signal channel. Next, for each summed channel the signal channel is divided by the reference channel to take out any variations in the spectrum due to changes in atmospheric transparency. There is still a slight curvature to the spectrum due to the variable transmission in wavelength of the interference filter. This curvature is found by scanning a bright white light source. The signal is then divided by the white light scan to correct for this wavelength dependence. The data are then in a form suitable for further analysis.

By looking at the reduced data (see Fig. 3) it is immediately apparent that the lithium lines, if present, are extremely weak and are hidden under the noise. To obtain a minimum upper bound on the lithium which might be present under

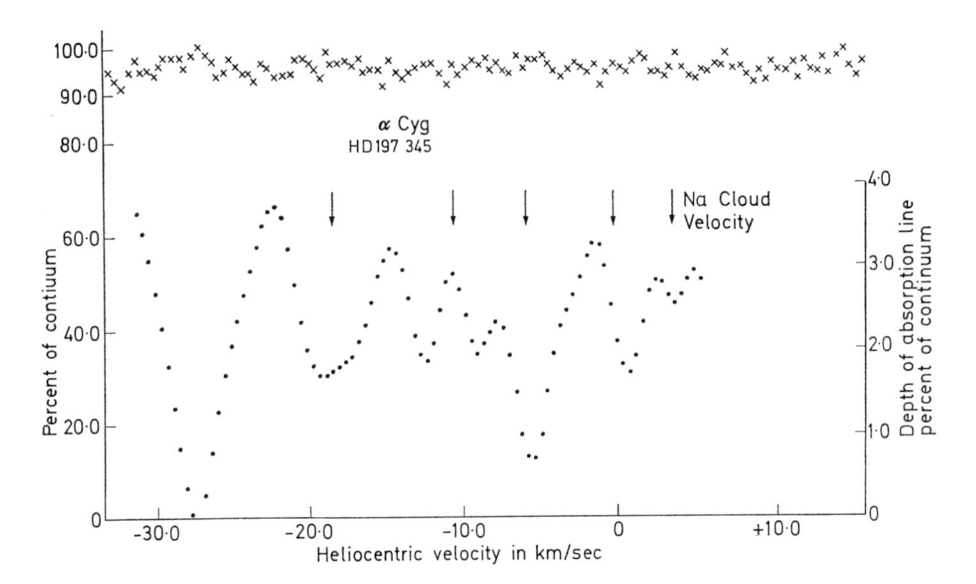

Fig. 3. Data and results for α Cyg. The crosses represent the corrected interstellar data. The dots represent the results of the fitting program. The ordinate on the left side of the figure gives the intensity of each data point in the scan. The ordinate on the right side gives the depth of the $^2S_{1/2}-^2P_{3/2}$ absorption feature

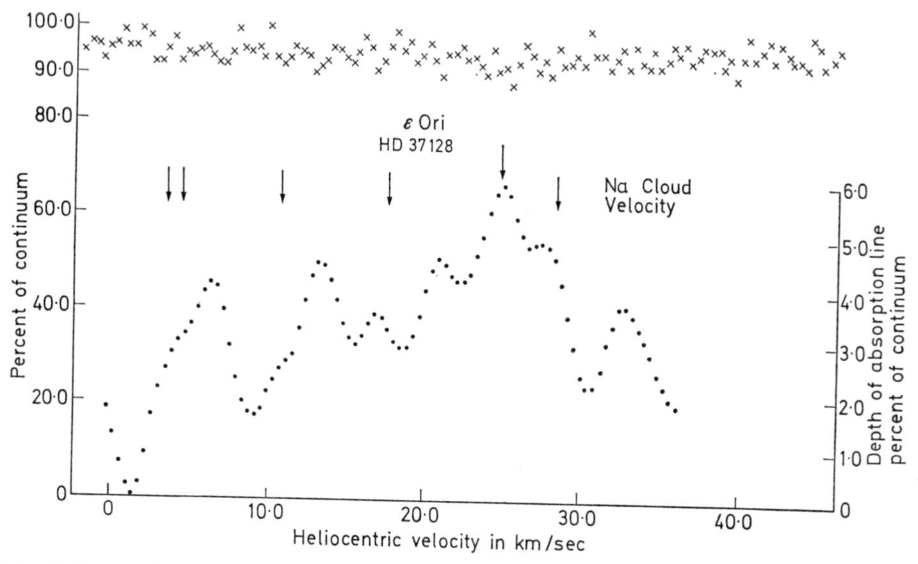

Fig. 4. Data and results for ε Ori. The crosses represent the corrected interstellar data. The dots represent the results of the fitting program. The ordinate on the left side of the figure gives the intensity of each data point in the scan. The ordinate on the right side gives the depth of the $^2S_{1/2}-^2P_{3/2}$ absorption feature

the noise, a fitting procedure was used. Since the line position was completely unknown, a lithium profile with a fixed line width was successfully fitted at each point in the spectrum, and an absorption strength was calculated for each point such that the variance was minimized. A Gaussian profile of the form

$$I(\lambda)=I_0\left\{1-S*\exp\left[-c\left(\frac{\lambda-\lambda_0}{W}\right)^2\right]-S/2*\exp\left[-c\left(\frac{\lambda-\lambda_0-\lambda_s}{W}\right)^2\right]\right\}$$

was used, where S is the depth of the absorption line, λ_0 is the apparent wavelength of the $^2S_{1/2}-^2P_{3/2}$ transition, λ_s is the fine structure splitting, W is the line width, C is a constant and I_0 is the continuum level. An estimate of the expected line widths was taken from HOBBS (1969b) line profiles of sodium and reduced for the different atomic weight of lithium, assuming a gas temperature of 100 K. The total equivalent width of the lithium feature is very insensitive to the line width chosen. The results of this fitting program are shown in Figs. 3 and 4.

IV. Theory

We wished to see if we had reached the goal of being able to detect lithium if it was found in some reasonable abundance such as an average T Tau abundance or chrondritic meteoric abundance, terrestrial abundance or perhaps the "cosmic" abundance quoted by ALLEN (1963). To carry out this program we have used the

Table 1. Theoretical expected equivalent widths of the lithium resonance line at varying star-cloud distances

1 Star	2 dis pc	3 Γ	4 N(Li) atoms per cm^2	5 N(Li)/N(LiI)	6 N(LiI) atoms per cm^2	7 W_λ mÅ
ε Ori, HD 37128	4	7.92×10^{-8}	5.5×10^{11}	1.30×10^{4}	4.2×10^{7}	1.7×10^{-2}
	8	1.99×10^{-8}		3.26×10^{3}	1.68×10^{8}	7.2×10^{-2}
	10	1.28×10^{-8}		2.12×10^{3}	2.5×10^{8}	0.104
	20	3.39×10^{-9}		5.55×10^{2}	9.9×10^{8}	0.42
	50	7.41×10^{-10}		1.21×10^{2}	4.5×10^{9}	1.89
	100	3.62×10^{-10}		5.94×10^{1}	9.3×10^{9}	3.92
α Cyg, HD 197345	4	1.29×10^{-9}		2.11×10^{2}	2.6×10^{9}	1.08
	8	4.92×10^{-10}		8.07×10^{1}	6.75×10^{9}	2.84
	10	4.00×10^{-10}		6.56×10^{1}	8.25×10^{9}	3.47
	20	2.77×10^{-10}		4.54×10^{1}	1.20×10^{10}	5.09
	50	2.43×10^{-10}		3.98×10^{1}	1.37×10^{10}	5.76
	100	2.38×10^{-10}		3.90×10^{1}	1.40×10^{10}	5.94
	4	1.29×10^{-9}	2.2×10^{11}	2.11×10^{2}	1.04×10^{9}	0.44
	8	4.92×10^{-10}		8.07×10^{1}	2.70×10^{9}	1.13
	10	4.00×10^{-10}		6.56×10^{1}	3.3×10^{9}	1.40
	20	3.39×10^{-9}		4.54×10^{1}	4.8×10^{9}	2.03
	50	7.41×10^{-10}		3.98×10^{1}	5.5×10^{9}	2.30
	100	3.62×10^{-10}		3.90×10^{1}	5.6×10^{9}	2.39

same model and many of the results of HERBIG's (1968) analysis of ξ Oph. In particular we have used a similar cloud structure to that used by HERBIG, i.e. thin, very dense interstellar clouds with most of the absorption coming from the HI regions. The electron densities were taken as 1 cm^{-3} in the clouds on the basis of HERBIG's analysis and similar density in the clouds studied in this report. Finally, we have used HERBIG's recombination coefficient for lithium in the HI regions.

Several assumptions must be made to calculate the ionization coefficient

$$\Gamma = \int_{\nu_0}^{\infty} \frac{4\pi J_\nu}{h\nu} a_1(\nu)\, d\nu,$$

where $a_1(\nu)$ is the atomic absorption cross-section from the ground state of the neutral atoms, J_ν is the intensity of the ionizing radiation and ν_0 is the frequency of the ionization edge for lithium. It is assumed that the ionizing radiation intensity comes from only two sources, the general interstellar radiation field and the radiation field from the background star. The interstellar radiation field is taken from LILLIE (1968) and the stellar radiation field in the case of α Cyg is taken from MIHALAS' (1966) Balmer line-blanketed model atmosphere for A stars and for ε Ori it is from MIHALAS and MORTON's (1965) model of a Bl V stellar atmosphere with line blanketing.

If the lithium in these clouds exists with ALLEN's (1963) "cosmic" abundance, [log N(Li)=3 on the scale where log N(H)=12] could we detect the lines?

Using the upper limits of the hydrogen column densities found by SAVAGE and CODE (1969) from their Orbiting Astronomical Observatory (OAO) observations toward ε Ori, an upper limit for lithium is 1.1×10^{12} atoms cm^{-2}. Also assume approximately half of all the material in the line of sight lies in a cloud moving with a heliocentric velocity of 25 $\mathrm{km \cdot s^{-1}}$. In the case of α Cyg, no OAO data were available on the hydrogen column densities in the direction of this star. However, assume that the hydrogen density is the same order of magnitude as that found for ε Ori. Consider that the density of material is distributed evenly between five clouds which are moving with the same heliocentric velocities as the sodium clouds found by HOBBS (1969a) and that each cloud has the same lithium column density as that of the single dense cloud in ε Ori. This assumption on the lithium column density in α Cyg seems appropriate considering the large equivalent width found in sodium absorption lines formed in these clouds. The results of this calculation are listed in Table 1 for various cloud star distances and also include lithium column densities for α Cyg which are less than that of ε Ori. Column 1 gives the star, column 2 gives the star-cloud distance in parsec, column 3 gives the ionization ratio coefficient, column 4 gives the total lithium column density, column 5 gives the ratio of total column density to neutral column density and column 7 gives the expected equivalent width of the $^2S_{1/2} - ^2P_{3/2}$ lithium line corresponding to the lithium density in column 6.

V. Results and Discussion

The results of the fitting program are shown graphically in Figs. 3 and 4. Besides the results of the fitting program, the corrected data and the position of the interstellar sodium clouds are shown. It may be reasonable to assume that if lithium is present, it resides in the same cloud complexes and moves with the same radial velocity as the sodium clouds. On this assumption we should look for maximum lithium absorption at radial velocities corresponding with those found for sodium by HOBBS (1969a). These results are shown in Table 2. Column 2 lists the maximum equivalent width found anywhere within the scan along with the radial velocity at which it was found. Column 3 lists the average equivalent width of all points within the scan, column 4 and column 5 list the radial velocities of HOBB's sodium clouds and the equivalent width of the lithium $^2S_{1/2} - ^2P_{3/2}$ feature which could be placed there, columns 6, 7, and 8 list the neutral lithium column densities which correspond to the equivalent widths found in columns 2, 3, and 5 respectively.

The results of the previous section show that if lithium were present in the interstellar clouds in a normal "cosmic" abundance then under the assumptions made for the cloud ionization and recombination model we could have detected the lithium resonance absorption lines if the star to cloud distance were greater than a certain distance (8 pc in the case of α Cyg and 100 pc in the case of ε Ori).

In the case of α Cyg it appears that no lithium was detected. At the position of the five strong sodium lines there is only one peak of the lithium fit which corresponds to one of the sodium absorption radial velocities. This correspondance is with one of the smaller peaks in the lithium fit and is smaller than the 1.5% r.m.s noise present in the data.

Table 2. Observed maximum equivalent widths of the lithium resonance line

1 Star	2 W_λ max in mÅ	3 W_λ Ave. mÅ	4 V_r km·s^{-1}	5 W_λ at V_r mÅ	6 N_2 No. of atoms per cm^2	7 N_3 No. of atoms per cm^2	8 N_5 No. of atoms per cm^2
α Cyg, HD 197345	2.54	1.54	−20.5	1.16			2.7×10^9
	at −23 km·s^{-1}		−11.9	1.85	6.0×10^9	3.7×10^9	4.4×10^9
			− 7.5	0.49			1.2×10^9
			− 2.5	1.43			3.4×10^9
			1.9	1.75			4.1×10^9
ε Ori, HD 37128	4.06	2.19	4.0	2.03			4.8×10^9
	at 25 km·s^{-1}		5.2	2.23			5.3×10^9
			11.7	2.03	9.6×10^9	5.2×10^9	4.8×10^9
			17.6	2.17			5.1×10^9
			25.2	4.06			9.6×10^9
			28.2	3.11			7.4×10^9

In the case of the ε Ori observation, the temptation becomes much stronger to say that lithium was detected. In the fitting program output, the maximum possible amount of lithium is found at the same radial velocity as the strongest sodium absorption line. It is also comforting that this amount of lithium is very close to what we might expect if a normal lithium abundance is assumed.

A closer examination of the statistics is necessary before any claim of detection can be made. The noise level on this data is 2.5% r.m.s. The central fitted absorption line depth is 6% of the continuum. A further calculation was made to determine the sensitivity of the depth of the absorption line as a function of the noise present in the data. The fitting program gives $S_v = f(y_1, y_2, \ldots, y_n)$, where S_v is the depth of the absorption line fitted at the radial velocity V and f is a function of the data points used to make the least squares fit. Considering that the data points y_i have a random noise with a variance σ_y^2, then we expect a variance σ_s^2 of S_v to be given by

$$\sigma_s^2 = \sum_i \left(\frac{\partial f}{\partial y_i}\right)^2 \sigma_y^2,$$

where the sum on i runs from 1 to the maximum number of data points used in the fitting procedure. Taking σ_y equal to the 2.5% r.m.s. noise level, S has a standard deviation $\sigma_s = 4.0\%$ r.m.s value of the continuum. A 4% standard deviation on the depth of the absorption line precludes a claim of the detection of the lithium resonance absorption when a maximum absorption depth using this model yields at best a 6% line depth.

VI. Conclusion

There is no doubt that the results of these observations are tantalizing. The line or lines are just on the limit of detectability. While no claim can be made for really detecting lithium in the interstellar media, the results on ε Ori are certainly suggestive and further observations should be undertaken. The upper limit on the lithium abundance in the direction of these two stars is taken to be equal to the value found in column 6 of Table 2. One might argue that a better upper limit might be that of the maximum value of column 8 of this table. However, we are assuming the worst possible case as our upper limit.

Further observations will require an instrumental resolution as high as was used in these observations due to the expected very narrow absorption lines. It is much easier to detect a line 4% deep that is 40 mÅ wide than it is to detect a line that appears to be 0.5% deep and 320 mÅ wide. The results on ε Ori are suggestive, although in light of HOBB's (1969a) interstellar sodium measurements there are several other stars which might be better candidates for interstellar lithium measurements.

Acknowledgements

I wish to thank the staff and personnel of Kitt Peak National Observatory, Tucson, Arizona, U.S.A., for their help and support of this project. I wish to thank Dr. WES TRAUB of the Smithsonian Astrophysical Observatory for his comments and suggestions on the first draft of this communication. I wish to acknowledge the financial support of the University of Wisconsin Alumni Research Foundation. I wish to thank F. E. BARMORE and W. D. EVANS of the University of Wisconsin for their help in setting up and carrying out the observations. I wish to acknowledge the help of Dr. F. L. ROESLER of the University of Wisconsin.

References

ALLEN, C. W.: Astrophysical Quantities. Second Edition. London: Athlone Press 1963.
BONSACK, W. H., GREENSTEIN, J. L.: Astrophys. J. **131**, 83 (1960.)
HERBIG, G. H.: Z. Astrophys. **68**, 243 (1968).
HOBBS, L. M.: Astrophys. J. **142**, 160 (1965).
HOBBS, L. M.: Astrophys. J. **157**, 135 (1969a).
HOBBS, L. M.: Astrophys. J. **157**, 165 (1969b).
HOBBS, L. M.: Astrophys. J. **158**, 461 (1969c).
LILLIE, C.: Ph. D. Thesis, University of Wisconsin (1968).
MACK, J. E., MCNUTT, D. P., ROESLER, F. L., CHABBAL, R.: Applied Optics **2**, 873 (1963).
MCNUTT, D. P.: J. Opt. Soc. Am. **55**, 288 (1965).
MIHALAS, D. M.: Astrophys. J. **144**, 454 (1966).
MIHALAS, D. M., MORTON, D. M.: Astrophys. J. **142**, 263 (1965).
ROESLER, F. L., MACK, J. E.: J. Physique **28**, C2—313 (1967).
SAVAGE, B. D., CODE, A. D.: In L. HOUZIAUX and H. E. BUTLER (eds.), Ultraviolet Stellar Spectra and Related Ground-Based Observations. IAU Symp. No. 36. Dordrecht-Holland: D. Reidel 1969.
SPITZER, L., Jr.: Astrophys. J. **109**, 548 (1949).

Ionized Gas in the Direction of the Galactic Center I: Kinematics and Physical Conditions in the Nuclear Disk

By P. G. Mezger, E. B. Churchwell, and T. A. Pauls
Max-Planck-Institut für Radioastronomie, Bonn, GFR

With 3 Figures

Abstract

A survey has been made of radio recombination line radiation from ionized gas in the direction of the galactic center. Interpretation of these observations suggests that the HII regions in the galactic center are ionized by recently formed massive stars and that the star formation rate in the region $150 \geqq R \geqq 50$ pc is extremely high. Star formation appears to be most active at positive galactic longitudes, where most of the molecular clouds and all discrete giant HII regions are observed. At negative longitudes star formation appears to have ceased, the region contains, apart from the extended HII region, only a few molecular clouds and two SN remnants.

I. The Gaseous "Nuclear" Disk

The two principal constituents of the central part of the Galaxy are a dense star cluster and a flat, rapidly rotating gaseous disk, usually referred to as "the nuclear disk". In another paper published in these Proceedings (Mezger, 1974, hereafter referred to as "Paper I") our present knowledge of the physical state and dynamics of the nuclear disk is summarized (see Paper I, Fig. 9). The gas layer extends out to a distance of about 800 pc from the center; thickness increases from about 80 pc in the center to about 250 pc at the edge of the disk (Rougoor 1964).[1]

The rotation curve for the outer part of the nuclear disk has been obtained from H λ21 cm line observations (Rougoor and Oort, 1960; Rougoor, 1964). In order to fit the observed rotation curve, a total mass within $R \leqq 800$ pc of about $10^{10}\, M_\odot$ is required (see, e. g. Sanders and Lowinger, 1972). The total mass of gas, estimated from H λ21 cm and a number of molecular lines, is $\simeq 10^7\, M_\odot$. The gravitational field is, therefore, dominated by the stellar mass distribution. In order to maintain the finite width of the gas layer, turbulent gas velocities in the central part of the nuclear disk must be of the order of $100\ \mathrm{km \cdot s^{-1}}$ (Rougoor and Oort, 1960).

Within $R \leqq 300$ pc, most of the gas appears to be in the form of molecules. Observations (reviewed in Paper I) indicate that the molecules are primarily found in dense clouds whose distributions (indicated by bars in Paper I, Fig. 9, lower part), are highly asymmetric with respect to the dynamical center of the Galaxy; most of

[1] For an adopted distance of the galactic center of 10 Kpc, an angular distance of 0°.57 corresponds to a linear (projected) distance of 100 pc.

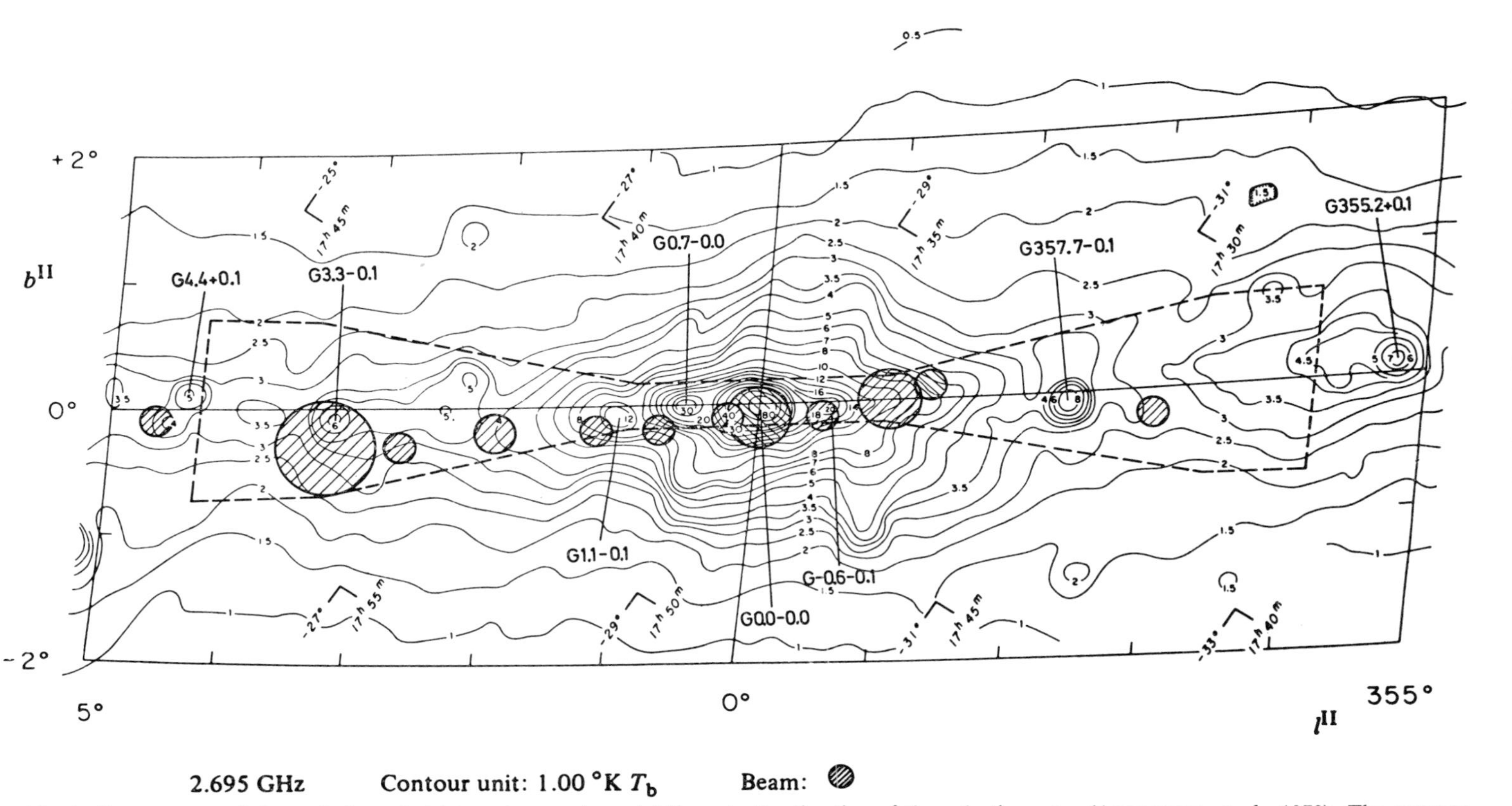

Fig. 1. Contour map of the main beam brightness temperature at λ11 cm in the direction of the galactic center (ALTENHOFF *et al.*, 1970). The antenna HPBW is 11 minutes of arc. The approximate extent of the gaseous nuclear disk is indicated by dashed lines. Hatched areas represent 100 µm IR sources observed by HOFFMAN *et al.* (1971 b). IR telescope HPBW 12 minutes of arc. *G*-numbers indicate radio sources; galactic co-ordinates are rounded off to one decimal

the clouds lie at positive galactic longitudes (ROBINSON and MCGEE, 1970; SCOVILLE *et al.*, 1972; SCOVILLE, 1972a; SOLOMON *et al.*, 1972). The observed velocities of the molecular clouds suggest the presence of expanding (or contracting) rings (SCOVILLE, 1972b, KAIFU *et al.*, 1972). Some dynamical parameters for the two limiting approximations of the expanding ring suggested by SCOVILLE (1972b) are given in Paper I, Table 6.

II. Radio and IR Continuum Observations

In the radio continuum, one observes strong emission in the direction of the galactic center. Fig. 1 shows a contour map for 11 cm continuum emission, observed by ALTENHOFF *et al.* (1970) with an angular resolution of 11 minutes of arc. The dashed line indicates the approximate extent of the gaseous nuclear disk. The shape of this disk is not well known and its outline in Fig. 1 should be considered only as a schematic representation. Strong IR λ100 μm continuum radiation has also been detected from an area $l \times b \simeq 4° \times 2°$ around the galactic center and from a number of discrete sources (HOFFMANN *et al.*, 1971a, b). Positions of the discrete IR sources are shown, superimposed on the 11 cm contour map (Fig. 1). When no size is given for the IR source, the circles drawn have diameters of 12 minutes of arc (the resolution of the IR observations).

Low-frequency radio observations of the galactic center region have been interpreted in terms of an extended thermal source embedded in a source of non-thermal radiation (MILLS, 1956; SMITH *et al.*, 1956; WESTERHOUT, 1958). The size (half power width = HPW) of the nonthermal source, $l \times b \simeq 5° \times 2°$, agrees approximately with the outer boundaries of the stellar spheroid shown in the schematic diagram Fig. 9, Paper I. At 11 cm, the non-thermal background radiation dominates and, therefore, most of the extended background radiation in the contour map, Fig. 1, is due to this non-thermal radiation. The contour maps, Fig. 2, on the other hand, have been observed by DOWNES *et al.* (1965) at 3.75 cm, at which wavelength the non-thermal radiation, owing to its steep spectrum ($T_b \propto \nu^{-2.7}$), has virtually disappeared, and the extended thermal radiation (which has a temperature spectral index $\beta \simeq -2.1$) dominates. HOLLINGER (1965) estimates the size of this extended thermal source to be $l \times b = 1\overset{\circ}{.}5 \times 0\overset{\circ}{.}5$. It is indicated in the schematic diagram Paper I, Fig. 9, by the hatched area. DOWNES (1970) estimates a maximum emission measure of the extended thermal source of 4×10^4 pc cm^{-6}.

III. Radio Recombination Line Observations of Discrete Sources

All discrete sources contained in the contour maps, Figs. 1 and 2, with peak flux densities $S_{5\,\mathrm{GHz}} \geqq 3.7$ f.u. (1 f.u. = 1 flux unit = 10^{-26} W · m^{-2} · Hz^{-1}) were included in three H109α recombination line surveys (MEZGER and HÖGLUND, 1967; REIFENSTEIN *et al.*, 1970; WILSON *et al.*, 1970). Weak recombination line emission in the direction of the center component G − 0.05 − 0.05[2] (Sgr A) is reported by ROBERTS and LOCKMAN (1970) and BALL *et al.* (1970).

[2] Galactic (G) coordinates rounded off to two decimals are used throughout this paper to designate sources.

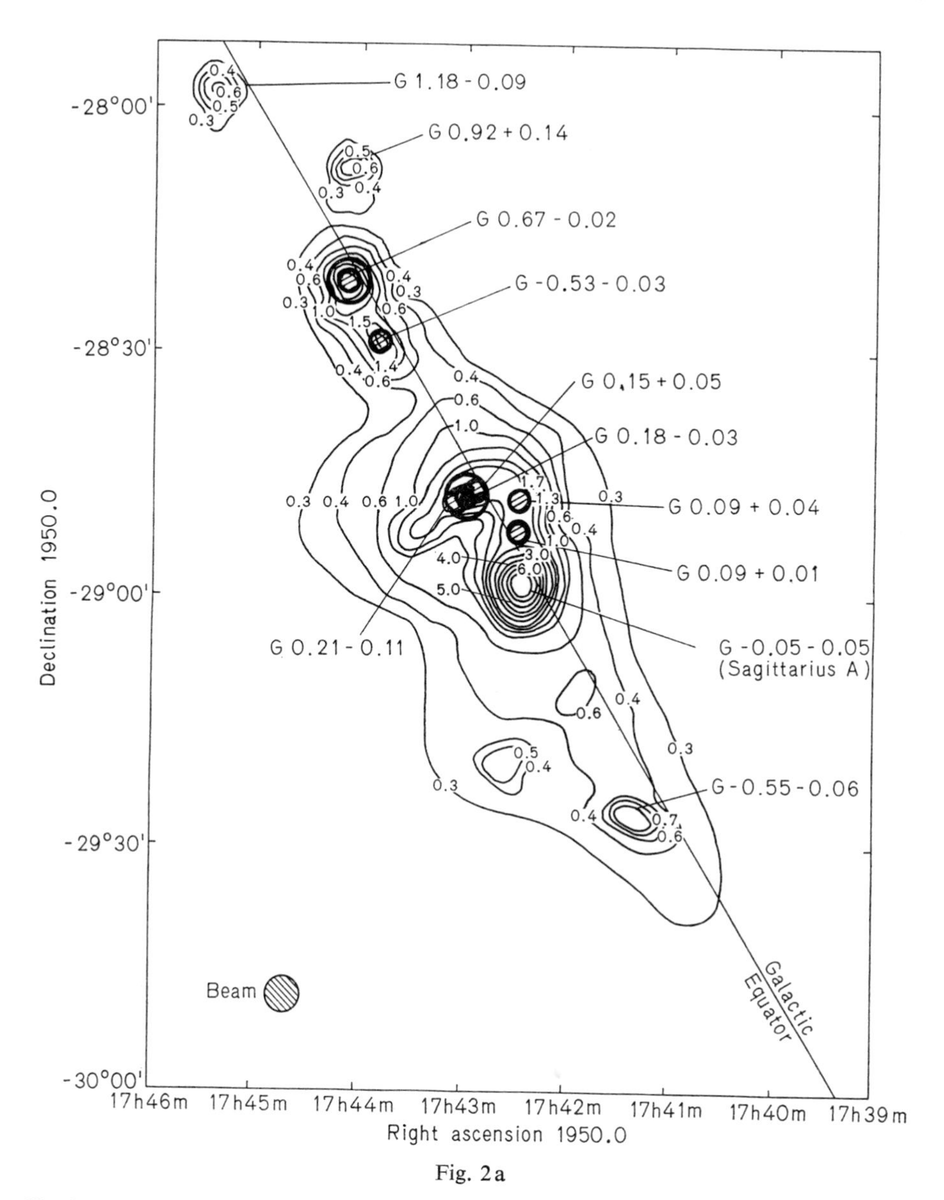

Fig. 2a

Fig. 2a and b. Contour map of the main beam brightness temperature at λ3.75 cm in the Sagittarius complex (DOWNES *et al.*, 1965). The antenna HPBW is 4 minutes of arc. The circles indicate the positions at which recombination line observations were made. The sizes of the circles indicate the HPBW of the telescope at the relevant wavelength. The discrete sources were observed for 109α (λ6 cm) and 85α (λ3 cm) emission (Fig. 2a); the extended thermal source was observed only at 109α (Fig. 2b)

We have recently made more observations of the discrete sources in the 109α and 85α recombination lines, at wavelengths 6 cm and 3 cm, respectively, using

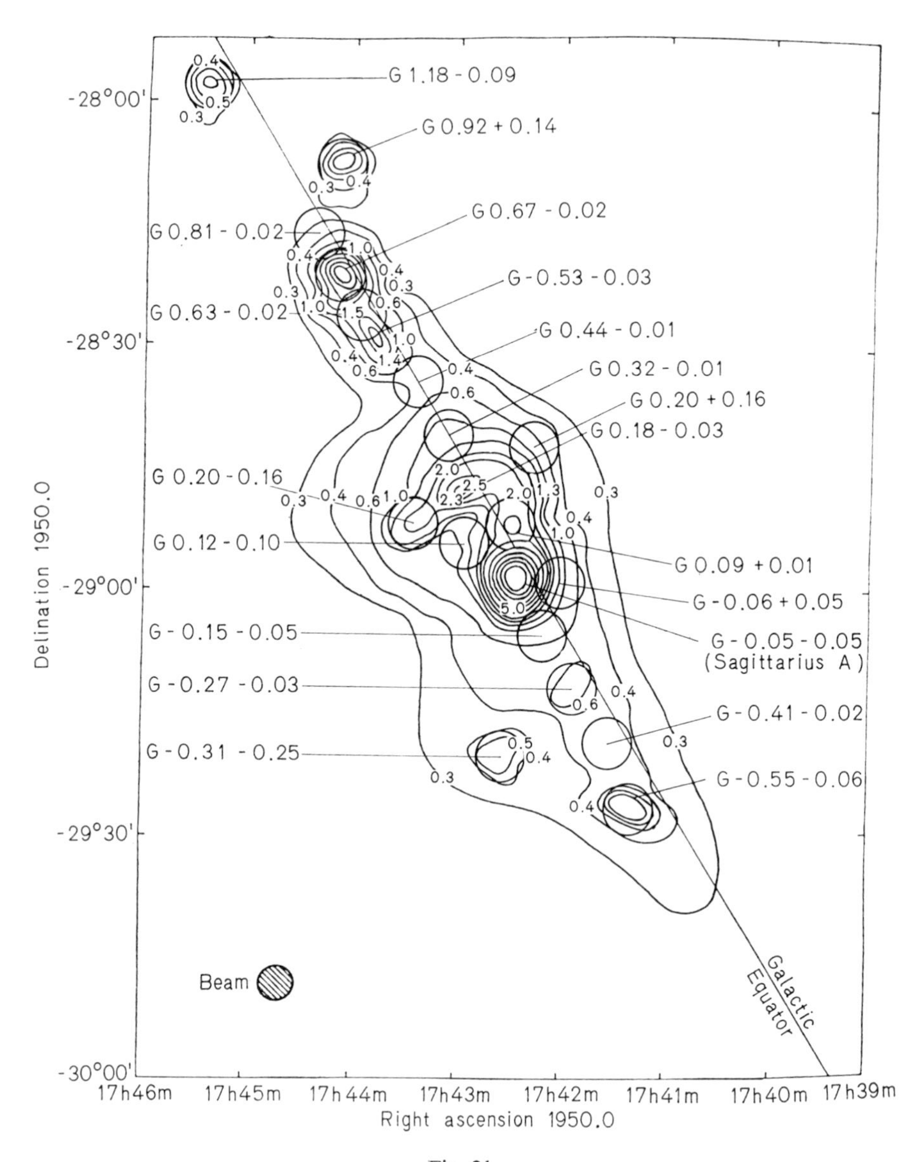

Fig. 2b

the NRAO[3] 140-ft telescope. The positions at which we searched for line emission are indicated in Fig. 2a by circles whose sizes correspond to the HPBW of the telescope.

All sources with $S_{5\,\mathrm{GHz}} \geqq 3.7$ f.u. and with projected positions inside the nuclear disk (Fig. 1) are listed in Table 1 together with their observed and derived properties. W49 A, a giant HII region located in the Sagittarius spiral arm, is given for

[3] The NRAO is operated by Associated Universities, Inc., under contract with the National Science Foundation.

comparison purposes. Columns 2 and 3 give the source HPW in α and δ. Column 4 gives the radio continuum flux density at $\nu = 5$ GHz, column 5 gives the continuum spectral index. The numbers in brackets in this column indicate the frequency interval (in GHz) in which the spectral index has been determined. Radial velocities, reduced to the Local Standard of Rest (LSR), are given in column 6. These velocities are usually weighted averages of values from more than one survey. Velocities given in brackets pertain to gaussian subcomponents, determined by a decomposition of observed non-gaussian line profiles. Lines observed in the direction of G−0.05−0.05 are weak and narrow and probably arise in the diffuse ionized gas along the line of sight (ROBERTS and LOCKMAN, 1970; BALL *et al.*, 1970).

Three criteria have been used to estimate the distances given in column 7: (1) the HI and molecular absorption lines in the spectrum of the source; (2) the radial velocity (from recombination lines); and (3) the identification with an optically visible object. We suggest that all sources in the longitude range $1.^{\circ}1 \geqq l \geqq -2.^{\circ}4$ are located in the nuclear disk. This suggestion is based on the similarity of the absorption spectra of all of these sources with that of G−0.05−0.05 (the non-thermal source, Sgr A, believed to be close to or at the center of the star cluster); also, the velocities of some of the sources are too large (either positive or negative) to fit the Schmidt model of galactic rotation outside the 3 kpc arm. For the southernmost source, G355.2+0.1, no recombination line has been detected; but it is believed to correspond to the local HII region NGC 6383. For the two northernmost sources, G3.3−0.1 and G4.4+0.1, H109α emission has been observed. The measured radial velocities are compatible with the Schmidt model of galactic rotation and two possible kinematic distances are given in column 7; in the following, we adopt for these HII regions the near distances. Thus, of the twelve sources in the direction of the nuclear disk, three, G355.2+0.1, G3.3−0.1, and G4.4+0.1 are not believed to be situated in the disk. It should be noted that our separation of sources located inside and outside the nuclear disk is in agreement with KERR and VALLAK (1967) and KERR and KNAPP (1970). LTE electron temperatures, computed from integrated H109α and H85α profiles, together with continuum measurements, are given in columns 8 and 9. Because of deviations from LTE, these temperatures usually represent lower limits.

Electron density N_e, emission measure E, and mass of ionized hydrogen (given in columns 10, 11 and 13) have been computed with the formulae given by SCHRAML and MEZGER (1969) for a spherical HII region of uniform density. The total number L_c of Lyman continuum photons per second required to maintain the ionization of the HII region (given in column 12) was computed with the following relation (see e.g. MEZGER, 1974, Eq. VI.18, hereafter referred to as Paper II)

$$\left[\frac{L_c}{\mathrm{s}^{-1}}\right] = 4.761 \cdot 10^{48}\, a(\nu_1\, T_e)^{-1} \left[\frac{\nu}{\mathrm{GHz}}\right]^{+0.1} \left[\frac{T_e}{{}^{\circ}\mathrm{K}}\right]^{-0.45} \left[\frac{S_\nu}{\mathrm{f.u.}}\right] \left[\frac{D}{\mathrm{kpc}}\right]^2 . \quad (1)$$

Here we substituted $a(\nu, T_e) = 1$, $T_e = 10^4$ K, and D and $S(\nu = 5$ GHz) from columns 4 and 7. Column 14 gives the IR flux of the sources, as observed by HOFFMAN *et al.* (1971a, b), in the wavelength range 75 μm–125 μm.

Of the nine sources which are believed to be actually located in the nuclear disk, five are HII regions with characteristics similar to those of giant spiral arm HII regions, such as W49 A.

Table 1. Characteristics of strong radio

Source	arc min θ_α	arc min θ_δ	$S_{5\,\mathrm{GHz}}$ f.u.	α	V_{LSR} $\mathrm{km\cdot s^{-1}}$	D kpc	$T_{e_{\mathrm{LTE}}}$ H85α	$T_{e_{\mathrm{LTE}}}$ H109α
G355.24+0.10	7.5	9.3	13.6	~0.0 (3–5)	no detection	$\leqq$ 3		
G357.66−0.09	4.4	3.4	15.7	−0.4 (1.4–5)	no detection	$\geqq$ 10		
G−0.53−0.06	5.7	6.7	15.4	−0.7 (1.4–5)	no detection	10		
G−0.05−0.05 (Sgr A)	3.7	3.7	190.0	−0.7 (2–15)	(+18; +4; −5)	10		
G0.07+0.01	7.0	5.0	80.0	−0.1 (5–15)	−56.2	10	7000	
G0.18−0.03	17.0	5.0	157.0	−0.1 (3–15)	−22.8; (+11.9; +43.1)	10	7600	
G0.53−0.03	4.4	5.2	35.5	−0.2 (3–15)	+45.7	10	10000	6300
G0.67−0.02 (Sgr B2)	2.9	3.6	47.8	−0.2 (3–15)	+63.1; (+36.1)	10	9000	6900
G0.87+0.10	6.4	1.6	7.9	+0.2 (3–8)	no detection	10		
G1.15−0.06	2.6	3.8	10.4	0.0 (1.4–8)	−21.7	10		6300
G3.27−0.10	8.2	8.0	11.0	~0.0 (3–5)	+4.0	2.2 or (17.8)		4200
G4.40+0.09	3.8	4.2	6.2	~0.0 (3–5)	+9.1	3.4 or (16.5)		5200
W49 A	1.5	2.0	49.8	–	–	13.8		7700

Notes:

1. Source sizes and flux densities at 5 GHz were taken from REIFENSTEIN *et al.* (1970). Spectral indices for all sources except G357.7−0.1, G355.2+0.1, G−0.6−0.1, G3.3−0.1, and G4.4+0.1 were taken from DOWNES and MAXWELL (1966). For these five sources spectral indices are taken from or inferred from ALTENHOFF *et al.* (1970). The numbers in parentheses indicate the frequency interval in GHz over which the spectral index is determined.

2. Radial velocities are weighted average values of our own observations and observations by REIFENSTEIN *et al.* (1970) and WILSON *et al.* (1970). Data on G−0.05−0.05 are from BALL *et al.* (1970) and ROBERTS and LOCKMAN (1970).

The other four sources show no recombination line emission. For three of the four, the spectral index differs sufficiently from the value expected for thermal emission ($\alpha = -0.1$ to -0.2) that a non-thermal radiation mechanism is indicated. G357.66−0.09 and G−0.55−0.06 have typical non-thermal spectra ($\alpha = -0.7$, ALTENHOFF *et al.*, 1970) and are believed to be supernova remnants. G−0.05−0.05

sources in the galactic plane between $-5° \leqq l \leqq 5°$

N_e cm^{-3}	E 10^5 pc cm^{-6}	L_c 10^{49} s^{-1}	$M_{HII}/M_\odot$ $\times 10^4$	F_{IR} (75 μm–125 μm) $\times 10^{-13}$ W cm^{-2}	Remarks
47	0.2	$\leqq 1$	0.07		Probably local since identified with NGC 6383 (REIFENSTEIN *et al.*, 1970). Calculations made on the assumption of 3 kpc for the distance. Distance given on the basis of HI absorption measurements of KERR and KNAPP (1970).
				6.4	
			}	24	Contains several compact radio components (EKERS and LYNDEN-BELL, 1971; DOWNES and MARTIN, 1971).
105	2.8	71	2.01		
76	2.3	139	5.47		
96	1.9	31	0.97		At least one compact radio component is associated with the source (EKERS and LYNDEN-BELL 1971).
201	5.6	42	0.63 }	10	Contains at least one associated compact radio component (EKERS and LYNDEN-BELL 1971); HOBBS *et al.*, 1971) find a compact component of 72 f.u. at λ 3 mm.
98	1.3	9	0.28	3.7	Contains one associated compact radio component (EKERS and LYNDEN-BELL, 1971).
52	0.2	0.5	0.03	1.5	Probably local (near distance used for calculations). On the basis of HI absorption KERR and KNAPP (1970) derive a distance of 6 kpc, but they indicate that this is uncertain.
90	0.5	0.6	0.02	0.4	Probably local (near distance used for calculations).
430	19	84	0.54	1.2	Contains compact radio components (WEBSTER, ALTENHOFF, and WINK, 1971). WYNN-WILLIAMS (1971) indicates six compact components.

3. For the calculations of N_e, E, L_c, and $M_{HII}/M_\odot$ it was assumed that $T_e = 10000$ K, $a(\nu, T_e) = 0.984$ and $N(He^+)/N(H^+) = 0.10$. The derived values do not depend very strongly on these assumptions.

4. All values given for W49 A were taken from REIFENSTEIN *et al.* (1970).

5. IR flux densities are taken from HOFFMAN, FREDERICK, and EMERY (1971a, b).

(Sgr A) has a typical non-thermal spectrum at cm-wavelengths ($\alpha = -0.7$); at longer wavelengths the spectrum gets considerably flatter ($\alpha = -0.25$ for 15 cm $\leqq \lambda \leqq$ 150 cm) and it eventually turns over at wavelengths longer than 150 cm. This behaviour can be explained by a model in which the non-thermal source Sgr A is surrounded by a thermal plasma with an emission measure $E = 4 \cdot 10^4$ pc $\cdot$ cm^{-6}

(for an assumed electron temperature of $T_e=6000$ K, BREZGUNOV *et al.*, 1971). G0.92+0.14 is peculiar: It has a positive spectral index ($\alpha=0.2$, DOWNES and MAXWELL, 1966), but no recombination line emission in the velocity range $300\ \mathrm{km \cdot s^{-1}} \geqq v_{\mathrm{LSR}} \geqq -300\ \mathrm{km \cdot s^{-1}}$. It is the only discrete source in the galactic center region which is located at positive latitudes; it is the only thermal source in this region which is not associated with a discrete IR λ100 μm source. It is possible that G0.92+0.14 does not belong to the nuclear disk; it may lie on the other side of the galactic center or it may even be an extragalactic object.

IV. Radio Recombination Line Observations of the Extended Thermal Source

We have recently extended our recombination line survey of the galactic center region to include the diffuse thermal source (apparent size $l \times b = 1\overset{\circ}{.}5 \times 0\overset{\circ}{.}5$). The H_2CO λ6 cm line is seen in absorption (at least in part) against this extended thermal source; from the similarity of the absorption spectra to that of G−0.05 −0.05, we conclude that this extended ionized gas is located in the inner part of the nuclear disk. In Fig. 2b the positions where we have searched for and found H109α emission are indicated by circles whose diameters correspond to the HPBW of the 140-ft telescope. This survey is not yet completed and a full account of the observations will be given elsewhere. The detection of recombination lines verifies the thermal nature of this source. However, the H109α lines emitted by the extended HII region are much wider than the lines emitted by the discrete HII regions; line widths range from 40 to 100 $\mathrm{km \cdot s^{-1}}$, compared to typical line widths of 20 to 40 $\mathrm{km \cdot s^{-1}}$ for the discrete sources (line widths in "spiral arm" HII regions are also typically 30 $\mathrm{km \cdot s^{-1}}$).

V. Comparison of the Dynamics of Ionized Gas and Molecular Clouds

Fig. 3 is a v_{LSR}, l-diagram for the ionized gas and molecular clouds (H_2CO λ6 cm absorption). The contours represent lines of constant optical depth of the H_2CO line; this quantity is an indicator of the column density of all neutral gas. Dotted contours correspond to gas which SCOVILLE *et al.* (1972) and SCOVILLE (1972a) attribute to the local arms and the 3 kpc arm. The asymmetry with respect to the galactic center is conspicuous. The two ellipses fitted by SCOVILLE (1972a, b), to the molecular clouds in this diagram are indicated as full and dashed curves labelled I and II. These ellipses correspond to expanding (or contracting) rings of gas (see also Paper I, Section VII). SCOVILLE's observations favour expansion for most of the clouds.

Points in Fig. 3 represent recombination line emission of the ionized gas. Dots refer to the discrete HII regions, triangles to the extended HII region (i.e. the diffuse thermal emission). Squares refer to the ionized gas in the immediate vicinity of the galactic center — with projected distances from the galactic center less than 50 pc. In the case of decomposed non-gaussian profiles, filled symbols refer to the main component, open symbols refer to subcomponents. Bars in the radial velocity co-ordinate indicate the width of the gaussian line (or line components, respectively).

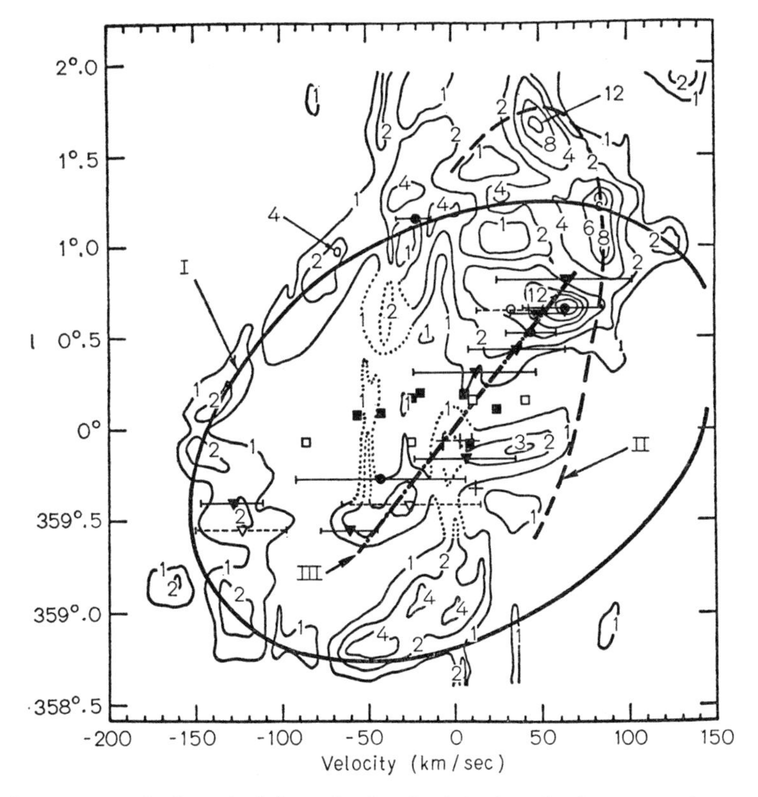

Fig. 3. Contour map (in l, v_{LSR}) of the molecular clouds in the galactic center region, as outlined by their HC_2O λ6 cm line in absorption. Contours indicate the optical depth, projected onto the galactic plane. The ellipses, I and II, represent two possible approximations to the observed cloud distribution (Scoville, 1972b). Dotted contours pertain to clouds in the local arm and in the 3 kpc arm. Points pertain to the ionized gas. Dots represent discrete giant HII regions, triangles represent the extended HII region, squares represent ionized gas with projected distances from the dynamical center of the Galaxy less than 50 pc. Filled symbols pertain to the main component, open symbols to subcomponents (in the case of composite line profiles.) Bars in v_{LSR} indicate line widths of the individual components.

The discrete HII regions, like the molecular clouds, show a strong asymmetry in their longitude distribution. In fact, all of the HII regions attributed, in Table 1, to the nuclear disk are located at positive galactic longitudes.

Apart from the ionized gas in the vicinity of the galactic center (represented by squares in Fig. 3), the ionized gas correlates well with the molecular clouds. However, only one discrete HII region, G1.15−0.06, and two lines emitted by the extended HII region at negative longitudes ($l \simeq -0^{\circ}.5$) appear to be correlated with molecular clouds which are a part of the expanding rings represented by the ellipses I and II. The majority of the ionized gas falls at projected distances from the galactic center of less than 150 pc (i.e. $|l| < 0^{\circ}.9$). Points representing discrete HII regions and the extended HII region between 50 and 150 pc projected distance from the center lie along a straight line, indicated in the diagram by the dash-dotted curve III. A straight line in this diagram is consistent with either a disk in solid-body rotation or a rotating ring which neither expands nor contracts. We believe that a rotating ring is the most likely distribution of the bulk of the ionized gas. The dense molecular clouds at $l \simeq 0.^\circ 7$, $v_{\mathrm{LSR}} \simeq 65\ \mathrm{km \cdot s^{-1}}$ (often referred to as Sgr B2 cloud) and at $l \simeq -0.^\circ 5$, $v_{\mathrm{LSR}} \simeq -50\ \mathrm{km \cdot s^{-1}}$ appear to be part of this rotating ring configuration. If the stellar mass distribution suggested by SANDERS and LOWINGER (1972) is correct, this rotating ring can obviously not be in dynamical equilibrium.

SCOVILLE (1972b) and KAIFU *et al.* (1972) suggested that the giant HII regions were located in the 'expanding' rings suggested by the molecule observations. The present observations indicate clearly that most of the giant HII regions together with the extended component are located internal to the rings ($R < 150$ pc) and have quite different dynamical properties.

The dynamics of the region within 50 pc of the galactic center is quite distinct from that of the outer regions. There does not appear to be much, if any, neutral gas. The ionized gas has radial velocities ranging up to $\pm 50\ \mathrm{km \cdot s^{-1}}$. The LTE electron temperature of this gas ranges from 5000 to 20000 K and possibly higher. Such high electron temperatures may indicate that collisional ionization and heating are important. Such a situation might be expected if massive infalling clouds collide with each other or with quiescent gas.

VI. Physical State of the Ionized Gas

Discrete HII regions and molecular clouds correlate well in position and radial velocity (Fig. 3). Physical conditions derived from some of these molecular clouds suggest, that they are in a state of gravitational contraction, which eventually will lead to the formation of stars (see Paper I, Sections IV and V). We, therefore, suggest that the discrete giant HII regions in the nuclear disk are ionized by recently formed O-stars. A total number of thirty-four O6-supergiants is required to account for the ionization of the five giant HII regions listed in Table 1 as being located in the nuclear disk ($\log\{L_c(\mathrm{O\,6\,I})\} = 49.93$; CHURCHWELL and WALMSLEY, 1972).

The extended HII region has linear dimensions of 300 pc × 150 pc in l and b. Turbulent velocities, as inferred from the observed width of recombination lines, range up to $100\ \mathrm{km \cdot s^{-1}}$. As mentioned above, turbulent velocities of this order of magnitude are necessary to maintain the finite width of the gas layer in the center of the nuclear disk. LTE electron temperatures, as inferred from the line-to-continuum ratio, are somewhat higher than those of the discrete HII regions listed in columns 10 and 11 of Table 1, but are still well below 10^4 K. The integrated flux density and peak emission measure of this extended HII region are estimated to be $S_{5\,\mathrm{GHz}} = 410$ f.u., and $E = 4 \times 10^4\ \mathrm{cm^{-6}}$ pc, respectively. Assuming a spheroidal distribution of the ionized gas of HPW's $\theta_l \times \theta_b = 61' \times 20'$, a constant electron density N_e and an electron temperature $T_e = 10^4$ K, we derive the following characteristics of the extended HII region:

$$N_e = 16\ \mathrm{cm^{-3}}, \quad M_{\mathrm{HII}}/\mathrm{M}_\odot = 7.3 \times 10^5 \quad \text{and} \quad L_c = 3.6 \times 10^{51}\ \mathrm{s^{-1}}.$$

Some of the discrete giant HII regions appear to be embedded in this extended HII region, whose density is an order of magnitude lower than that of the discrete HII regions. The extended HII region, therefore, could be composed of a larger number of expanded low-density regions. A total number of forty-three O6-supergiants would be required to account for its ionization. However, some other mechanism of ionization can not be excluded.

VII. Correlation between 100 μm IR Emission and Thermal Radio Emission

Various authors have suggested that the far IR emission from the galactic center is due to thermal radiation of dust grains (LEQUEUX, 1970; OKUDA and WICKRAMASINGHE, 1970; KRISHNA SWAMY, 1971). POTTASCH (1971), on the other hand, showed that line radiation could only account for the observed IR flux density if the abundances of C, N, O, Si and Fe in the galactic center region are a factor of ten higher than in the solar neighbourhood. Fig. 1 shows that the positions of discrete IR sources correlate well with positions of some of the discrete radio sources. HARPER and LOW (1971) showed that there is a correlation between radio flux density [or, according to Eq. (1), the total number of Lyman continuum photons emitted per second by the ionizing star(s)] and IR flux density. Because of the high albedo of dust grains in the UV, it was suggested that trapped Lyman-alpha photons are the prime source of heating of the dust grains. In this case, one would expect the following relation between total IR flux density and radio flux density (Paper II, eq. VI.31).

$$\left[\frac{F_{\mathrm{IR}}}{W\,\mathrm{cm}^{-2}}\right] = 1.30 \times 10^{-15} \left[\frac{\nu}{10\,\mathrm{GHz}}\right]^{0.1} \left[\frac{S_\nu}{\mathrm{f.u.}}\right]. \tag{2}$$

The empirical relation between the observed IR and radio flux densities (HARPER and LOW, 1971) gives, on the average, an IR flux density five times that predicted by Eq. (2). If the IR radiation is actually thermal radiation from heated dust grains, this observation implies that only 1/5 of the heating of the dust grains is provided by absorption of Lyman alpha photons, and about 4/5 by absorption of other photons. In this section we investigate whether a similar correlation holds between IR and radio flux of thermal sources in the galactic center region.

HOFFMAN *et al.* (1971 a) mapped the IR emission in the wavelength band 75 μm to 125 μm in an area $l \times b = 3\overset{\circ}{.}6 \times 2^\circ$ around the galactic center. According to their evaluation, 36% of the total IR flux density is provided by four discrete sources and 64% of the flux density comes from an extended, smooth background emission. However, the radio observations (Section II), which have higher angular resolution than the IR observations, clearly indicate that the discrete sources are superimposed on an extended source whose dimensions correspond to those of the $22 \times 10^{-4}\,\mathrm{erg} \cdot \mathrm{s}^{-1} \cdot \mu\mathrm{m}^{-1} \cdot \mathrm{sterad}^{-1} \cdot \mathrm{cm}^{-2}$ contour in HOFFMAN *et al.*'s diagram. The IR contours are consistent with the presence of such an underlying extended component. Thus, we decompose the IR flux into three components: I) A smooth background emission extending over $l \times b = 3\overset{\circ}{.}6 \times 2^\circ$, II) A more intense and less extended emission which coincides with the extended HII region, i.e. $l \times b = 1\overset{\circ}{.}5 \times 0\overset{\circ}{.}5$, III) Four point sources. The results are given in Table 2. The total IR

Table 2. IR and radio flux densities of galactic center sources (IR flux densities according to our re-evaluation of observations by HOFFMAN *et al.*, 1971a; see text)

IR Component	Source	Radio flux $S_{5\,\mathrm{GHz}}$ f.u.			IR flux F_{IR} (75 μm–125 μm) $\times 10^{-13}$ W · cm^{-2}	
	total IR emission ($l \times b = 3\overset{\circ}{.}6 \times 2^\circ$)					127
I	smooth background ($l \times b = 3\overset{\circ}{.}6 \times 2^\circ$)	—				101
II	extended HII region ($l \times b = 1\overset{\circ}{.}5 \times 0\overset{\circ}{.}5$)	410.0				21
III	G−0.55−0.06	15.4			0.2	5
	G−0.05−0.05	190.0			3.3	
	G 0.07+0.01	80.0	237.0			
	G 0.18−0.03	157.0				
	G 0.53−0.03	35.5	83.3	155.3	1.5	
	G 0.67−0.02	47.8				
	mm-component	72.0				
	G 0.87+0.10	7.9			—	
	G 1.15−0.06	10.4			0.2	

flux density agrees within 5% with the value given by HOFFMAN *et al.* Component II accounts for 16%, the discrete sources together (=Component III) for only 4% of the total IR flux density.

A comparison of the radio and IR flux densities given in Table 2 shows that relation (2) is in fact fullfilled for the discrete HII regions in the nuclear disk, whereas the diffuse IR emission associated with the extended HII region ($l \times b \simeq 1\overset{\circ}{.}5 \times 0\overset{\circ}{.}5$) is about five times higher than the value predicted by Eq. (2) This may be partly the result of an incorrect separation of the IR emission from components II and III; here, we want mainly to stress the point that the total IR emission associated with the observed ionized hydrogen in the galactic center (i.e. component II and III) is less than the empirical ratio determined by HARPER and LOW (1971) for typical spiral arm HII regions.

However, 80% of the total IR λ100 μm radiation observed in the area $l \times b \simeq 3\overset{\circ}{.}6 \times 2^\circ$ comes from regions outside the discrete or extended HII regions observed in the radio range. SANDERS and LOWINGER (1972) suggested that this radiation comes from dust particles which are heated by the integrated star light from the central stellar cluster. However, with this mechanism it appears to be difficult to explain grain temperatures of the order of 100 K.

On the assumption that thermal emission from dust grains with grain temperatures of $T_g \simeq 100$ K is actually the prime mechanism that produces the observed λ100 μm IR emission, POTTASCH (1973) has estimated a total mass of $M_{\mathrm{Dust}} \simeq 3 \cdot 10^3$ $M_\odot$ involved in this emission. If the gas-to-dust ratio in the galactic center region is similar to the corresponding ratio in the solar vicinity, we would expect a total gas mass of $M_{Gas} \simeq 3 \cdot 10^5\ M_\odot$ to be associated with the emitting dust grains. This compares reasonably well with the observed mass of ionized gas ($\simeq 8 \cdot 10^5\ M_\odot$); the puzzling fact is the difference in distribution between the extended IR component (I) which extends to 200 pc from the plane and all the observed gas (i.e. ionized or neutral) which is within 50 pc of the plane.

There are several possible explanations for this marked difference in the distribution. The heated dust grains may be associated with a tenuous ($N_e \ll 10\ \text{cm}^{-3}$) and/or hot ($T_e \gg 10^4$ K) plasma, which would be hard to observe in the radio range. Or the distribution of dust is determined by some different mechanism such as radiation pressure.

VIII. Comparison with Nuclei of External Galaxies

Cameron (1968) has suggested that the thermal radio sources in the nuclear region of the Galaxy are analogous to the hot spots 'HS' (Morgan, 1958, 1959) and the amorphous nuclei 'AN' (Sersic and Pastoriza, 1965, 1967) observed in the nuclei of some late-type spirals. From the photographs published by Morgan, it appears that the dimensions of the HS's are characteristically 1/5 to 1/10 those of the nuclear bulge and thus of the order of 200–400 pc. Thus, one such region would appear to correspond to the entire thermal complex in the galactic center (i.e. both extended HII region and giant discrete HII regions) rather than to only one of the discrete sources as suggested by Cameron (1968). Spectra of some of the nuclei with HS or AN have been obtained by Pastoriza (1967) and by Weedman (1970). All show emission lines characteristic of low excitation HII regions. ([OIII] is rarely observed, the intensity ratio, $H\alpha$/[NII] is between 2 and 3). Hence the spectra are very like those of HII regions in the spiral arms of external galaxies (Burbidge and Burbidge, 1962), and similar to those of HII regions in the Galaxy (Johnson, 1953). The spectra of the HS and AN nuclei are unlike the spectra of the nuclei of Seyfert galaxies (higher excitation) and of early type and elliptical galaxies ($H\alpha/[\text{NII}] \leqq 1$).

Line widths in the spectra of HS and AN nuclei are of the order of 100–200 km $\cdot$ s^{-1} (Weedman, 1970). This corresponds well to the line width expected for the integrated spectrum of the extended HII region in the center of our Galaxy on the basis of observed line widths and the radial velocity distribution of radio recombination lines.

IX. Summary and Conclusion

Most of the gas within 800 pc of the galactic center is concentrated to a flat disk – "the nuclear disk". The properties of the gas vary markedly with distance from the center. In the outer part of the nuclear disk most of the neutral gas appears to be in the form of atomic hydrogen of low space density; it rotates in dynamical equilibrium on nearly circular orbits. However, with decreasing distance from the dynamical center the space density of the gas increases. At distances $R < 400$ pc, dense molecular clouds are observed; these obviously do not move on circular orbits. In fact, a large fraction of the molecular clouds appears to belong to a system of expanding rings, with radii between 200 and 300 pc. There is marked asymmetry; most of the molecular clouds lie at positive galactic longitudes.

Within 150 pc of the center, much of the gas is ionized. There are 9 radio continuum sources with flux densities $S_{5\,GHz} \geqq 3.7$ f.u. which appear to be associated with the nuclear disk. Five of these sources are thermal, four are non-thermal. All the thermal sources are located at positive galactic longitudes. Two of the four non-thermal sources have characteristics similar to those of spiral arm SN remnants and are located at negative longitudes. The other two non-thermal sources are Sgr A (G−0.05−0.05)—possibly the galactic center—and a peculiar source (G0.92+0.14) which is at positive galactic longitude.

The thermal sources do not share the expansion motions of the majority of the molecular clouds. However, they correlate in position and radial velocity with 'non-expanding' dense molecular clouds. Since these clouds appear to be gravitationally unstable, we have suggested that the associated HII regions are ionized by recently formed O-stars. Thirty-four O6 supergiants are required to account for the ionization of these discrete HII regions.

Apart from the discrete HII regions, there is an extended HII region of size 300×150 pc which shares the dynamic properties of the discrete HII regions but which appears to have a rather symmetrical distribution about the galactic center. The density is an order of magnitude lower than that in the discrete HII regions. If it is thermal, a total of forty-three O6 supergiants would be required to account for its ionization. The complex of extended and discrete thermal sources appears to be similar to "hot spots" seen in the nuclei of other galaxies.

Within 50 pc of the galactic center, all of the gas appears to be ionized and it shows high velocities (-50 km $\cdot$ s^{-1}) in the line of sight.

The presence of giant HII regions in the galactic center indicates that massive stars are being formed at a high rate. However, it appears that star formation is occuring predominantly at positive galactic longitudes. A similar burst of star formation may have occured at negative galactic longitudes some time ago, but it has now ceased; here we observe two SN remnants but fewer molecular clouds and no giant HII regions.

Within a spheroid of major and minor axes 300×125 pc, the total stellar mass is $M_* = 3 \cdot 10^9$ $M_\odot$, based on the mass distribution in the model of SANDERS and LOWINGER (1972). The total amount of gas contained in the molecular clouds within $R \leqq 300$ pc is estimated by SCOVILLE (1972b) to be 3×10^6 $M_\odot$. We estimate a total mass of ionized gas in this region of $M_{HII} \leqq 10^6$ $M_\odot$. Thus, the gas-to-star mass ratio within $R \leqq 300$ pc is $M_{gas}/M_* \simeq 10^{-3}$ and it appears possible that this amount of gas is provided by mass loss from evolved stars in the central stellar cluster, as suggested by SPITZER and SASLAW (1966). Stellar densities, on the other hand, appear to be too low for stellar collisions to contribute significantly to the mass loss.

It is of interest in this context that the gas in the central region, $R \leqq 300$ pc, of the nuclear disk shows anomalies in its chemical composition. The C^{12}/C^{13}-ratio in the molecular cloud associated with the thermal component G0.67−0.02 appears to be less than the terrestrial value of 90, which also prevails elsewhere in the interstellar gas (FOMALONT and WELIACHEW, 1973; ZUCKERMAN, 1971). Our observations of the He 109α and the He^+ 173α-lines (CHURCHWELL and MEZGER, 1973) indicate an underabundance of helium in the giant HII regions in the nuclear disk.

Acknowledgement

It is our pleasure to acknowledge stimulating discussions with W. ALTENHOFF and D. DOWNES, MPIfR; P. STRITTMATTER, University of Arizona; and P. PISHMISH, University of Mexico. We thank F. KERR (University of Maryland) and J. H. OORT (Sterrewacht te Leiden) for critical remarks. Our special thanks goes to L. F. SMITH, Institut d'Astrophysique, Liège, for her critical reading of this manuscript and her help in preparing Section VIII of this paper.

Discussion

KIPPENHAHN, R.:
Can you derive the gravitating mass of the galactic center from the velocity curves ot the rotating disc you observe?

MEZGER, P. G.:
No, because the rotation of that ionized gas is obviously *not* an equilibrium rotation.

References

ALTENHOFF, W. J., DOWNES, D., GOAD, L., MAXWELL, A., RINEHART, R.: Astron. Astrophys., Suppl. **1**, 319 (1970).
BALL, J. A., GOTTLIEB, C. A., LILLEY, A. E., RADFORD, H. E.: Astrophys. J. **162**, L203 (1970).
BREZGUNOV, V. N., DAGHESAMANSKY, R. D., UDAL'TSOV, V. A.: Astrophys. Letters, **9**, 117 (1971).
BURBIDGE, E. M., BURBIDGE, G. R.: Astrophys. J. **135**, 694 (1962).
CAMERON, M. J.: Observatory **88**, 254 (1968).
CHURCHWELL, E. B., MEZGER, P. G.: Nature **242**, 319 (1973).
CHURCHWELL, E. B., WALMSLEY, C. M.: Astron. Astrophys., **23**, 117 (1972).
DOWNES, D.: Ph. D. Thesis, Harvard University (1970) (unpublished).
DOWNES, D., MARTIN, A. H. M.: Nature **233**, 112 (1971).
DOWNES, D., MAXWELL, A., MEEKS, M. L.: Nature **208**, 1189 (1965).
DOWNES, D., MAXWELL, A.: Astrophys. J. **146**, 653 (1966).
EKERS, R. D., LYNDEN-BELL, D.: Astrophys. Letters **9**, 189 (1971).
FOMALONT, E. B., WELIACHEV, L. N.: Astrophys. J. **181**, 781 (1973).
HARPER, D. A., LOW, F. J.: Astrophys. J. **165**, L9 (1971).
HOBBS, R. W., MODALI, S. B., MARAN, S. P.: Astrophys. J. **165**, L87 (1971).
HOFFMANN, W. F., FREDERICK, C. L., EMERY, R. J.: Astrophys. J. **164**, L23 (1971a).
HOFFMANN, W. F., FREDERICK, C. L., EMERY, R. J.: Astrophys. J. **170**, L89 (1971b).
HOLLINGER, J. P.: Astrophys. J. **142**, 609 (1965).
JOHNSON, H. M.: Astrophys. J. **118**, 370 (1953).
KAIFU, N., KATO, T., IGUCHI, T.: Nature Phys. Science **238**, 105 (1972).
KERR, F. J., KNAPP, G. R.: Australian J. Phys., Astrophys. Suppl. **18**, 9 (1970).
KERR, F. J., VALLAK, R.: Australian J. Phys., Astrophys. Suppl. **3**, 1 (1967).
KRISHNA SWAMY, K. S.: Astrophys. J. **167**, 63 (1971).
LEQUEUX, J.: Astrophys. J. **159**, 459 (1970).
MEZGER, P. G., HÖGLUND, B.: Astrophys. J. **147**, 490 (1967).

MEZGER, P. G.: In L. N. MAVRIDIS (ed.), Stars and the Milky Way System, p. 88. Berlin-Heidelberg-New York: Springer 1974.
MILLS, B. Y.: Observatory **76**, 65 (1956).
MORGAN, W. W.: Publ. Astron. Soc. Pacific **70**, 364 (1958).
MORGAN, W. W.: Publ. Astron. Soc. Pacific **71**, 394 (1959).
OKUDA, H., WICKRAMASINGHE, N. C.: Nature Phys. Science **226**, 134 (1970).
PASTORIZA, M. G.: Observatory **87**, 225 (1967).
POTTASCH, S. R.: Astron. Astrophys. **13**, 152 (1971).
POTTASCH, S. R.: In L. N. MAVRIDIS (ed.), Stars and the Milky Way System, p. 209. Berlin-Heidelberg-New York: Springer 1974.
REIFENSTEIN, E. C. III., WILSON, T. L., BURKE, B. F., MEZGER, P. G., ALTENHOFF, W. J.: Astron. Astrophys. **4**, 357 (1970).
ROBERTS, M. S., LOCKMAN, F. J.: Astrophys. J. **161**, 877 (1970).
ROBINSON, B. J., MCGEE, R. X.: Australian J. Phys. **23**, 405 (1970).
ROUGOOR, G. W., OORT, J. H.: Proc. Nat. Acad. Sci. **46**, 1 (1960).
ROUGOOR, G. W.: Bull. Astron. Inst. Neth. **17**, 381 (1964).
SANDERS, R. H., LOWINGER, Th.: Astron. J. **77**, 292 (1972).
SCHRAML, J., MEZGER, P. G.: Astrophys. J. **156**, 269 (1969).
SCOVILLE, N. Z.: Ph. D. Thesis, Columbia University, New York (1972a) (unpublished).
SCOVILLE, N. Z.: Astrophys. J. **175**, L127 (1972b).
SCOVILLE, N. Z., SOLOMON, P. M., THADDEUS, P.: Astrophys. J. **172**, 335 (1972).
SERSIC, J. L., PASTORIZA, M.: Publ. Astron. Soc. Pacific **77**, 287 (1965).
SERSIC, J. L., PASTORIZA, M.: Publ. Astron. Soc. Pacific **79**, 152 (1967).
SMITH, F. G., O'BRIEN, P. A., BALDWIN, J. E.: Monthly Notices Roy. Astron. Soc. **116**, 282 (1956).
SOLOMON, P. M., SCOVILLE, N. Z., PENZIAS, A. A., WILSON, R. W., JAFFERTS, K. B.: Astrophys. J., **178**, 125 (1972).
SPITZER, L., Jr., SASLAW, W. G.: Astrophys. J. **143**, 400 (1966).
WEBSTER, W. J., Jr., ALTENHOFF, W. J., WINK, J. E.: Astron. J. **76**, 677 (1971).
WEEDMAN, D. W.: Astrophys. J. **159**, 405 (1970).
WESTERHOUT, G.: Bull. Astron. Inst. Neth. **14**, 215 (1958).
WILSON, T. L., MEZGER, P. G., GARDNER, F. F., MILNE, D. K.: Astron. Astrophys. **6**, 364 (1970).
WYNN-WILLIAMS, C. G.: Monthly Notices Roy. Astron. Soc. **151**, 397 (1971).
ZUCKERMAN, B.: In L. SNYDER (ed.), Molecules in Space. Proc. Charlottesville Symp. (1971) (in press).

Results of Lunar Occultations of the Galactic Center Region in HI, OH and H_2CO Lines and in the Nearby Continua

By AA. SANDQVIST
Stockholm Observatory, Saltsjöbaden, Sweden

With 5 Figures

Abstract

A discussion of recent results of lunar occultations of the galactic center and the structure of the $+40\ \mathrm{km \cdot s^{-1}}$ absorption feature is presented.

Eleven lunar occultations of the galactic center region have been observed with the 140-foot radio telescope of the National Radio Astronomy Observatory[1], Green Bank, West Virginia, in the 21-cm HI line, the 18-cm OH lines, the 6-cm H_2CO line, and in the nearby continua. The observations and the resulting line contour maps have been presented by SANDQVIST (1971, 1973), of which the latter paper will be referred to as "Paper I".

The most interesting results have been concerned with the structure and internal motions of absorbing clouds in front of the galactic center. The present paper deals principally with a cloud of molecules with a radial velocity centered on $+40\ \mathrm{km \cdot s^{-1}}$, because this cloud is the most compact and appaers to be closest to the center itself.

First, let us look at the continuum. The first figure of Paper I illustrates one occultation situation on a 21-cm radio continuum contour map of a two-degree region of the galactic center. Sgr A is seen to be structureless in that 14′ arc-resolution map and to be immersed in a large general background. The complete two-dimensional brightness distribution of Sgr A at 21 cm, with an effective resolution of about 25″ arc, is now presented here in Fig. 1. It has been reconstructed using the complex Fourier transforms of the restored continuum curves from occultations at different position angles. The contour map was then produced using an interferometer reduction process developed by HÖGBOM (1973). The general background continuum has been subtracted out and is not included in Fig. 1.

In addition to the compact source and main peak of Sgr A, the occultations revealed a complex structure resembling somewhat the 15″ arc-resolution 2.2 μm map of BECKLIN and NEUGEBAUER (1968). There are a few exceptions to this resemblance, of which one of the more noteworthy is a secondary source (which I shall call "component S") which is about 2′ arc away from the main peak of Sgr A. The positions of Sgr A and component S are given in Table 1. The size of the Sgr A complex, to the limits of detectability, is 4.′5 arc in galactic longitude by 3.′5 arc in latitude. The integrated flux density of the complex is $220 \times 10^{-26}\ \mathrm{W.m^{-2} \cdot Hz^{-1}} \pm 10\%$.

[1] Operated by Associated Universities, Inc., under contract with the National Science Foundation.

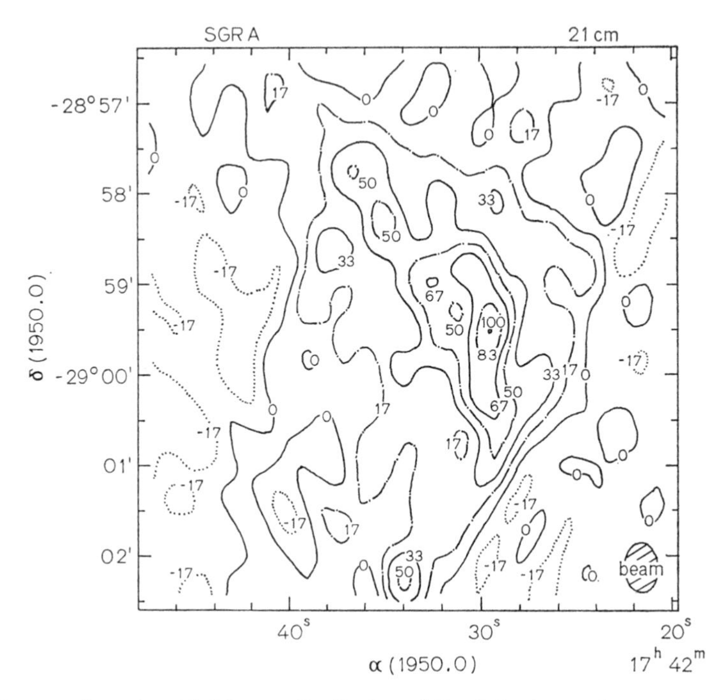

Fig. 1. The two-dimensional brightness distribution of Sgr A in the continuum at a wavelength of 21 cm. The general background continuum is not included. When multiplied by 10^{-22}, the contour values give brightness in $W \cdot m^{-2} \cdot Hz^{-1} \cdot sterad^{-2}$. The effective halfpower beamwidth is indicated in the lower right hand corner

Table 1. Positions derived from the occultation series

	α (1950.0)	δ (1950.0)
Sgr A (21 cm)	$17^h\ 42^m\ 29^s.7\ (\pm 1^s)$	$-28°\ 59'\ 30''\ (\pm 15'')$
Component *S* (21 cm)	$17^h\ 42^m\ 37^s.9\ (\pm 1^s)$	$-28°\ 58'\ 37''\ (\pm 15'')$
OH Centroid	$17^h\ 42^m\ 38^s.6\ (\pm 1^s)$	$-28°\ 58'\ 45''\ (\pm 15'')$

Now we return to the $+40\ km \cdot s^{-1}$ absorption feature. In the third figure of Paper I, this feature is shown in the 1667- and 1665-MHz OH absorption profiles. There has been a great deal of discussion in the literature about the apparent close similarity in the shapes of these two lines for the $+40\ km \cdot s^{-1}$ feature. The lack of a clear saturation effect has been interpreted in terms of many small clumpy condensations, each of high optical depth, and almost none lying in the same line of sight (ROBINSON *et al.* 1964; PALMER and ZUCKERMAN, 1967). This study has given somewhat different results (Fig. 2), in which the line ratio varied from about 1.8, the equilibrium value, in the wings of the feature to a minimum value of 1.2 at $+48\ km \cdot s^{-1}$, indicating a clear saturation effect. (I believe the difference has resulted from more careful efforts to determine the true baseline.) The hypothesis of small condensations is, therefore, no longer needed in this respect. The occultation

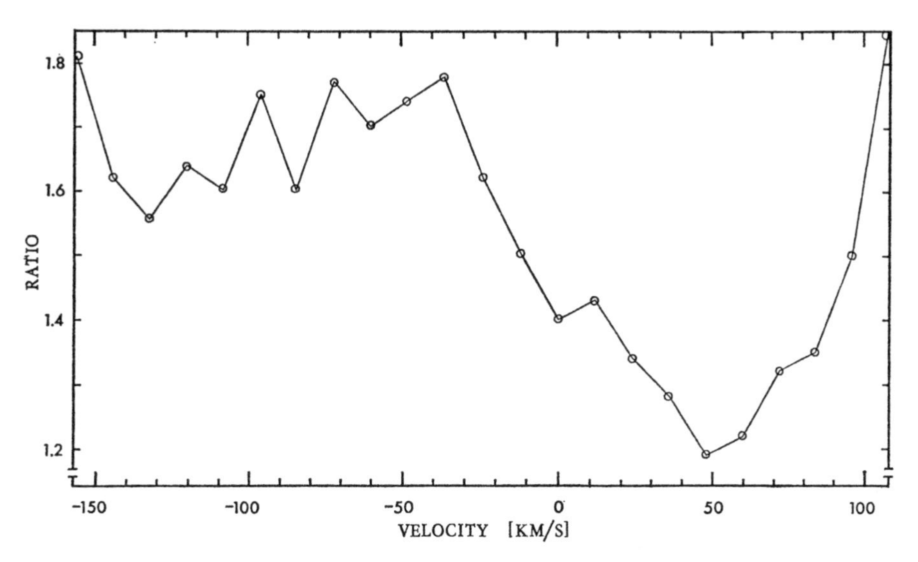

Fig. 2. The observed ratio of the OH-line intensities at 1667 and 1665 MHz, using the absorption profile observed towards Sgr A presented in Paper I. The velocity is with respect to the local standard of rest (LSR)

observations also clearly indicated a saturation type of behaviour, as can be seen in Fig. 3. These curves show that the outer edges of the cloud, as revealed at the ends of immersions and beginnings of emersions, are much closer to the equilibrium ratio of 1.8, while the regions of the cloud closer to its centroid evidently are dense enough to cause line saturation, as indicated by the abnormal line ratio of 1.2.

An intercomparison of the occultation observations has shown that the $+40\ \mathrm{km \cdot s^{-1}}$ features of the HI, OH and H_2CO lines come from the same region of the sky (SANDQVIST 1971). Since the feature is strongest in the 1667-MHz OH line, this line is best suited to study the extent and the velocity structure of the cloud. For each occultation two curved lines, representing the lunar limb, are obtained which give the limits of detectability of this cloud, one line for the immersion phase, the other for emersion. Some of these are shown in Fig. 4, which also includes the radial velocity of the feature at the moment of disappearance or reappearance. The angular dimensions, to the limits of detectability, are 6′ by 4′ arc in galactic longitude and latitude, respectively. The velocity gradient appears to be linear across the cloud, with a value of $8\ \mathrm{km \cdot s^{-1}}$ per minute of arc. This same velocity gradient was detected in the other OH, HI and H_2CO lines as well. The pattern suggests rotation, in the same direction as the galactic rotation, about an axis parallel to the galactic axis. The OH centroid, the position with maximum optical depth for the $+40\ \mathrm{km \cdot s^{-1}}$ feature, is also shown in Fig. 4, and its coordinates are given in Table 1. It is interesting that the OH centroid has the same velocity as that for which the line saturation effect is a maximum, namely $+48\ \mathrm{km \cdot s^{-1}}$.

Comparisons with other molecules, the infrared emission and the radio continuum are shown in Fig. 5. The $+40\ \mathrm{km \cdot s^{-1}}$ cloud and the Sgr A complex are superimposed almost symmetrically. The CO emission peak reported by PENZIAS

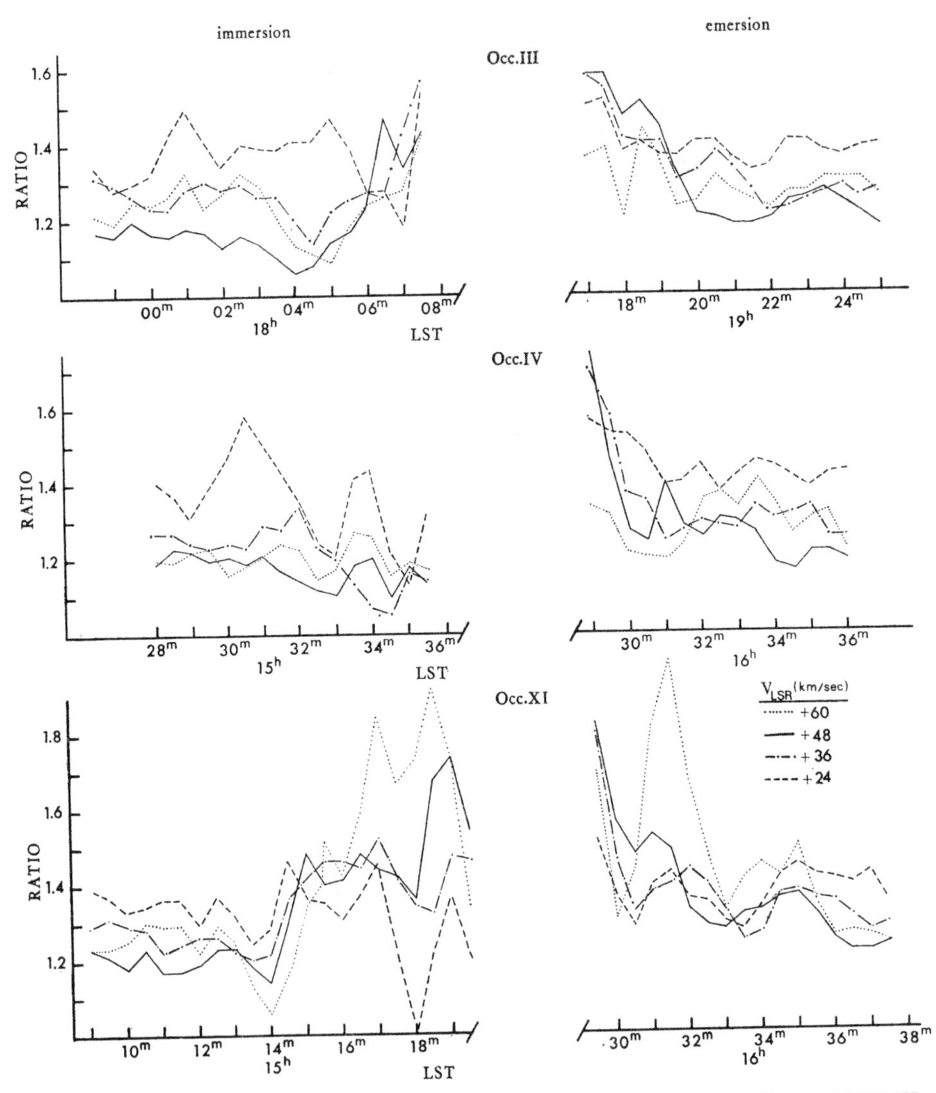

Fig. 3. The variation of the observed ratio of the OH-line intensities at 1667 and 1665 MHz during several occultations. "LST" indicates local sidereal time. Ratios have been plotted for four different velocities within the range of the $+40$ km · s^{-1} feature

et al. (1971) is close to the OH centroid as is also the continuum component *S*. On the other hand, NH_3 emission (CHEUNG *et al.* 1968) and HCN emission (SNYDER and BUHL 1971) peak in the other marked position. The peak velocities for the two groups are different, each agreeing well with the velocity expected for a rotating cloud as described before. In the case of the infrared, the 2.2 μm map of BECKLIN and NEUGEBAUER (1968) shows strong emission at the position of Sgr A but there is almost an anticorrelation with the distribution of the molecules. Perhaps the cloud is too dusty in its central region to transmit the infrared radiation

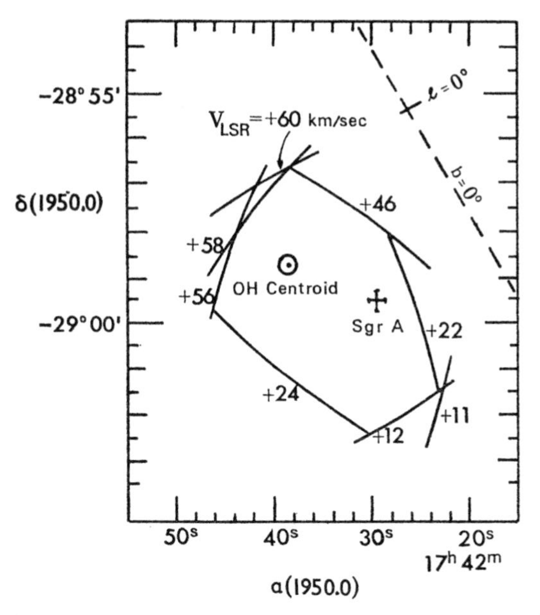

Fig. 4. The limits of detectability and velocity structure of the $+40 \mathrm{~km} \cdot \mathrm{s}^{-1}$ cloud as observed in the 1667-MHz OH line with an effective angular resolution of about 7 seconds of arc. The positions of the OH centroid and the main peak of Sgr A are also indicated

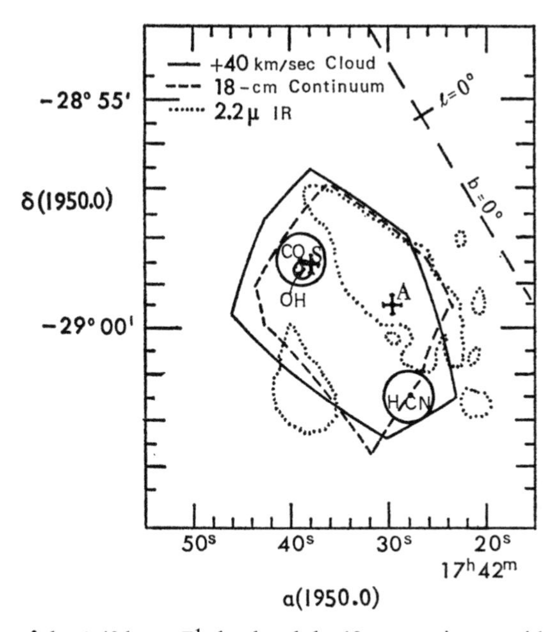

Fig. 5. Comparison of the $+40 \mathrm{~km} \cdot \mathrm{s}^{-1}$ cloud and the 18-cm continuum with emission from CO and HCN, and with the 2.2 μm infrared (IR) emission. "A" and "S" mark the 21-cm peak positions of Sgr A and component S, respectively, "OH" indicates the centroid of the $+40 \mathrm{~km} \cdot \mathrm{s}^{-1}$ cloud. The dimensions of the circles and crosses are measures of the uncertainties in the positions

from the galactic center. This cloud contains a large concentration of molecules, presumably a large amount of dust to shield the molecules and absorb the infrared, it is in rapid rotation, and it appears to be very close in distance to the galactic center. Its physical relationship to the continuum source Sgr A is not known. At 10 kpc, it has linear dimensions of 18×12 pc, and a consideration of its rotational and gravitational energies gives a minimum mass of about $4 \times 10^5\ M_\odot$. If the cloud is at the galactic center, its mass probably has to be much greater. This is because of the high external forces present in the galactic center that would tend to disrupt the cloud (OORT, private communication).

ZUCKERMAN *et al.* (1970) and SCOVILLE *et al.* (1972) have mapped the region giving rise to the $+40\ \mathrm{km \cdot s^{-1}}$ H_2CO absorption feature towards Sgr A with an angular resolution of 6.'5 arc. Both groups of authors obtain a velocity gradient of about $2\ \mathrm{km \cdot s^{-1}}$ per minute of arc across the cloud. This is much less than the $8\ \mathrm{km \cdot s^{-1}}$ per minute of arc which was obtained from the occultation results. But let me emphasize the very high angular resolution obtained by the occulations, namely about 7″ arc a factor of 55 better! The smearing, caused by a 6.'5 arc beam, would significantly lower any velocity gradient.

The lower angular resolution, obtained by SCOVILLE *et al.* (1972), may also be the cause of the disagreement of 3.'5 arc between the positions of maximum optical depth of their H_2CO cloud on the one hand and the CO and OH cloud on the other. Furthermore, their velocity for maximum optical depth is $+46\ \mathrm{km \cdot s^{-1}}$ which, according to the velocity structure obtained from the occultation results, would place this position very close to the OH centroid. The remarkable similarity of the high-frequency-resolution OH and H_2CO absorption profiles, obtained by SANDQVIST (1970, 1971) also speaks strongly for a similar distribution of OH and H_2CO in this cloud.

The velocity gradient across the $+40\ \mathrm{km \cdot s^{-1}}$ cloud is about 8 times that of the nuclear disk of neutral hydrogen. It seems, therefore, unlikely that the two phenomena are related in a simple manner. Furthermore, there is the uncomfortably high positive radial velocity of the cloud at the position of the main peak of Sgr A, namely about $+37\ \mathrm{km \cdot s^{-1}}$ possibly indicating the rapid movement of this part of the cloud in towards Sgr A.

It is interesting that LALLEMAND *et al.* (1960) have observed a rapid rotation of the nucleus of M31. They find a radius of the nucleus of 7.5 pc, a velocity gradient of $12\ \mathrm{km \cdot s^{-1}}$ per pc and a mass of $10^7\ M_\odot$. The radius of the $+40\ \mathrm{km \cdot s^{-1}}$ cloud, if at 10 kpc, is 9 pc and the velocity gradient is about $3\ \mathrm{km \cdot s^{-1}}$ per pc. While the similarity in size, velocity structure and position between this stellar object in M31 and this predominantly gas and dust object in our own Galaxy may be fortuitous, it certainly can invite further study of both objects.

Acknowledgement

I should like to thank Professors F. J. KERR, J. H. OORT and J. HÖGBOM for interesting discussions. This work was begun while I was at the Astronomy Program at the University of Maryland.

References

BECKLIN, E. E., NEUGEBAUER, G.: Astrophys. J. **151**, 145 (1968).
CHEUNG, A. C., RANK, D. M., TOWNES, C. H., THORNTON, D. D., WELCH, W. J.: Phys. Rev. Letters **21**, 1701 (1968).
HÖGBOM, J.: In preparation (1973).
LALLEMAND, A., DUCHESNE, M., WALKER, M. F.: Publ. Astron. Soc. Pacific **72**, 76 (1960).
PALMER, P., ZUCKERMAN, B.: Astrophys. J. **148**, 727 (1967).
PENZIAS, A. A., JEFFERTS, K. B., WILSON, R. W.: Astrophys. J. **165**, 229 (1971).
ROBINSON, B. J., GARDNER, F. F., DAMME, K. J. VAN, BOLTON, J. G.: Nature **202**, 989 (1964).
SANDQVIST, AA.: Astron. J. **75**, 135 (1970).
SANDQVIST, AA.: Ph. D. Thesis, University of Maryland (1971).
SANDQVIST, AA.: Astron. Astrophys., Suppl. **9**, 391 (1973).
SCOVILLE, N. Z., SOLOMON, P. M., THADDEUS, P.: Astrophys. J. **172**, 335 (1972).
SNYDER, L. E., BUHL, D.: Astrophys. J. **163**, L47 (1971).
ZUCKERMAN, B., BUHL, D., PALMER, P., SNYDER, L. E.: Astrophys. J. **160**, 485 (1970).

A Model for the Galactic Center

By V. de Sabbata, P. Fortini and C. Gualdi
Istituto di Fisica, Universita di Bologna, Bologna, Italy

Abstract

A model is proposed to explain the main features of γ, infrared and radio emission from the galactic center on the basis of hydrogen ionization in a very strong magnetic field and the resulting proton synchrotron radiation.

Moreover the model provides a solution for the problem of injection of cosmic rays in the Galaxy with the correct number and energy of protons, and also, at least in part, for the emission of gravitational waves.

The model consists of a cluster of dense stars (neutron stars) located in the center of our galaxy, where the radius of the cluster is $\sim 10^{17}$ cm, the number of stars $\sim 1.2 \cdot 10^{11}$ of which 1/10 with very strong magnetic field (with magnetic moment $\sim 10^{33}$ erg Gauss^{-1}) and with the interstellar gas density inside the cluster of $\sim 10^{8}$ hydrogen atoms cm^{-3}, of which 1% ionized. Each star has a mass of $\sim 1\, M_{\odot}$. We notice that in general people consider a cluster with 10^{6}–$10^{8}\, M_{\odot}$ but our confidence in missing mass caused us to push the number up to 10^{11}. This is an upper limit because of the relativistic instability occurring at $\tau \sim 3\, R_{\mathrm{Schwarzschild}}$.

Diffusion of Elements in the Galaxy

By Ewa Basinska and Wilhelmina Iwanowska
Polish Academy of Sciences, Institute of Astronomy, Laboratory of Astrophysics, Toruń, Poland

With 9 Figures

Abstract

Effects of gravitational separation of elements in the interstellar gas above the galactic plane are estimated for a very simple evolutionary model of the Galaxy. They show an increase of He/H and Fe/H contents in the solar vicinity by factors 1.15 and 4.0 respectively during the time interval of 14×10^9 years.

I. Introduction

Some years ago several spectroscopists (Morgan and Mayall, 1957; McClure, 1969; Spinrad and Taylor, 1971; Arp, 1965) stated that stars populating nuclei of galaxies, including our own, show strong metallic lines and molecular bands in their spectra. This means that the galactic nucleus and the galactic plane both are rich in heavy elements in contrast to the galactic halo.

This statement, as well as some results of our own work on the stellar populations problem, led us to consider, if and to what extent the process of gravitational diffusion might have contributed to the observed distribution of elements in our Galaxy (Iwanowska, 1970). Preliminary calculations of the diffusion velocity of some heavy elements, relatively to hydrogen, for the present conditions of the interstellar gas in the solar vicinity resulted in values of the order of few $\mathrm{km \cdot s^{-1}}$. This looked promising and one of us (E. B.) started the calculations of the changes of chemical composition caused by the process of the diffusion of elements in a column of gas above the galactic plane in the solar vicinity. These calculations were made by solving the equations of continuity for a very simple model of the Galaxy.

II. Assumptions and Calculations

The calculations were based on the formulae given for the diffusion velocity by Chapman and Cowling (1952) for a two-component mixture of gases, namely

$$v_1 - v_2 = -D_{12}\left[\frac{1}{c_1 c_2}\frac{\partial c_1}{\partial z} + \frac{m_2 - m_1}{c_1 m_1 + c_2 m_2}\frac{1}{p}\frac{\partial p}{\partial z} - \frac{m_1 m_2 (F_1 - F_2)}{kT(c_1 m_1 + c_2 m_2)} + \right.$$
$$\left. + \frac{\alpha_{12}}{T}\frac{\partial T}{\partial z}\right], \tag{1}$$

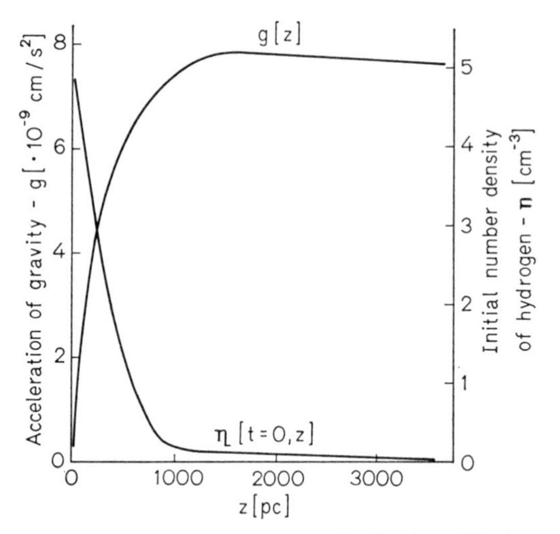

Fig. 1. Acceleration of gravity $g(z)$ and initial density distribution of hydrogen $n(t=0, z)$

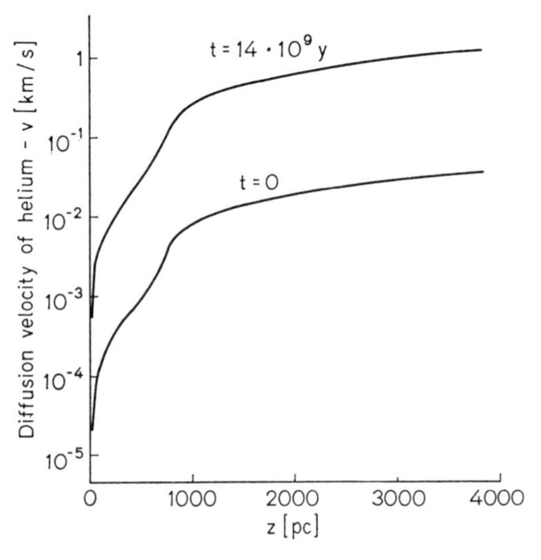

Fig. 2. Diffusion velocity of helium atoms $v(z)$ at the beginning ($t=0$) and presently ($t=14\times10^9$ years)

where v_1 and v_2 the z-velocity components of the particles of the two gas constituents; m_1, m_2 the masses; $c_1=n_1/n$, $c_2=n_2/n$ the relative number densities; $n=n_1+n_2$ the total number density; α_{12} the thermal coefficient of diffusion; F_1, F_2 the external forces per unit mass;

$$D_{12}=\frac{3}{8n\sigma_{12}^2}\left[\frac{kT(m_1+m_2)}{2\pi m_1 m_2}\right]^{1/2} \quad \text{and} \quad \sigma_{12}=\tfrac{1}{2}(\sigma_1+\sigma_2),$$

where σ_1, σ_2 the diameters of the particles.

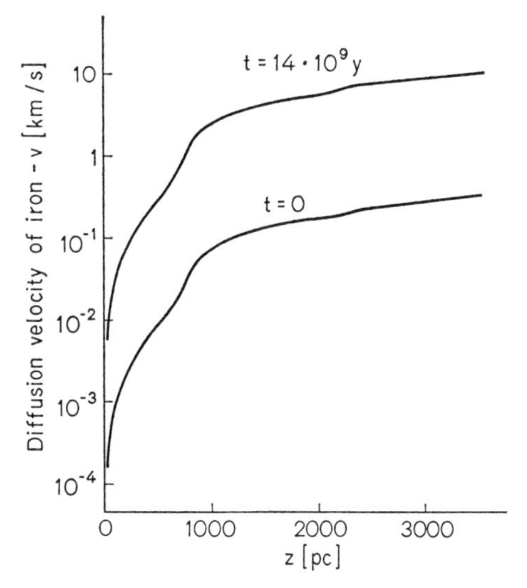

Fig. 3. Diffusion velocity of iron atoms $v(z)$ at the beginning ($t=0$) and presently ($t=14\times10^9$ years)

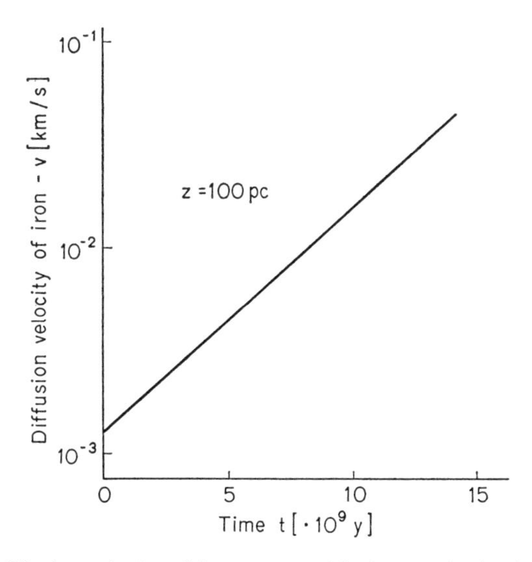

Fig. 4. Increase of diffusion velocity of iron atoms with time at the level $z=100$ pc above the galactic plane

As the main constituent (1) the was considered neutral hydrogen, since it represents a major part of the mass of interstellar matter. The overall temperature was taken equal to $T=100$ K, the initial gas density in the galactic plane equal to 7.4 atoms of hydrogen per cm^3 with solar abundances of other elements (ALLEN, 1963). The initial distribution of the interstellar gas with z-height over the galactic plane in pc was taken from the model given by PEREK (1962). This model is based on the dynamical potential of the Galaxy similar to that given by OORT (1965), derived from the observed stellar motions. It depends on the distribution of the total mass – stellar, interstellar and unknown. We assume in our simplified model that all this mass was gaseous at the beginning and that its density distribution does not change with time, except that gas is converted gradually into stars, its density diminishing with time at the rate

$$n(z,t)=n(z,0)\exp(-t/\tau), \tag{2}$$

with $\tau=4.006\times10^9$ years taken after TRURAN and CAMERON (1971). Thus, we keep PEREK's gravity acceleration $g(z)$ constant with time.

In the formula (1) we disregarded all other terms than that containing the pressure gradient which was expressed by the acceleration of gravity assuming hydrostatic equilibrium. Thus formula (1) goes into

$$v_1-v_2=D_{12}(m_2-m_1)\,g/kT \quad \text{with } D_{12}=\frac{3}{8n\sigma_{12}^2}\left[\frac{kT(m_1+m_2)}{2\pi m_1 m_2}\right]^{1/2}. \tag{3}$$

First, the density distribution of hydrogen $n_1=n_H$ had to be calculated as it varies with time owing to star formation according to formula (2). Next, the velocity of diffusion for each heavy element separately was calculated as a function of z and t, assuming $v_1=0$.

Then, the distribution of this element with z, $n_i(z,t)$, as it varies with time owing to two processes – star formation and diffusion – was calculated by solving the equation of continuity

$$\partial\rho/\partial t+v\,\mathrm{grad}\,\rho+\rho\,\mathrm{div}\,v=0. \tag{4}$$

Taking $\rho_i=n_i m_i$ and $n_i(z,t)=n_i(z,0)\exp(-t/\tau)$, one gets

$$\frac{\partial n_i(z,t)}{\partial t}=-\frac{n_i(z,t)}{\tau}+v_i(z,t)\frac{\partial n_i(z,t)}{\partial z}+n_i(z,t)\frac{\partial v_i(z,t)}{\partial z}. \tag{5}$$

In fact, the computer program was arranged for the ratio $u_i(z,t)$ of the heavy element content to that of hydrogen

$$u_i(z,t)=n_i(z,t)/n_H(z,t).$$

III. Results

The main results shown in Figs. 1–9 for helium and iron are as follows.

1. The relative heavy element content $u_i(z,t)$ in the galactic plane ($z=0$) increased during the time interval preceding the formation of the Sun, assumed as $t=10^{10}$ years by the factors (f_i):

Element:	He	Fe
f_i:	1.1	2.0

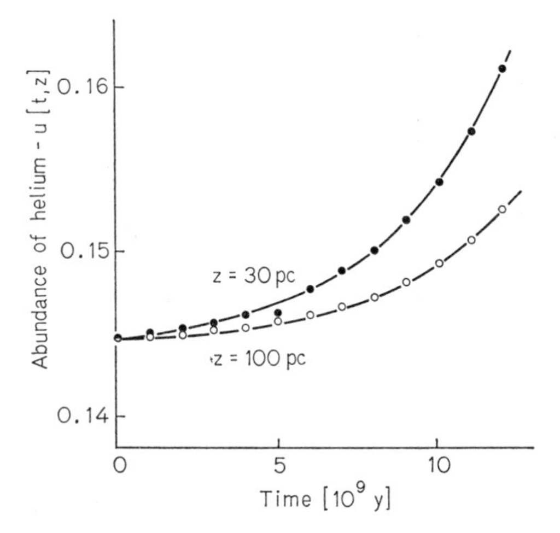

Fig. 5. Changes of the helium to hydrogen abundance ratio with time at the levels $z=30$ pc and $z=100$ pc

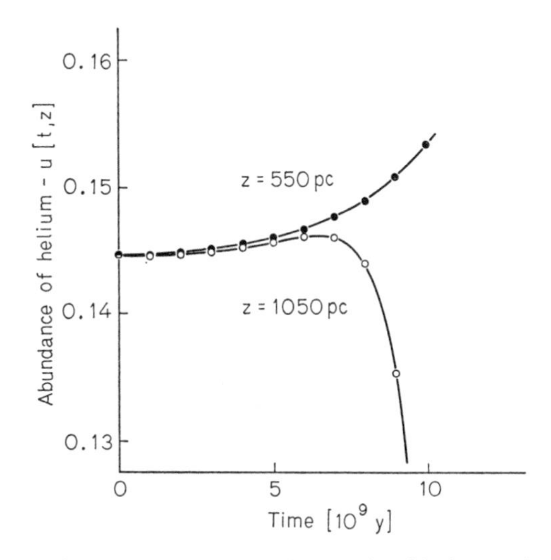

Fig. 6. Changes of the helium to hydrogen abundance ratio with time at the levels $z=550$ pc and $z=1050$ pc

Over some level z_i^0 the relative heavy element content decreases with time, z_i^0 decreasing with time and with increasing atomic weight.

2. The diffusion processes are accelerating with time owing to the decrease of the gas density caused by star formation. Thus, stars younger than 10^8 years were formed out of material enriched in heavy elements by much greater factors f_i as given below:

Element:	He	Fe
f_i:	1.3	4.0

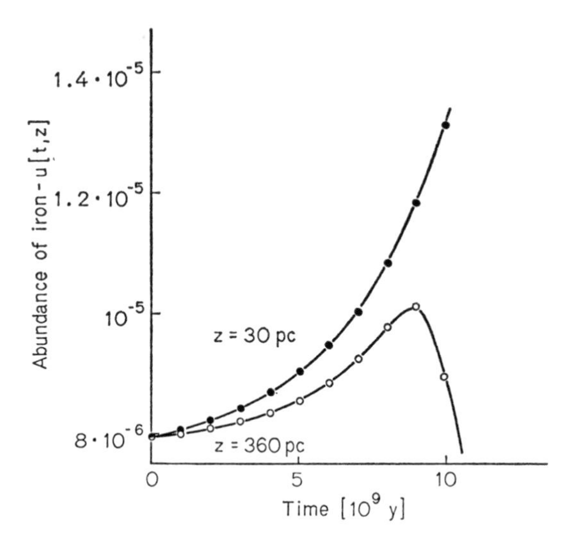

Fig. 7. Changes of the iron to hydrogen abundance ratio with time at the levels $z=30$ pc and $z=360$ pc

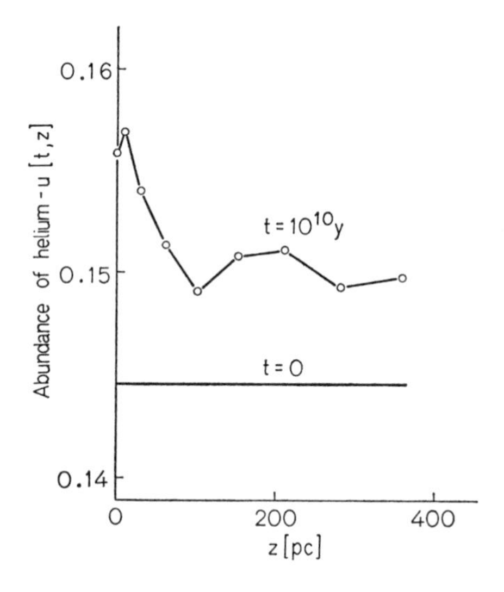

Fig. 8. Helium to hydrogen abundance ratio as a function of z at $t=0$ and $t=10^{10}$ years

3. Through the process of diffusion the distribution of heavy elements may cease to be a monotonic function of z. Local maxima and minima of heavy element contents may be formed depending on the shapes of the two relevant functions, $g(z)$ and $n_i(z, t)$, as shown in Figs. 8 and 9.

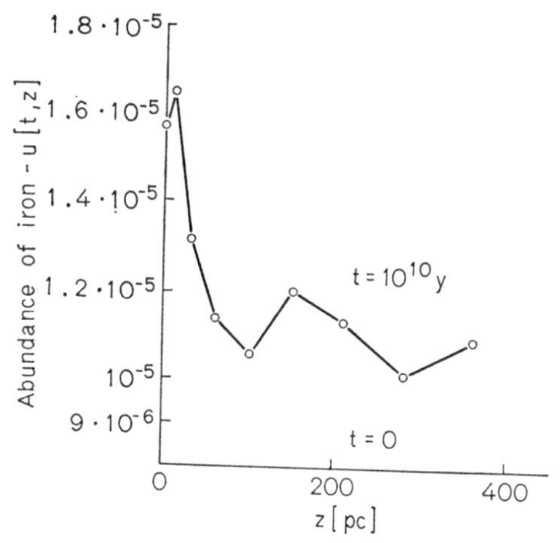

Fig. 9. Iron to hydrogen abundance ratio as a function of z at $t=0$ and $t=10^{10}$ years

IV. Discussion

Thus, the contribution of the process of diffusion to the progressive concentration of heavy elements towards the galactic plane, estimated for our simplified model and the supposed age of our Galaxy, is quite significant. Of course, this model is subject to criticism for its over-simplicity.

1. We have neglected motions mixing the gas, as are for example turbulent motions of clouds with estimated r.m.s. velocities of the order of 5–10 $\mathrm{km \cdot s^{-1}}$, what gives in one z-coordinate less than 3–6 $\mathrm{km \cdot s^{-1}}$. We do not think that it is necessary for the effectiveness of diffusion that its velocity exceeds the velocity of turbulence. We consider turbulent motions as random motions, regarding turbulence as a kind of noise superposed upon the systematic motion of diffusion, since the mean time between collisions of two clouds (about 10^3 years) is less by many orders of magnitude than the assumed age of the Galaxy (14×10^9 years). Also the mean free path of an atom moving with such velocity is a small fraction of a parsec, what means that the distribution of elements and any discontinuities in it are washed out in the limits much less than one parsec.

2. We have chosen an one phase density distribution of interstellar gas, contrary to the presently supported two-phase model consisting of cold dense clouds imbedded in a hot intercloud medium. In our future work such model will be discussed.

3. We assumed that the overall mass distribution of matter and consequently the distribution of gravitational acceleration $g(z)$ does not change with time, the only changes consisting in the loss of gas for star formation.

With this simplification we got the last billions of years as the time at which diffusion of elements was really effective. In a more realistic model one can foresee the phase of early contraction as another time favourable for diffusion, unless too strong turbulence would be present at that time.

In fact, we are proceeding to calculate the effect of diffusion for a more realistic evolutionary model of the Galaxy taking into account the effects mentioned above and others. This paper has the purpose to estimate the order of magnitude of the diffusion effects which probably will not drastically change by further refinements.

One remark is to be done concerning the observed distribution of elements. The content of elements in super metal rich stars, analysed up to now by several spectroscopists, shows an increase of overabundance with increasing atomic weight (KUROCZKIN 1972). This is what should be expected if gravitational diffusion of elements was at work in the interstellar matter before and during formation of stars.

References

ALLEN, C. W.: Astrophysical Quantities. Second edition. London: The Athlone Press 1963.
ARP, H.: Astrophys. J. **7**, 141 (1965).
CHAPMAN, S., COWLING, T. G.: The Mathematical Theory of Non-Uniform Gases. Cambridge: Cambridge University Press 1952.
IWANOWSKA, W.: Trans. I.A.U. XIV B, 201 (1970).
KUROCZKIN, D.: Private communication (1972).
MCCLURE, R. D.: Astrophys. J. **74**, 50 (1969).
MORGAN, W. W., MAYALL, N. U.: Publ. Astron. Soc. Pacific **69**, 291 (1957).
OORT, J. H.: In A. BLAAUW and M. SCHMIDT (eds.), Galactic Structure, p. 470. Chicago: Chicago University Press 1965.
PEREK, L.: Adv. Astron. Astrophys. **1**, 165 (1962).
SPINRAD, H., TAYLOR, B. J.: Astrophys. J. **163**, 303 (1971).
TRURAN, J. W., CAMERON, A. G. W.: Astrophys. Space Sci. **14**, 1 (1971).

Discussion

KIPPENHAHN, R.:
I wonder whether turbulent motion is not more important. Although as you said the turbulent motion is random while the diffusion goes systematically in one direction, does not the ratio of turbulent velocity to diffusion velocity give a measure how well matter will be mixed? and is not this ratio sometimes much bigger than one?

IWANOWSKA, W.:
Turbulent motions should smear out the effects of diffusion over Δz comparable to the mean free path of atoms between collisions. This Δz is at present in the galactic plane much less than 1 pc assuming 4 $\text{km} \cdot \text{s}^{-1}$ for z-component of turbulence.

OORT, J.:
There is now good evidence that hydrogen clouds exist at large distances (up to at least 2 kpc) from the galactic plane. This gas must have come from the galactic layer. The mechanism which has carried the clouds up there must have disturbed the galactic layer and must probably have mixed this thoroughly and have mixed out differences in composition within this layer.

IWANOWSKA, W.:
What percent of the total gas mass do you estimate to be involved into this mixing mechanism?

OORT, J.:
About 15%.

IWANOWSKA, W.:
Then we have to consider this fraction of the gas mass as being involved into turbulent motions with high velocities.

WILLIAMS, P. M.:
Have you considered the effects of radiation pressure? Since we see interstellar lines, there must be some radiation force on the interstellar medium, systematically away from the galactic centre and disk. Does it compare with the gravitational force?

IWANOWSKA, W.:
We have not taken into account the effect of radiation pressure, neither other effects, like the gravitational contraction of the Galaxy, or the fact that some mass fraction of the gas is ionized and subject to the galactic magnetic field. This is just a first approximation solved in order to see what might be the order of magnitude of the effects of diffusion. Further refinements are going to be made.

Narrow-Band Observations of Super-Metal-Rich Stars

By P. M. Williams
Institute of Astronomy, Cambridge, United Kingdom

Abstract

The iron abundances of 300 evolved stars have been estimated from narrow-band photometry analysed using synthetic spectra computed from model atmospheres. About 3% of the stars appear to have iron abundances significantly greater than that of the Hyades, and are regarded as SMR. Using the sodium and calcium abundances determined similarly, it is demonstrated that the apparent super-metallicity is not caused by the stars' having unusually large microturbulence or different $T(\tau)$ from normale stars. Some of the SMR stars appear from their space motions and evolutionary status to be old.

I. Introduction

It has become increasingly apparent that the older stars in the galactic disk are not all metal poor, but that they show a broader abundance distribution than the younger stars. Owing to the implications for galactic evolution, the analyses of super-metal-rich (SMR) stars by Spinrad and Taylor (1969) have attracted a lot of discussion, mostly on the interpretation of these stars' strong absorption lines and heavy line blanketing in terms of high metal abundances. The Spinrad and Taylor abundances were determined from low-resolution spectrum scans of strong lines, and are not always in accord with those derived from conventional spectroscopic analyses. However, a third independent set of metal abundances tend to support the Spinrad and Taylor results for the stars observed in common, and can also be used for testing the metal-rich interpretation of these stars' spectra.

The narrow-band photometric observations were made with the spectrograph at the coudé focus of the Cambridge 92-cm reflector (Griffin and Redman 1960, Griffin 1961), and were analysed using theoretical indices formed from syntheses of the spectral regions measured. The synthetic spectra were computed from each of a network of metal atmospheres covering the ranges $4000 \leqq T_e \leqq 5600$ K, $0.9 \leqq \log(g) \leqq 3.7$, and $-1.0 \leqq [\mathrm{Fe/H}] \leqq +0.5$. The temperatures were derived from broad-band photometry, and the gravities from independent luminosity estimates and theoretical mass-luminosity relations. The results and details of the analyses have been published (Williams 1971a, 1972).

The four Hyades cluster giants were found to have a mean abundance [Fe/H]=0.22, in good agreement with other analyses, while a number of stars were found to be significantly more iron-rich. Those having both [Fe/H]>0.3 and [Na/H]>0.3 are collected in Table 1. The sodium and calcium abundances are also from narrow-band photometry (WILLIAMS 1971b, REGO *et al.* 1972). Several of the other stars found to have high [Fe/H] do not have high [Na/H], or were not observed in that program, and might also be SMR. It is noteworthy that the two K1 III stars in Table 1 have gravities more appropriate to luminosity class IV stars, so that the stars in Table 1 fall into two fairly well-defined groups in the HR diagram – the hot subgiants having $T_e \sim 5000$ K, and the K2 giants having $T_e \sim 4500$ K.

II. Possible Sources of Error

Besides accidental errors which could affect any of the abundances derived, two sources of systematic error have been invoked to explain away the SMR stars. The first possible explanation of these stars' high derived metal abundances is that the microturbulence was underestimated. The high metal abundances found

Table 1. Metal abundances in probable SMR stars

Star	HD	Spectrum	[Fe/H]	[Na/H]	[Ca/H]
δ And	3627	K3 III	+0.36	+0.41	+0.27
49 Per	25975	K1 III	+0.50	+0.48	
	72184	K2 III	+0.46	+0.31	
μ Leo	85503	K2 III	+0.46	+0.47	+0.22
	102328	K3 III	+0.48	+0.39	
	145148	K0 IV	+0.45	+0.46	
β Oph	161096	K2 III	+0.40	+0.44	+0.09
31 Aql	182572	K0 IV	+0.50	+0.50	
27 Cyg	191026	K1 III	+0.50	+0.50	

Table 2. Overestimates in metal abundances from underestimates of the atmospheric temperature gradients or microturbulence

T_e	log (g)	Δ[Fe/H]	Δ[Na/H]	Δ[Ca/H]	
4500	2.7	+0.27	−0.01	+0.61	ξ
		+0.21	+0.49	+0.50	$T(\tau)$
5000	3.7	+0.52	+0.02	+0.41	ξ
		+0.43	+0.53	+0.42	$T(\tau)$

The first and third rows give the abundance changes arising from the microturbulence change discussed in the text, and the second and fourth the abundance changes from the $T(\tau)$ change discussed.

for two high-luminosity stars not included in Table 1 (β Cam and ε Leo) are probably spurious on this account. The theoretical indices for the models corresponding to the stars in Table 1 were all computed assuming a microturbulence of $v_t = 2.5$ km · s^{-1}. The effect of a microturbulence underestimate on the derived abundances was determined by recomputing some of the indices using $v_t = 4$ km · s^{-1} and then deriving the apparent overabundances that solar composition stars having $v_t = 4$ km · s^{-1} would show when analysed with the theoretical indices computed using $v_t = 2.5$ km · s^{-1}. These spurious overabundances are given in Table 2 for two specimen models representing the two groups of SMR stars noted above.

The three sets of abundances differ widely in their sensitivity to microturbulence errors owing to the different strengths of the lines measured by the narrow-band indices. Hence, the derived [Na/Fe] and [Ca/Fe] abundances are sensitive to the stars' microturbulent velocities. Real variation in [Na/Fe] and [Ca/Fe] from star to star complicate the issue, but on average, we would expect stars whose microturbulent velocities had been underestimated to show patterns of overabundances as in Table 2. None of the stars in Table 1 shows such a pattern, so that it is unlikely that they owe their high metal abundances to underestimates of the microturbulence.

The other possible source of systematic error in the abundances derived for strong-lined stars was put forward by STROM *et al.* (1971). They pointed out that increased line-blanketing in such stars would steepen the temperature gradients in their atmospheres, leading to overestimates of the abundances derived from neutral lines of most elements. The increased blanketing could be caused by a high metal abundance, greater microturbulence, or increased CN absorption. This effect will lead to some overestimates of the abundances in genuinely metal-rich stars, but cannot simulate large overabundances in normal composition stars unless they have high microturbulent velocities or enhanced C/Fe or N/Fe, possibly as a result of their evolution. We already have grounds for believing that the stars in Table 1 do not have, on average, higher microturbulent velocities than solar composition stars. Now, CN is not a significant source of blanketing in the $T_e \sim 5000$ K stars, so that only the atmospheres of the cooler stars are likely to have steeper $T(\tau)$ from extra CN blanketing. Therefore, the effect of a steepened temperature structure on the theoretical narrow-band indices and the derived abundances has been investigated following STROM *et al.* The changes in the derived abundances for an arbitrary steepening of the models' $T(\tau)$ for $\tau < 0.1$ amounting to 10% near the surface are also given in Table 2. Although less drastic than the change used by STROM *et al.*, it is fairly realistic since they started with an unblanketed model while the present models used scaled solar $T(\tau)$ relations. As can be seen from Table 2, if the high [Fe/H] abundances given in Table 1 for the $T_e \sim 4500$ K stars arose from steeper $T(\tau)$ in their atmospheres they would show even greater overabundances for Na and Ca, which are not observed. Even if a combination of higher microturbulence and steeper $T(\tau)$ increased both the Na and Fe derived abundances, this would be detectable by an even greater overabundance of Ca, not found in any metal-rich star observed for Ca. Hence, it seems that the stars in Table 1 are genuinely SMR unless accidental errors have affected individual abundances.

III. Status of the SMR Stars

The stars in Table 1 form no exception to the observation that the SMR stars are among the oldest objects in the Galaxy (SPINRAD and TAYLOR 1969, TORRES-PEIMBERT and SPINRAD 1971). In particular, 31 Aql, 49 Per and HD 145148 all fall below the M 67 subgiant branch in the HR diagram, suggestion that they are older than the cluster, and also have high space velocities. It is more difficult to be certain about the ages of the cooler SMR stars owing to the "funnelling" effect in the HR diagram, but one of them (HD 72184) also has a large space motion and an eccentric galactic orbit. The SMR stars do not seem to comprise a separate group among the kinematically old stars, but are the metal-rich end of a broad abundance distribution which can be explained most simply in terms of poor mixing of the interstellar medium at the time these objects were formed.

In conclusion, it is worth recalling that although narrow-band or scanner abundance investigations will never supplant conventional spectroscopic analyses using high-dispersion spectrograms, they have a number of advantages in addition to their speed. Most important is the homogeneity of the data, particularly when one is comparing the compositions of a number of stars. Many observers have found their measures of the equivalent widths of spectral lines from spectrograms differ from those measured by other observers of the same star, so that inter-comparison of the compositions of two stars analysed by different observers is often uncertain. On the other hand, narrow-band observations can be made of a large number of stars using the same equipment, and, in effect, the same impersonal determination of the continuum height. Nevertheless, detailed analyses of some of the interesting stars observed photometrically are very necessary.

Acknowledgements

I am very grateful to Professor R. D. REDMAN for the hospitality of the Institute of Astronomy (formerly the Cambridge Observatories), and to St. John's College, Cambridge, for a Fellowship.

References

GRIFFIN, R. F.: Monthly Notices Roy. Astron. Soc. **122**, 181 (1961).
GRIFFIN, R. F., REDMAN, R. O.: Monthly Notices Roy. Astron. Soc. **120**, 287 (1960).
REGO, M. E., WILLIAMS, P. M., PEAT, D. W.: Monthly Notices Roy. Astron. Soc. **160**, 129 (1972).
SPINRAD, H., TAYLOR, B. J.: Astrophys. J. **157**, 1279 (1969).
STROM, S. E., STROM, K. M., CARBON, D. F.: Astron. Astrophys. **12**, 177 (1971).
TORRES-PEIMBERT, S., SPINRAD, H.: Bol. Obs. Tonantzintla Tacubaya **6**, 15 (1971).
WILLIAMS, P. M.: Monthly Notices Roy. Astron. Soc. **153**, 171 (1971a).
WILLIAMS, P. M.: Monthly Notices Roy. Astron. Soc. **155**, 215 (1971b).
WILLIAMS, P. M.: Monthly Notices Roy. Astron. Soc. **158**, 361 (1972).

Fine Analytic Abundance Determination of Magellanic Cloud A-Supergiants and its Importance for the Discrimination of Theories for the Chemical Evolution of the Galaxy

By B. Wolf
European Southern Observatory, Santiago, Chile

Abstract

Model atmosphere analyses have been done for the two brightest stars of the Magellanic Clouds: HD 33579 in the Large Magellanic Cloud and HD 7583 in the Small Magellanic Cloud. The results are presented and discussed in connection with the chemical evolution of the Galaxy.

The Magellanic Clouds differ from the Galaxy in mass, densities, gas content, dynamics and so on. A totally different evolution history must be expected. Thus, for a discrimination of competing theories of the chemical evolution of the Galaxy, fine analytic abundance determinations of some stars of the Magellanic Clouds are desirable.

I have obtained some high dispersion spectra (12.3 Å/mm in the blue and 20 and 30 Å/mm in the red region at H_α) of the two brightest stars in the Magellanic Clouds with the coudé-spectrograph of the 1.5 m telescope of the European Southern Observatory on La Silla.

The two stars are:

HD 33579 in the LMC (A3Ia-O, $M_v = -9.53$ [determined with the modul $(m-M) = 18.5$] Wolf 1972a) and
HD 7583 in the SMC (AOIa-O, $M_v = -9.26$ [determined with the modul $(m-M) = 19.2$] Wolf 1972b).

These supergiants are extreme population I objects, for which the evolution time, in which 10% of the total amount of hydrogen is converted to helium, can be estimated to be less than half a million years. Thus, the derived abundances can be regarded as showing the results of the chemical evolution in these stellar systems until now.

For the interpretation of the observations a grid of model atmospheres was computed, assuming radiative equilibrium, LTE with the Planck function as the source function, hydrostatic equilibrium, no line blanketing, and plane parallel layers. The final model parameters T_e and $\log g$ were selected in such a way, that they give the observed colour indices $(U-B)$ and $(B-V)$, the observed Balmer jump, the MgI/MgII- and FeI/FeII-ionisation equilibria, and the observed equivalent widths of the Balmer lines (see Table 2).

The empirical curves of growth of both stars have a very steep shape, with broadening parameters much too large for supergiants with pure radiation damping. This can be the result of a depth dependent microturbulence; and, in deed,

Table 1. Characteristic data of the model atmospheres of HD 33579 (LMC) and HD 7583 (SMC)

	HD 33579 (LMC)	HD 7583 (SMC)
Temperature	$T_e = 8130 \pm 300$ $\theta_e = 0.620 \pm 0.020$	$T_e = 8960 \pm 300$ $\theta_e = 0.562 \pm 0.020$
Gravitational acceleration	$\log g = 0.70 \pm 0.20$	$\log g = 0.87 \pm 0.20$
Electron pressure	$\log P_e = 0.25$ at $\bar{\tau} = 0.7$	$\log P_e = 0.27$
Gas pressure	$\log P_g = 0.26$ at $\bar{\tau} = 0.7$	$\log P_g = 0.61$
Microturbulence	6–26 $\text{km} \cdot \text{s}^{-1}$ decreasing with the optical depth	3.5–15.6 $\text{km} \cdot \text{s}^{-1}$
Macroscopic motions	differences for different lines up to 12 $\text{km} \cdot \text{s}^{-1}$	no difference within the probable error

Table 2. Comparison of some measured and calculated features of HD 33579 and HD 7583

	HD 33579		HD 7583	
	observed	calculated	observed	calculated
U–B	−0.29	−0.28	−0.36	−0.42
B–V	0.14	0.13	0.05	0.08
	corrected with: $E_{B-V} = 0.05$		$E_{B-V} = 0.07$	
Balmer jump	$D = 0.30$	$D = 0.36$	$D = 0.34$	$D = 0.32$
W_λ of the Balmer lines				
H_γ	1405	1640	1223	1352
H_δ	1467	1562	1357	1331
H_8	1385	1499	1360	1248
H_9	1479	1479	1492	1232
H_{10}	1425	1463	1558	1221
Ionisation equilibria				
Mg	$\log \varepsilon_{\text{MgI}} = 7.25$ $\log \varepsilon_{\text{MgII}} = 7.46$		$\log \varepsilon_{\text{MgI}} = 7.57$ $\log \varepsilon_{\text{MgII}} = 7.88$	
Fe	$\log \varepsilon_{\text{FeI}} = 7.69$ $\log \varepsilon_{\text{FeII}} = 7.38$		$\log \varepsilon_{\text{FeI}} = 7.70$ $\log \varepsilon_{\text{FeII}} = 7.25$	

with a depth dependent microturbulence, which I have incorporated in the model atmospheres, the empirical and theoretical curves of growth agree resonably. The characteristic data of the model atmospheres for these two stars are given in Table 1. With these models, observable features were calculated, and are compared with observations in Table 2. I deduced from these models, the data of Table 3 and the abundances of Table 4.

The observed oxygen abundance is determined from the OI-triplet at λ 3947.50 in HD 33579 and HD 7583 and is uncertain in both cases. Also uncertain on account of the observations are the abundances of Co and Ni in HD 33579 and Al, Ni, Sr, and Ba in HD 7583. As a whole, the abundances of

Table 3: Some calculated data of HD 33579 and HD 7583

	HD 33579	HD 7583
Bolometric correction	B.C. = 0.06	B.C. = 0.19
Absolute bolometric magnitude	$M_{bol} = -9.59$	$M_{bol} = -9.45$
Luminosity	$L/L_{\odot} \approx 530000$	$L/L_{\odot} \approx 470000$
Radius	$R/R_{\odot} \approx 360$	$R/R_{\odot} \approx 280$
Mass	$M/M_{\odot} \approx 25$	$M/M_{\odot} \approx 25$
Evolution time	$t_E \approx 300000$ years	$t_E \approx 320000$ years

Table 4: Abundances of HD 33579 and HD 7583 together with the abundances for the galactic supergiants α Cyg and η Leo and for the Sun

Element	HD 33579 A3Ia-O (WOLF, 1972a)	HD 7583 A0Ia-O (WOLF, 1972b)	α Cyg A2Ia (GROTH, 1960)	η Leo A2Ia (WOLF, 1971)	Sun[a] (GOLDBERG *et al.*, 1960)
H	12.00	12.00	12.00	12.00	12.00
O	8.14:	8.95:	9.36	8.83	8.96
Mg	7.36	7.72	7.81	7.65	7.36
Al	6.03	6.44	6.59	6.45	6.20
Si	7.64	7.78	7.88	8.03	7.45
Ca	6.16	6.17	6.47	6.16	6.15
Ti	4.66	4.65	5.13	4.99	4.68
V	3.74	—	3.88	3.97	3.70
Cr	5.59	5.49	5.67	5.61	5.36
Mn	5.06	—	5.57	4.65	4.90
Fe	7.43	7.48	7.62	7.78	7.60[b]
Co	4.72:	—	3.73	3.65:	4.64
Ni	5.71:	5.17:	4.82	5.18:	5.91
Sr	—	3.38:	3.11	—	2.60
Ba	—	3.28:	—	3.63	2.10

[a] Abundances with corrections according to ZWAAN (1962).
[b] Fe—abundance from GARZ *et al.* (1969).

those metals, which are determined with good accuracy in both stars (Mg, Si, Ca, Ti, Cr, Fe), agree even better with the values of the Sun than with those of the galactic supergiants α Cyg and η Leo.

For the average of these metals we find:

$$\Delta \log \bar{\varepsilon} = \log \bar{\varepsilon}\,(\text{HD } 33579) - \log \bar{\varepsilon}_{\odot} = 0.01;$$
$$\Delta \log \bar{\varepsilon} = \log \bar{\varepsilon}\,(\text{HD } 7583) - \log \bar{\varepsilon}_{\odot} = 0.12.$$

These differences are certainly not significant. The corresponding difference for the galactic supergiants with the Sun are:

$$\Delta \log \bar{\varepsilon} = \log \bar{\varepsilon}\,(\alpha \text{ Cyg}) - \log \bar{\varepsilon}_{\odot} = 0.33;$$
$$\Delta \log \bar{\varepsilon} = \log \bar{\varepsilon}\,(\eta \text{ Leo}) - \log \bar{\varepsilon}_{\odot} = 0.26;$$

This equality of metal abundances in the atmospheres of these very young A-supergiants in the Magellanic Clouds with the abundances of the Sun and of the galactic A-supergiants within a factor of two (which we regard as the accuracy of this fine analytic abundance determination) is surprising.

If the element distribution, now observed in population I-objects, has its origin in a variety of processes in stars (BURBIDGE *et al.*, 1957), one would expect a priori a different distribution in stars, which were formed in so different systems as the Magellanic Clouds and the Galaxy. A possible way out would be, that the Magellanic Clouds were detached from the Galaxy after the bulk of the population I material had been formed. But there is little observational evidence for such an event.

References

BURBIDGE, E. M., BURBIDGE, G. R., FOWLER, W. A., HOYLE, F.: Rev. Mod. Phys. **29**, 547 (1957).
GARZ, T., HOLWEGER, H., KOCK, M., RICHTER, J.: Astron. Astrophys. **2**, 446 (1969).
GOLDBERG, L., MÜLLER, E. A., ALLER, L. H.: Astrophys. J. Suppl **5**, 1 (1960).
GROTH, H. G.: Z. Astrophys. **51**, 206 (1960).
WOLF, B.: Astron. Astrophys. **10**, 383 (1971).
WOLF, B.: Astron. Astrophys. **20**, 275 (1972a).
WOLF, B.: Astron. Astrophys. **28**, 335 (1973).
ZWAAN, C.: Bull. Astron. Inst. Neth. **16**, 225 (1962).

High Resolution Ultraviolet Stellar Spectra Obtained with the Orbiting Stellar Spectrophotometer S 59

By K. A. VAN DER HUCHT and H. J. LAMERS
Space Research Laboratory, Astronomical Institute Utrecht, The Netherlands

With 3 Figures

Abstract

The experiment S 59 will observe the ultraviolet spectra of about 200 bright O, B, and A stars in three wavelength bands, 100 Å wide and centered around 2110 Å, 2540 Å and 2820 Å, with a spectral resolution of 1.8 Å. A study of the first obtained spectra has shown that:

1. the Mg II resonance lines at 2800 Å in early-B stars are considerably stronger than was predicted by theory,

2. the spectrum of γ^2 Velorum (WC8+O8) shows a wide C IV emission line at 2530 Å, which indicates mass loss by the star,

3. the interstellar lines of Mg^0, Mg^+, Fe^+ and Mn^+ in the spectrum of ζ Puppis (O4f) indicate that compared to hydrogen these elements are deficient by a factor 30 relative to the Sun.

I. Introduction

The Orbiting Stellar Spectrophotometer, S59, on board the ESRO satellite TD-1A, observes the ultraviolet spectrum of bright stars. The instrument was designed and constructed by the Space Research Laboratory of the Astronomical Institute at Utrecht and the Institute of Applied Physics, TNO-TH (TPD) at Delft. It consists of a gimballed telescope-spectrometer combination with its own star-tracking facility. The instrument has been described briefly by HOEKSTRA *et al.* (1972).

The spectra of about 200 bright stars will be observed in three wavelength bands, 100 Å wide and centered around 2110 Å, 2540 Å and 2820 Å, with a spectral resolution of 1.8 Å. At present (August, 1972) the spectra of about 100 stars have been obtained and some preliminary results will be reported.

II. Magnesium II Resonance Lines

The Mg II resonance lines have been studied in the spectra of 25 stars. A number of representative line profiles are shown in Fig. 1. These lines, being the ultraviolet counterparts of the Ca II H and K lines, are probably formed in the upper layers of the stellar atmosphere and they are particularly sensitive to the temperature stratification of chromospheres and non-LTE effects.

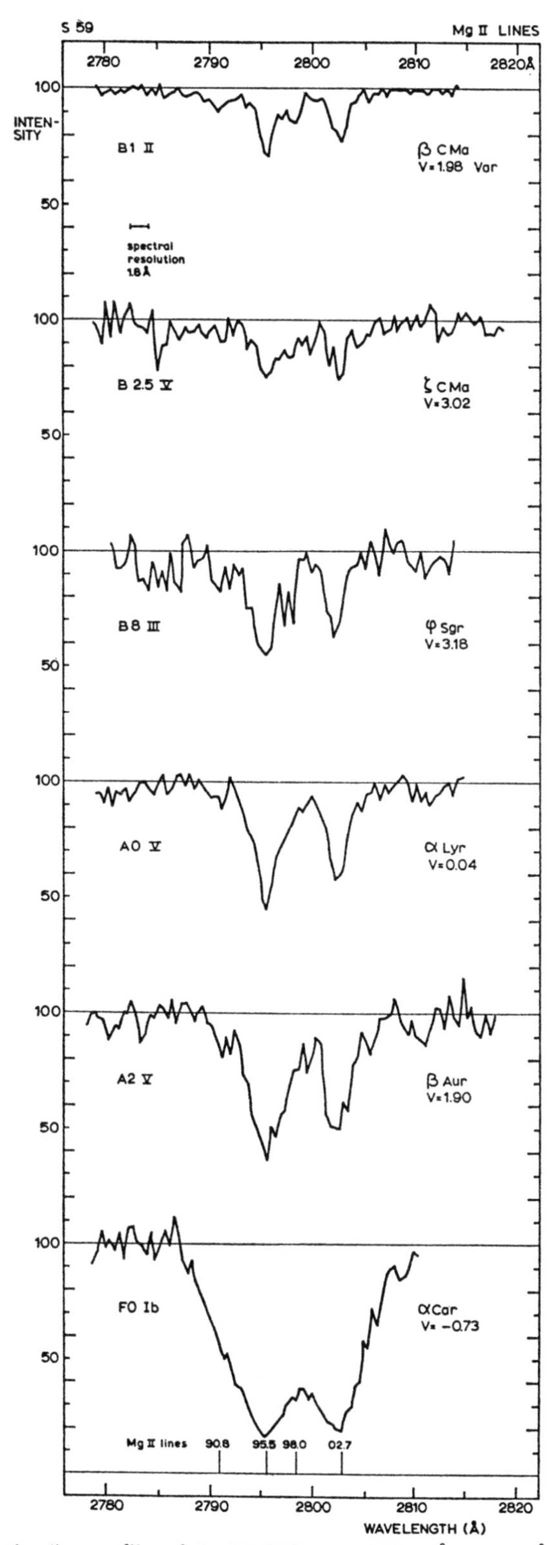

Fig. 1. Representative line profiles of the Mg II lines at 2795.5 Å, 2802.7 Å ($3S^2-3p^2P^0$) and at 2790.8 Å, 2798.0 Å ($3p^2P^0-3d^2D$). The intensities and wavelengths are calibrated provisionally

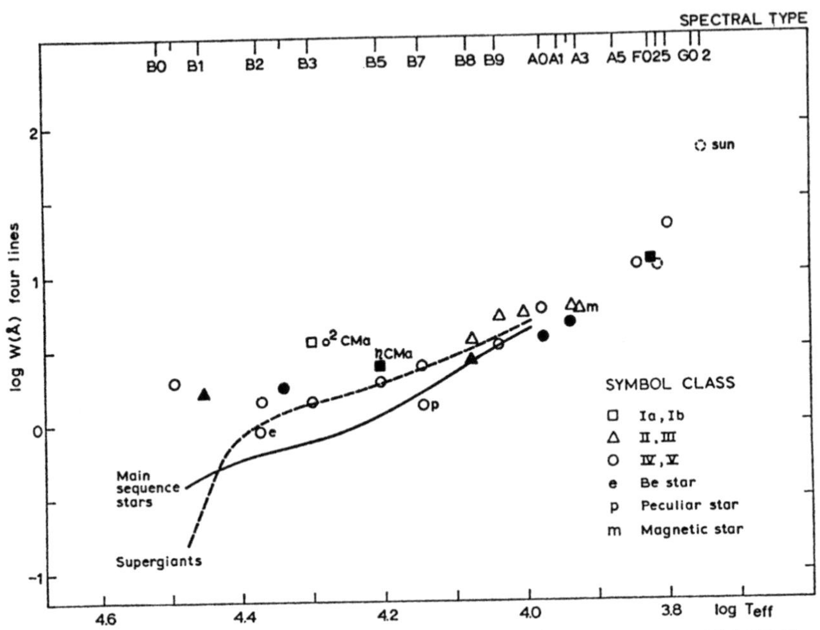

Fig. 2. The equivalent widths of the Mg II lines λ 2790.8, λ 2795.5, λ 2798.0 and λ 2802.7 versus spectral type. The LTE-predictions for main-sequence stars and supergiants are indicated. The observations of γ Lyr (KONDO *et al.*, 1972) and the Sun (KATCHALOV and YEKOVLEVA, 1962) are indicated by interrupted circles. Open symbols refer to one single observation (accuracy about 30 percent), filled symbols refer to more than one observation of the same star (accuracy about 20 percent)

The equivalent width of the four Mg II lines at 2795.5 Å, 2802.7 Å ($3s^2S-3p^2P^0$) and 2790.8 Å, 2798.0 Å ($3p^2P^0-3d^2D$), corrected for the interstellar component, are plotted versus spectral type or effective temperature in Fig. 2. The predicted equivalent widths for main sequence stars and supergiants, assuming an abundance of $N(\text{Mg})/N(\text{H})=3.10^{-5}$ and using hydrogen-line blanketed model atmospheres (KINGLESMITH, 1971) are shown for comparison.

The following conclusions on the Mg II lines can be drawn from Fig. 2:

a) The lines are sensitive temperature indicators at the spectral types A and F.

b) The Be and Bp stars show weaker lines than the normal stars.

c) In late-B stars the equivalent width agrees reasonably with the predictions, whereas in middle and early-B stars the lines are too strong by a factor of 1.4 (B5–B6) to 4 (B0–B1).

The most probable explanation for the discrepancy between LTE predictions and observations in B-stars is the occurence of non-LTE effects.

The population density of the ground level of Mg^+ depends strongly on the radiative ionization rate, due to photons in the Lyman continuum. As the studies of non-LTE model atmospheres of B-type stars by AUER and MIHALAS (1970) and MIHALAS and AUER (1970) have shown that the flux in the Lyman continuum is weakened by the overpopulation of the $n=1$ level of hydrogen, the ionization of Mg^+ will be decreased and the equivalent width of the Mg II resonance lines will be increased. Detailed computations by SAKHIBULIN (1972) have indicated that this non-LTE effect may explain the observed Mg II lines in B-stars.

III. Emission in γ^2 Velorum

The star γ^2 Vel (BS 3207, WC8+O8, $V=1.82$) was observed by S59 35 times, on May 17, 18, and 19, 1972. The spectrum in the 2500 Å band is dominated by a strong emission feature at 2530.3 Å, identified with C IV 2529.97 Å ($4f^2F^0-5g^2G$). Several other C IV emission lines have been observed in this star, indicating the spectrum to be formed by recombination.

The complex structure of the profile is presented in Fig. 3. The half half-width of the line is 9.5 Å, corresponding to a velocity of 1100 km·s^{-1}. The measured wavelengths of the various features of this profile are listed in Table 1. As the system consists of a spectroscopic binary, the emission components are attributed to the spectrum of the WC star and the absorption lines to the O8 star. The preliminary identifications are given in Table 1, columns 2, 3, 4, and 5. When in the near future other O-stars are observed by S59, it will be possible to correct the presented spectrum of γ^2 Vel for the contribution of the O-component, which will result in a more definite interpretation of the emission profile.

The C IV emission line shows a diffuse red wing, which extends to 2560 Å, indicating that material is moving away from us with velocities up to 3500 km·s^{-1}. From the known orbital period (GANESH and BAPPU, 1967) we found that the S59

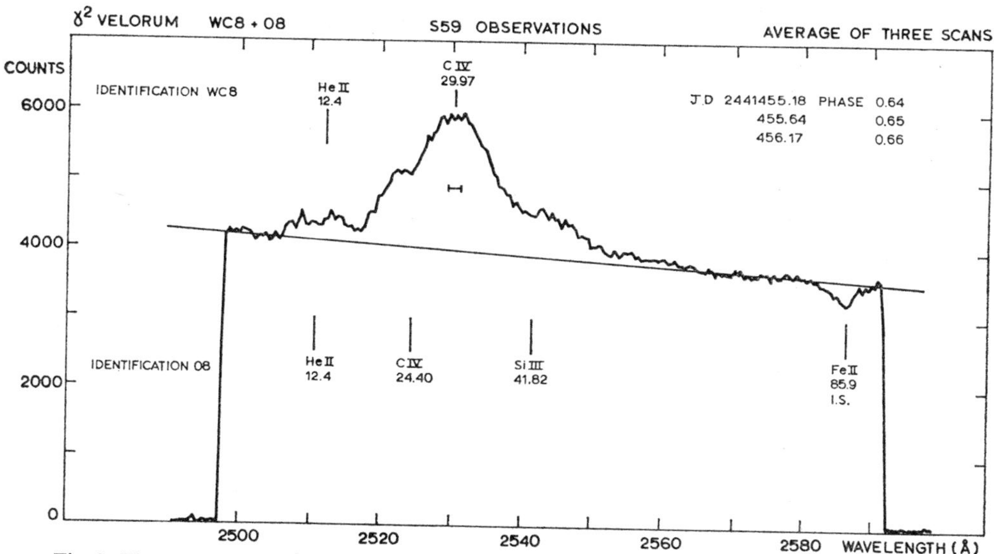

Fig. 3. The spectrum of γ^2 Velorum (WC8+O8) as observed by S59 in the 2500 Å channel, with a preliminary calibrated wavelength scale. The proposed identifications of the spectral features are indicated. The λ 2529.97 C IV emission line has a equivalent width $W_\lambda=8.5$ Å, while the line flux is 0.48 times the continuum flux at the central wavelength. |——| Indicates the spectral resolution of 1.8 Å

Table 1. Wavelengths and preliminary identifications of features in the spectrum of γ^2 Vel

Measured[a]	Proposed identification			
wavelength (Å)	spectrum	(lab) Å	mult	origin[b]
2510.6 a	He II	2511.25	$P\delta$	O
2512.1 e	He II	2511.25	$P\delta$	WC
2524.3 a	C IV	2524.40	14	O
2530.3 e	C IV	2529.97	15	WC
2541.1 a	Si III	2541.82	8	O
2586.2 a	Fe II	2585.9	1	IS

[a] a=absorption; e=emission.
[b] WC=WC-star; O=O8-star; IS=interstellar.

observations were made during phase 0.65 in which the WC component is nearly in front of the O-star. We suggest that the red wing is produced by mass transfer from the WC to the O star. This is the more probable, because Hanbury Brown *et al.* (1970) found the λ 4650 C III/C IV emission to occur near the critical Roche lobe around the Wolf-Rayet star. A similar red wing has been observed by Code and Bless (1964) in the line C IV 5805 Å, in approximately the same phase 0.61.

IV. Interstellar Lines

Several interstellar absorption lines have been detected in the spectrum of the O4f star ζ Pup (BS 3165, $V=2.25$). The lines are listed in Table 2. The doublet ratio of the interstellar Mg II resonance lines is 1.2, indicating that the lines are saturated. This doublet ratio, together with the equivalent widths indicates a velocity dispersion of 15 km $\cdot$ s^{-1} and a column density of $N_{\mathrm{Mg}^+} \sim 10^{13.7}$ cm^{-2} (cf. Strömgren, 1948). Assuming the same velocity dispersion, the equivalent widths of the other interstellar lines indicate column densities of $N_{\mathrm{Mg}^0} \sim 10^{12}$, $N_{\mathrm{Fe}^+} \sim 10^{13.4}$, $N_{\mathrm{Mn}^+} \sim 10^{12.3}$ cm^{-2}.
If we assume that the interstellar material is mainly singly ionized, then the relative column densities of $\mathrm{Mg}^+ : \mathrm{Fe}^+ : \mathrm{Mn}^+ = 2:1:0.1$, which agrees reasonably with the presently adopted solar abundances.

Table 2. Interstellar absorption lines in the spectrum of ζ Pup

Observed line		Proposed identification			
λ(Å)	W_λ(m Å)	spectrum	lab (Å)	$f_{\cdot \mathrm{l.u.}}$	ref.
2852	160	Mg I	2852.1	1.81	a
2803	410	Mg II	2802.7	0.313	a
2796	480	Mg II	2795.5	0.627	a
2586	70	Fe II	2585.9	0.043	b
2576	40	Mn II	2576.1	0.3	b

[a] Wiese *et al.* (1969).
[b] Warner (1967).

JENKINS (1971) derived a column density of $N_H{}^\circ \sim 10^{19.8}$ cm^{-2} from $L\alpha$ observations in the spectrum of ζ Pup. If we compare this value with the above mentioned column density of Mg^+, we find $\log(N_{Mg}/N_H) = -6.1$, which is a -1.6 less than normally accepted for the Sun. This recalls the well known deficiency of interstellar Ca by about a factour 100 (e.g. HERBIG, 1968).

Details of these investigations will be described by: LAMERS *et al.* (1973): Mg II resonance lines; VAN DER HUCHT and LAMERS (1973): C IV emission in γ^2 Vel; de BOER *et al.* (1972): UV interstellar absorption lines.

Acknowledgement

We wish to thank our colleagues of the Space Research Laboratory who contributed to the realization of the experiment. In particular T. KAMPERMAN and R. HOEKSTRA who were largely responsible for the project and L. DE FEITER and his collaborators for continuous help in data reduction. The optical part of the experiment was designed and constructed by the Institute of Applied Physics, TNO-TH (TPD), under the direction of A. HAMMERSCHLAG and W. WERNER.

References

AUER, L. H., MIHALAS, D.: Astrophys. J. **160**, 233 (1970).
DE BOER, K. S., HOEKSTRA, R., HUCHT, K. A. VAN DER, KAMPERMAN, T. M., LAMERS, H. J., POTTASCH, S. R.: Astron. Astrophys. **21**, 447 (1972).
CODE, A. D., BLESS, R. C.: Astrophys. J. **139**, 787 (1964).
GANESH, K. L., BAPPU, M. K. V.: Kodaikanal Obs. Bul. Ser. A. No. **183**, 177 (1967).
HANBURY BROWN, R., DAVIS, J., HERBISON-EVANS, D., ALLEN, L. R.: Monthly Notices Roy. Astron. Soc. **148**, 103 (1970).
HERBIG, G.: Z. Astrophys. **68**, 243 (1968).
HOEKSTRA, R., HUCHT, K. A. VAN DER, KAMPERMAN, T. M., LAMERS, H. J., HAMMERSCHLAG, A., WERNER, W.: Nature, Phys. Sc. **236**, 121 (1972).
HUCHT, K. A. VAN DER, LAMERS, H. J.: Astrophys. J. **181**, 537 (1973)
JENKINS, E. B.: Astrophys. J. **169**, 25 (1971).
KATCHALOV, V., YAKOVLEVA, A.: Izv. Krymsk Astrofiz. Obs. **27**, 5 (1962).
KINGLESMITH, D. A.: Hydrogen Line Blanketed Model Stellar Atmospheres. NASA SP-3065 (1971).
KONDO, Y., GIULI, R. T., MODISETTE, J. L., RYDGREN, A. E.: Astrophys. J. **176**, 153 (1972).
LAMERS, H. J., HUCHT, K. A. VAN DER, SNIJDERS, M. A. J., SAKHIBULIN, N.: Astron. Astrophys. **25**, 105 (1973).
MIHALAS, D., AUER, L. H.: Astrophys. J. **160**, 1161 (1970).
SAKHIBULIN, N.: Private communication (1972).
STRÖMGREN, B.: Astrophys. J. **108**, 242 (1948).
WARNER, B.: Mem. Roy. astr. Soc. **70**, 165 (1967).
WIESE, W. L., SMITH, M. W., MILES, B. M.: Atomic Transition Probabilities II, NSRDS-NBS 22 (1969).

Discussion

RIJSBERGEN, R. VAN:
How big is the noise over the whole spectrum in the star you discussed?

LAMERS, H.:
In the spectra of the weakest program stars, the noise of the one single observation is about 10 per cent. In case of brighter stars and after averaging over many spectral scans the noise is reduced considerably to about 3 per cent for the weakest stars.

Infrared Astronomy

(General Lecture)

By J. BORGMAN
Kapteyn Observatory, Roden, The Netherlands

With 10 Figures

Abstract

This is a review paper on selected topics of infrared astronomy. After a discussion of problems arising in connection with the interpretation of observations a sort description is given of characteristic infrared features of stars, H II regions, planetary nebulae and extragalactic sources. A discussion of infrared observations of the galactic center leads to the tentative suggestion of a dense cluster of 3×10^6 late-type giants behind 200 magnitudes of visual extinction.

I. Introduction

Observations in the infrared during the last five years have dramatically reconfirmed that advances in the physical sciences are often related to the exploitation and refinement of new techniques. A large number of papers with numerous observations have been published and already a number of excellent review papers are available. An extensive bibliography and description of earlier work can be found in a paper by NEUGEBAUER *et al.* (1971). More recently WOOLF (1973) has reviewed the observations which are understood as circumstellar infrared emission. On the same subject but concentrating on infrared excesses in early-type stars is a review by PECKER (1972). A paper by SPINRAD and WING (1969) deals with the spectra of stars in the near infrared. The general review by WEBBINK and JEFFERS (1969), though of course not including the important work of the last few years, is of interest to those who want a broad introduction.

Some of the review papers cited are fairly recent. In view of this fact and in order to avoid too much overlap with other papers at this conference I have tried to limit myself to a general introduction, emphasizing on observations by describing characteristic objects and in particular those observations which have not been included in earlier reviews. Completeness has not been attempted and would be far beyond the scope of this paper. The important work on high resolution spectroscopy in the near infrared will not be reviewed in this paper, as the conference has a separate paper on multiplex spectrometry by Mrs. CONNES.

Some Remarks on Techniques and Practical Problems

Though this conference has a separate paper on infrared techniques it appears appropriate to make a few brief comments which are relevant to the quantitative significance and the interpretation of the observations.

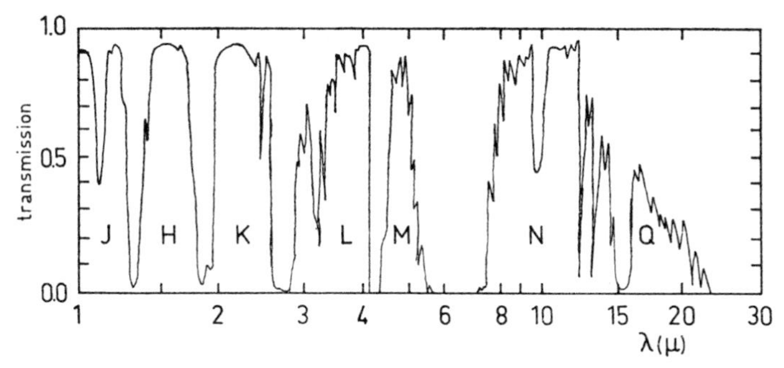

Fig. 1. Atmospheric transmission at 2000 m elevation and 2 mm of precipitable water. (Adapted from T. DE GROOT in Sterrengids, 1971)

The Earth's atmosphere is opaque in a large portion of the infrared. Ground-based observations are possible only in a number of atmospheric windows (Fig. 1). Much infrared work is being done by means of broad-band photometry, using filters which cover the entire width of the windows. In this way the magnitude system of Table 1 (JOHNSON, 1966; LOW, 1968) has been defined. It should be noted that the quoted effective wavelength of the *Q*-band, 22 μm is not typical for wide band ground-based photometry (cf. Fig. 1).

The reproduction of the infrared magnitude system causes some problems which are reminiscent of the problems encountered when transforming the *U*-magnitude of the *U,B,V* system. The original observations which define the infrared magnitude system have been made with response functions (JOHNSON, 1965) which cut rather far into the enclosing absorption bands. As a consequence the response depends on the physical parameters which define these absorption bands, notably the barometric pressure and the amount of precipitable water. This has a negative effect on the internal and external reproducibility of the magnitude

Table 1. JOHNSON's magnitude system[a]

Filter band	λ_{eff} (μm)	$\Delta\lambda$ (μm)	0 mag ($\mathrm{W \cdot cm^{-2} \cdot \mu m^{-1}}$)	0 mag ($\mathrm{W \cdot m^{-2} \cdot Hz^{-1}}$)
U	0.36	0.066	4.35×10^{-12}	1.88×10^{-23}
B	0.44	0.098	7.20×10^{-12}	4.44×10^{-23}
V	0.55	0.087	3.92×10^{-12}	3.81×10^{-23}
R	0.70	0.21	1.76×10^{-12}	3.01×10^{-23}
I	0.90	0.23	8.30×10^{-13}	2.43×10^{-23}
J	1.25	0.30	3.40×10^{-13}	1.77×10^{-23}
H	1.60	0.30	1.28×10^{-13}	1.09×10^{-23}
K	2.20	0.58	3.90×10^{-14}	6.30×10^{-24}
L	3.40	0.70	8.10×10^{-15}	3.10×10^{-24}
M	5.00	1.13	2.20×10^{-15}	1.80×10^{-24}
N	10.20	4.33	1.23×10^{-16}	4.30×10^{-25}
Q	22.00	7.5	7.70×10^{-18}	1.02×10^{-25}

[a] Adapted from JOHNSON (1966) and LOW (1968).

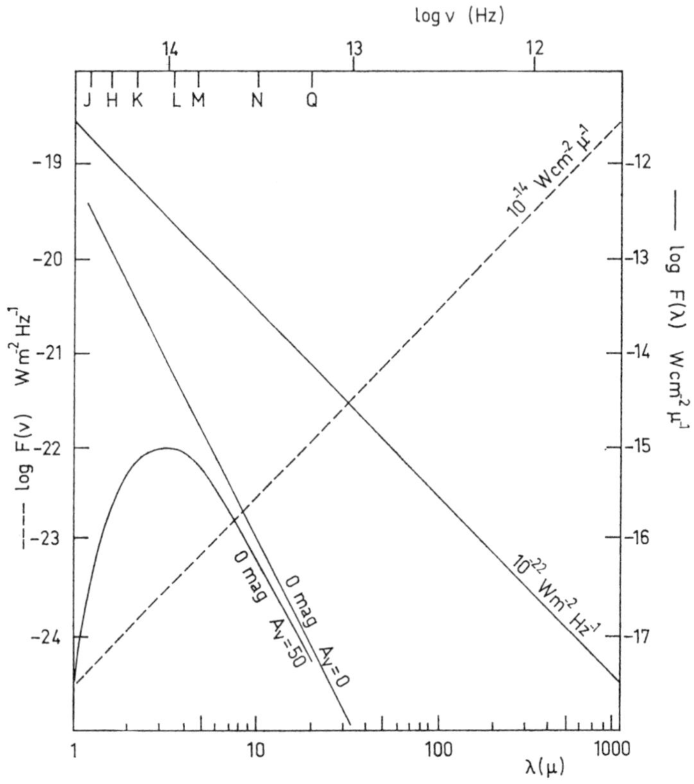

Fig. 2. Mutual conversion of flux density $F(\lambda)$ in $\mathrm{W \cdot cm^{-2} \cdot \mu m^{-1}}$ and $F(\nu)$ in $\mathrm{W \cdot m^{-2} \cdot Hz^{-1}}$. The curves in the lower left corner refer to $F(\lambda)$ for a star of 0 mag at all wavelengths, at 0 mag visual extinction and 50 mag visual extinction, respectively

system, especially in the case of the *M*-band and *Q*-band where the atmospheric influences on the passband are highly asymmetrical. The reproducibility and internal accuracy of a magnitude system will benefit from choosing the passband limits well within the atmospheric windows. Some observers tend to use more than one filter within the 10 μm and 20 μm windows; the increased spectral resolution is very useful to detect broad-band features in the spectrum (cf. COHEN and WOOLF, 1971).

Nowadays most authors have abandoned magnitudes and give the data in absolute flux density $F(\lambda)$ (units $\mathrm{W \cdot cm^{-2} \cdot \mu m^{-1}}$, often used for stars) or $F(\nu)$ (units $\mathrm{W \cdot m^{-2} \cdot Hz^{-1}}$, often used for radio sources). The mutual conversion of these units is given in Fig. 2 which further includes a calibration of the magnitude system and an example of the distortion of the energy distribution by interstellar reddening. More universal is the presentation of results in terms of the parameter $\lambda F(\lambda)$ in units of $\mathrm{W \cdot m^{-2}}$, a practice which, to my knowledge, has been initiated by astronomers at the University of Minnesota. This will be done throughout the present paper. As the properties of the parameter $\lambda F(\lambda)$ are possibly not generally appreciated we will briefly discuss some of them. It is evident that $\lambda F(\lambda) = \nu F(\nu)$ which makes it convenient to transform $\lambda F(\lambda)$ either to $F(\lambda)$ or $F(\nu)$ Furthermore,

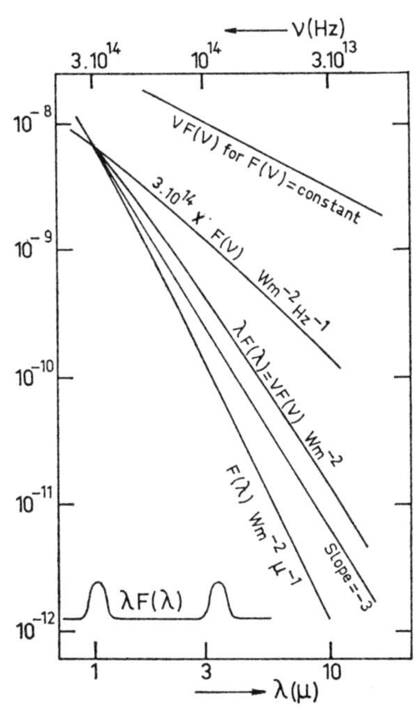

Fig. 3. $F(\lambda)$, $F(\nu)$ and $\nu F(\nu)$ for a star of 0 mag at all wavelengths. The line $\nu F(\nu)$ for $F(\nu)$= constant represents the slope for a free-free continuum. The two idential features in the "spectrum" at the bottom contain equal amounts of energy

the energy contained in spectral features at different wavelengths can be readily compared in a $\lambda F(\lambda)$ versus $\log \lambda$ diagram:

$$\int \lambda F(\lambda)\, d(\log \lambda) = \int F(\lambda)\, d\lambda.$$

The Rayleigh-Jeans law of black body radiation at long wavelengths has a slope of -3 in a $\log \lambda F(\lambda)$ versus $\log \lambda$ diagram instead of the -4 in the $\log F(\lambda)$ versus $\log \lambda$ and $+2$ in the $\log F(\nu)$ versus $\log \nu$ diagram, respectively. This is illustrated in Fig. 3.

Observations expressed in flux density units are more convenient to analyse than magnitudes but it is always wise to study the calibration procedure carefully. The transformation of broad-band measurements to monochromatic flux densities and colour temperatures is not a trivial problem. Laboratory calibrations as well as transformations to celestial sources having a different colour temperature should be regarded with caution, especially if sets of observations obtained with different instruments or at different times are to be studied for effects which are not much in excess of the internal errors. These considerations apply in particular to airborne and balloonborne experiments where additional problems like beam-size transformations, stabilization and pointing are restricting the significance of observations, specifically if these observations are to be used in an analysis together with groundbased optical and radio observations. It is suggested that observers include at least examples of raw data and the reduction procedures in their papers, helping readers to get an insight in the significance of the derived quantities.

Infrared radiation is detected by a beam-switching technique. This means that the recorded response is proportional to the difference of the radiation received from two regions in the sky, separated by the so-called chopper throw. An extended source, even a bright one, may escape detection if its dimensions are considerably larger than the chopper throw as has been demonstrated by the work of HOFFMANN and FREDERICK (1969) and AUMANN and LOW (1970) on the galactic center region. This effect tends to a systematic underestimate of flux densities and dimensions of extended sources; point sources will be similarly effected if stabilization or pointing are inadequate.

Fortunately, many of the discoveries in infrared astronomy are based on pronounced effects and the interpretation would not change substantially if more precise and reliable reduction procedures could have been applied. However, it is likely that significant progress in the near future will depend on better bandpass definition, better absolute calibration, more extensive use of relative calibration with respect to a net of well observed celestial standard sources and, in the case of airborne or balloonborne equipment, better pointing and stabilization. In addition it is becoming clear that improvements in spatial resolution are possible and can be used profitably to study spatial detail to the level of resolution that is compatible with optical and radio observations. Significant improvements in spectral resolution are soon expected to provide an answer to the question of the contributions of line emission to the large far infrared fluxes which are observed in several sources (cf. section IVb on planetary nebulae).

II. Distribution of Infrared Sources over the Sky

Using a 60-inch telescope and an array of PbS detectors at the prime focus, NEUGEBAUER and LEIGHTON (1969) carried out a survey of the northern sky at declination between $-33°$ and $+80°$. Their catalogue consists of 5500 sources brighter than $K=3^m$. Positive identifications include a large number of known late-type variables; approximately 1/3 of the sources are visible to the naked eye.

The Dearborn Catalogue of Faint Stars (LEE *et al.*, 1947) contains over 44.000 red objects; the catalogue is based on an objective prism survey of the sky north of declination $-4\overset{\circ}{.}5$. A comparison with the 2.2 μm survey indicates that 93% of the 2.2 μm sources in the declination zone where the two surveys overlap can be identified in the Dearborn Catalogue (GRASDALEN and GAUSTAD, 1971).

ACKERMANN (1970) conducted a search for red objects in Cygnus by means of objective prism exposures with a 25-cm Schmidt telescope. This technique seems to be quite successful if the colour temperature of the objects is not too low. PRICE (1968) made a partial survey of the southern sky ($-30°>\delta>-52°$) at 2.2 μm.

A survey at 5 μm of the southern sky is being prepared by a group at the Kapteyn Laboratory in collaboration with the European Southern Observatory. A 50-cm telescope with a small secondary mirror chopping at 400 Hz and two InSb detectors will be used. It is expected that this survey will be capable to detect stars with low-temperature circumstellar shells. LOW (1970) reports trial runs of a survey at 5 μm and 10 μm. The initial results indicate no unexpected findings.

Very promising results have been obtained by HOFFMANN *et al.* (1971a) who have operated a 30-cm balloonborne stabilized telescope equipped with a 80–135 μm bandpass photometer. They report 72 sources (58 of which had not been previously detected) in a survey covering 750 square degrees, mostly in the galactic plane. The results of this survey will be discussed later in this paper. Similar work reported separately at this conference is being done by the space research group of the Kapteyn Laboratory.

III. Radiation from Stars and Starlike Objects

We will divide the material in four categories:

a) O, B and A stars and related objects; interstellar extinction.

b) F, G, K and early M stars and related objects.

c) Late M, S and C stars.

d) Some selected stars.

a) O-, B- and A-Type Stars and Related Objects; Interstellar Extinction

A review of the near infrared intrinsic colours of normal stars has been compiled by JOHNSON (1966); revised values are given in a later paper on interstellar extinction (JOHNSON, 1968). Relative magnitudes of some stars have been plotted in Fig. 4.

It has been noted that emission-line stars, as a rule, show excess radiation at 3.5 μm and beyond (JOHNSON, 1967; WOOLF *et al.*, 1970). These observations have been interpreted as radiation by free-free emission from circumstellar gas envelopes. The infrared excesses are a significant fraction of the infrared energy that would be radiated by a normal star of the same spectral type; compared to the total energy the observed infrared excess in Be stars is typically less than 1%. Using the data of WOOLF *et al.* (1970) we have plotted the virtually unreddened B2IIIe star ζ Tau in Fig. 5; the infrared excess at K, L and M can be appreciated by shifting the curve labelled "O mag" downward to coincide with the data of ζ Tau at V, R and I.

Observations and interpretations of infrared excesses were reviewed at a recently held symposium of the ASP on infrared excesses in stars, held in Santa Cruz, which was summarized in Nature (Vol. 238, p. 72). In that summary observations are reported of the 10 μm excess radiation in and around some early-type stars in Orion I out to several seconds of arc (cf. STEIN and GILLETT, 1971) their spectrum of θ^1 C Ori has been reproduced in Fig. 5; meanwhile a 0.1 μm resolution spectrum of the trapezium has been reported by WOOLF (1973) in a discussion of observations by GILLETT and STEIN. The 10 μm feature, common in the spectra of many late-type stars, is attributed to stretching and bending resonances of silicates in dust grains. It is likely that this broad emission band has its peak at 9.7 μm, practically coinciding with the 9.6 μm absorption band of ozone in the terrestrial atmosphere.

Photometry in J, H, K and L of stars in the association VI Cyg has been reported by VOELCKER and ELSÄSSER (1973). They find no evidence of anomalous infrared colours for stars brighter than $V=15\overset{m}{.}6$ and they conclude that these stars show no evidence of low temperature shells. A similar conclusion, relevant to lower temperature shells was reached by STEIN and GILLETT (1971) on the

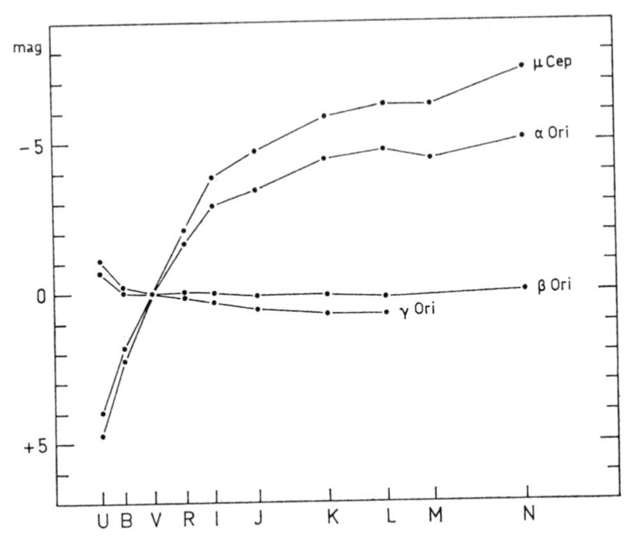

Fig. 4. Relative magnitudes for some "normal" stars

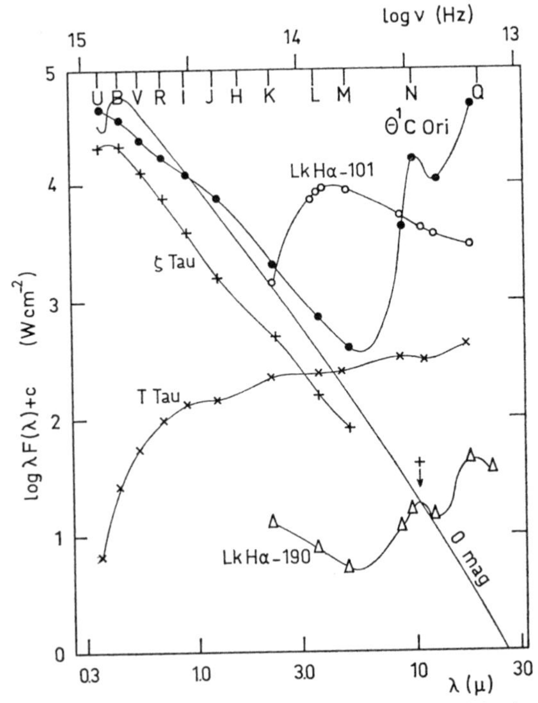

Fig. 5. Relative log $\lambda F(\lambda)$ versus λ; data are from sources mentioned in the text

basis of 10 μm photometry of VI Cyg No. 12, though it should be noted that the steep increase of interstellar extinction towards shorter wavelengths tends to mask the silicate bump.

Observations of stars in NGC 2264 (STROM *et al.*, 1972a) in H, K and L show several stars with excess infrared radiation. The authors suggest that their data demonstrate a very high (>10) value of R within some of the postulated dust shells. BREGER and DYCK (1973) combining the infrared observations of NGC 2264 with polarization measurements in the visible conclude that the data of some of the stars remind of Be-stars, requiring a flattened radiating gas envelop.

COHEN and WOOLF (1971) have reported on the remarkable infrared radiation from the sources LkHα-101 and LkHα-190 (cf. HERBIG, 1971). These are emission line objects with strong radiation in the infrared. BD+40°4124 (COHEN, 1972; STROM *et al.*, 1972b) at the center of a group of emission-line objects (including LkHα-224 and LkHα-225) has similar characteristics though the infrared energy distribution varies considerably from one object to the other.

Interstellar extinction in the infrared is small but not negligible. Adopting VAN DE HULST'S (1949) curve No. 15 for dielectric grain scattering one finds $R=3.05$; the extinction at 2.2 µm is 9% of A_v and at 10 µm it is 1% of A_v. JOHNSON (1968) has carefully reviewed the evidence for a variation of the extinction law and he finds values of R between 3 and 6, based on the "cluster diameter method" and the "variable extinction method". The variation of the extinction law seemed to be confirmed by infrared observations of early- and some late- type stars (cf. the study of LEE (1968) of the Orion region). However, these observations could also be explained by infrared excess radiation. The independent evidence of the "variable extinction method" is rather weak and, moreover, not entirely independent, whereas a critical examination of the definition of cluster diameter indicates that the results of the "cluster diameter method" do not support a variation of R (HARRIS, 1973).

De VRIES (1972) has demonstrated that on the basis of near infrared photometry of early-type stars there is no evidence for a considerable variation of the reddening law if the sample is restricted to normal stars, specifically excluding emission line objects. There is a tendency now to believe again in a uniform reddening law in the interstellar field. Recent literature, analyzing infrared photometry for evidence of a departure from $R=3$ includes a study by VOELCKER and ELSÄSSER (1973) of VI Cyg ($R=3.25$) and an investigation of Sco-OB1 ($R=3$) by SCHILD *et al.* (1971). KOORNNEEF (private communication), analyzing L photometry of stars in a highly reddened cluster in Ara finds neither evidence for shell radiation nor a departure from $R\approx 3$ (BORGMAN *et al.*, 1970).

Though the number of early type stars with infrared observations (including 10 and 20 µm) is quite limited it is possible to arrive at a tentative interpretation: the excess radiation, if present, in the range 1–5 µm is due to free-free emission from a gas envelop. In some cases there is evidence for an extended cool dust cloud, radiating at 10 µm and 20 µm; this cloud could be the remnant of the material from which the star was formed (cf. PECKER, 1972).

b) F, G, K and Early M Stars and Related Objects

Relative infrared magnitudes of normal F-M stars are included in JOHNSON'S (1966) tables of infrared magnitudes (see Fig. 4).

When passing to cooler stars in the HR-diagram the frequency of the 10 µm excess radiation, characterized as the silicate bump, increases. One would except a

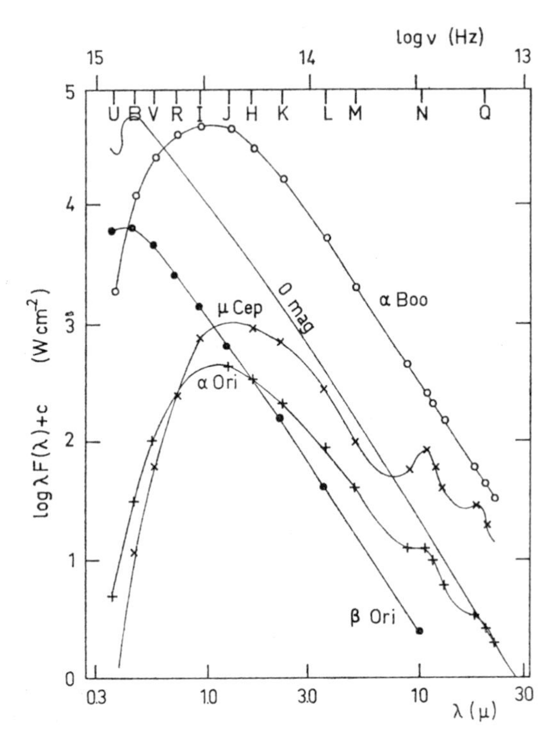

Fig. 6. Relative log $\lambda F(\lambda)$ versus λ. (Data are from WOOLF, 1973, and JOHNSON, 1968)

close correlation with mass loss parameters but there are striking exceptions as described by GILLETT *et al.* (1970). Out of four F and G supergiants with mass loss evidence they find only 89 Her (F2Ia) to show a significant 10 μm excess. GEHRZ (1972), from an investigation of several late-type stars finds that "normal" stars show excess radiation at 10 μm if they are cooler or more luminous than the trajectory GoIa$^+$, G8Ia, M1Iab, M5Ib, M6III in the HR-diagram. μ Cep (M2Ia), formerly providing seemingly important evidence for an anomalous reddening law in Cepheus (JOHNSON, 1968) is one of the "normal" stars with a pronounced excess radiation at 10 μm, much more pronounced than in α Ori (M2Iab) which was used by JOHNSON as a less reddened comparison star (Fig. 6).

Large infrared excesses have been observed in T Tauri stars. Recent observations include V 1057 Cyg (COHEN and WOOLF, 1971).

c) Late M, S and C Stars

Generally the infrared photometry of M- and S-type stars is characterized by a double bump (10 μm and 20 μm), characteristic for absorption by silicates; carbon stars seem to show only the bump at 10 μm. The magnitude of the bump varies from a few tenths to a full magnitude (HACKWELL, 1971). For some of the stars in this category there is good evidence for a low temperature shell of carbon (cf. FORREST *et al.* (1971) on the carbon-rich hydrogen-deficient supergiant R CrB).

d) Some Selected Stars

We will conclude this brief discussion of infrared properties of stars by a discussion of some selected objects which do not fall readily into one of the categories described before.

(i) *NML Cyg*. This star is one of the reddest objects in the 2.2 μm sky survey. The observations both in the infrared and at the 1612 MHz line of OH have recently been discussed by HYLAND *et al.* (1972), in connection with a larger study of similar objects. NML Cyg is a late M-type star with emission features (HERBIG and ZAPPALA, 1970). The infrared energy distribution has been plotted in Fig. 7. Recently, DAVIES *et al.* (1972) have described their detailed 1612 MHz observations which result in a large number of components within an area of 3.3 arc sec × 2.2 arc sec. These authors present a model of NML Cyg, in which the oxygen-rich M-type supergiant has an infrared radiating envelop, surrounded by an expanding and rotating ring with OH-emission; they favour a pre-main sequence stage for their model of NML Cyg. NML Cyg is visually fainter than 18th mag, which is due to interstellar extinction. The object is probably similar to VY CMa which has been discussed by HERBIG (1970); cf. HYLAND *et al.* (1972).

(ii) *IRC 10216*. This star, visually a fuzzy object fainter than 18th mag, was discovered in the 2.2 μm sky survey and subsequently observed at longer wavelengths (BECKLIN *et al.*, 1969). The energy distribution is plotted in Fig. 7. It was

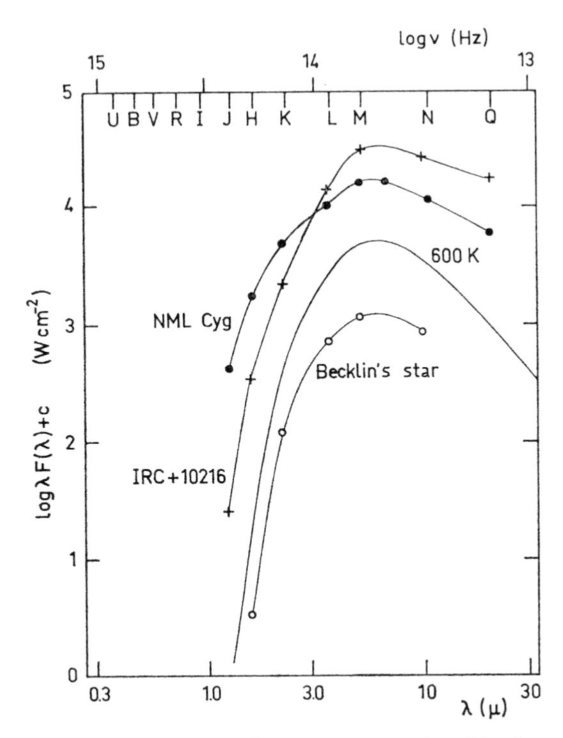

Fig. 7. Relative log $\lambda F(\lambda)$ versus λ; data are from sources mentioned in the text. For comparison a 600 K black body has been drawn

suggested that the object could be a Mira variable. Later observations by LOCKWOOD (1970) revealed the absence of the TiO and VO bands, whereas MILLER (1970) detected three CN bands in the spectrum, indicating the presence of a carbon star (cf. HERBIG and ZAPPALA, 1970). TOOMBS *et al.* (1972) have determined infrared diameters from lunar occultations. The brightness of the object allows high time resolution; the reappearance took from 1 sec (2.2 μm) to 3 sec (10 μm). The authors propose a model in which the central carbon star is surrounded by a dust shell of two components, the inner part being optically thick, diameter 0.4 seconds of arc at 600 K, the outer part being optically thin, diameter 2 seconds of arc and temperature 375 K.

(iii) *Becklin's Object*. This is an extremely red pointlike object, approximately 1 minute of arc NW of the trapezium in the Orion nebula and apparently embedded in the KLEINMANN-LOW extended source. Recently PENSTON *et al.* (1971) have suggested, on the basis of circumstatial evidence, that BECKLIN's star is a normal F-type supergiant, reddened by approximately 80 magnitudes in V. The suggestion of a F-type supergiant follows from excluding early-type supergiants (there is no compact HII region around the object) and later-type supergiants (there are no CO and H_2O absorption bands). The large extinction is in qualitative agreement with the Orion formaldehyde cloud observations, discussed by THADDEUS *et al.* (1971). Further support for the hypothesis of a starlike object with large extinction in front of it comes from the detection of a broad 9.7 μm emission feature of the trapezium region which is seen in absorption against Becklin's object (WOOLF, 1973, reporting observations of GILLETT, FORREST and COHEN). Another absorption band at 3.1 μm is attributed to interstellar ice. PENSTON *et al.* (1971) show that absorption and radiation of the stellar energy by the KLEINMANN-LOW nebula is in good agreement with reasonable assumptions on the diameter, temperature and mass of the postulated dust cloud. A comparison of the depth of the 9.7 μm feature (1.6 mag) and the adopted extinction ($A_v = 80$ mag) suggests a ratio A_v/9.7 μm absorption = 50, a result which we will use later in a discussion of the galactic center.

IV. Other Galactic Sources

Unlike stars, some extended sources radiate sufficient energy to be observable beyond 20 μm from balloonplatforms, aircrafts or rockets. We will describe a number of galactic objects on which observations have recently been accumulating; some objects, e.g. HII regions, will be more extensively covered at this conference by POTTASCH.

a) HII Regions

Several HII regions have been observed in the infrared, notably in the partial sky survey at 100 μm by HOFFMANN *et al.* (1971a). They report observations of 72 sources, 46 of which are identified with HII regions. Earlier, HARPER and LOW (1971) measured fluxes (45–750 μm) of eight HII regions; in an analysis of this material, supplemented by earlier data on three sources in the galactic center region they find evidence for a constant ratio between the total infrared flux and the 2 cm radio flux densities of SCHRAML and MEZGER (1969). HARPER and LOW favour an explanation of their data in terms of thermal radiation from dust grains, which are heated by L_α photons, resulting in a linear relation between the 2 cm free-free flux density and the total flux. The total flux is said to be approximately a factor of 5 too high, if it is assumed that all Lyman-continuum photons are degraded to L_α photons before absorption by the grains. Therefore, is would be necessary that a substantial fraction of the primary Lyman-continuum photons contribute directly to the thermal input of the dust cloud. This simple model seems not to fit to the data of HOFFMANN *et al.* (1971 a). They find a variation of the ratio of 2 cm flux density to 100 μm flux density over three orders of magnitude. A critical analysis of this situation will be given by POTTASCH at this conference. The ratio of total infrared flux to 100 μm flux density appears to be between 2 and 4, a result which we will use in the interpretation of the galactic center data.

New data at 345 μm on the twin HII region M17 have recently been published by JOYCE *et al.* (1972). The interpretation of their data presents a difficult problem, since their beam size (1.4 arc min) is small in comparison with the dimensions of the HII region (SCHRAML and MEZGER, 1969), the infrared source at 10 μm (KLEINMANN, reported by LOW, 1971) and the infrared source as seen by a total infrared flux measurement (HARPER and LOW, 1971; LOW and AUMANN, 1970). However, it is apparent that the infrared radiation at longer wavelengths is far too large to be understood as part of a black body model which fits the 10 and 20 μm radiation; the problem is even more severe if the possible size of the grains is taken into account.

b) Planetary Nebulae

Planetary nebulae show a large excess radiation in the infrared. The radiation at 10 μm is about 100 times larger than expected from free-free emission, fitted at 1.65 μm. The first discovered and one of the best observed planetary nebulae is NGC 7027. Its low resolution energy distribution (GILLETT *et al.*, 1967) can be interpreted as a continuum radiation of graphite grains which are heated by L_α photons (KRISHNA SWAMY and O'DELL, 1968). Meanwhile, it has been shown that several planetaries exhibit line emission, notably in the SIV line at 10.52 μm (RANK *et al.*, 1970); they found this line in NGC 7027 not substantially broader than the instrumental resolution (0.3 cm^{-1}, corresponding to 3 μm). However, their high resolution spectra do not support the suggestion by GOLDBERG (1968) that the infrared excess should be due to a collection of weak lines. Recently spectroscopic observations of a number of planetaries have been published (GILLETT *et al.*, 1972). They find that in some planetaries a significant fraction of the infrared excess is caused by emission lines. WOOLF (1973), reporting spectroscopic observations of NGC 7027 by GILLETT, FORREST and STEIN has demonstrated the existence of considerably more structure than the SIV line. Their spectrum, covering 8–14 μm at a resolution of 100 mμ shows at least four broad features, possibly including SIV, which could be only a minor feature at this resolution. The energy in the broad emission features is at least 60% of the total radiation or more if the continuum level has been overestimated.

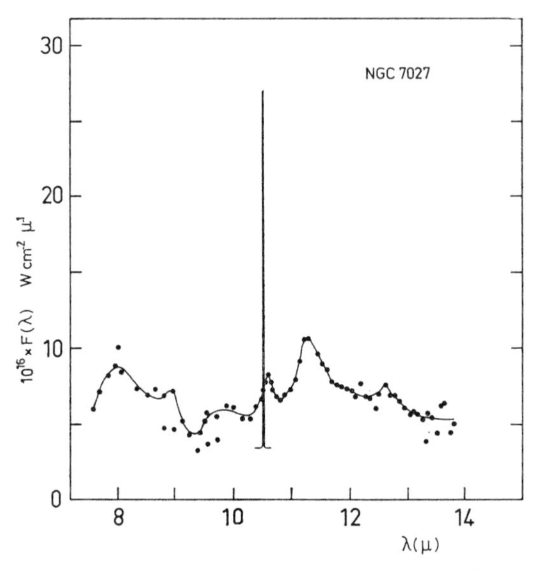

Fig. 8. The spectrum of the planetary nebula NGC 7027; points plotted are adapted from WOOLF (1973); the sharply peaked S IV emission profile is from RANK *et al.* (1970)

c) The Galactic Center Region

The infrared radiation from the galactic center region is strong and complex. Using a balloonborne 30 cm telescope and a 100 μm photometer with a beam size of 12 minutes of arc HOFFMANN *et al.* (1971 b) mapped an extended source of $2° \times 4°$, showing outstanding concentrations at Sgr A and Sgr B2.

Earlier, BECKLIN and NEUGEBAUER (1968, 1969) observed the galactic center region from 1.7–20 μm. They found evidence for an extended source of 1–5 μm radiation, with the maximum brightness centered on Sgr A; within 20 seconds of arc of the brightness peak at 2.2 μm they detected a pointlike source and an extended source, approximately 16 seconds of arc in diameter. The pointlike source has an energy distribution which resembles the extended source observed at 1–5 μm. Probably this source is a late-type reddened supergiant at the distance of the galactic center. BECKLIN and NEUGEBAUER assigned to it an upper limit of the 10 μm flux density of 10^{-25} W·m^{-2}·Hz^{-1}; RIEKE and LOW (1971) identified the pointsource at a 10 μm map of the galactic nucleus with a flux density concentration which is approximately 3 times the upper limit quoted. Assuming this difference to be significant the source must be variable. The extended 16 seconds of arc source of BECKLIN and NEUGEBAUER must probably be identified with one or more concentrations within the map of RIEKE and LOW; their 10 μm map shows radiation coming from a 30 seconds of arc source with four concentrations.

Adapting the presentation of NEUGEBAUER *et al.* (1971), the infrared picture of the galactic center region can be summarized as follows:

a) an extended $2° \times 4°$ region of 100 μm radiation with Sgr A and Sgr B2 as outstanding concentrations. Sgr A is a non-thermal radio source, Sgr B2 is a thermal HII region. The observed luminosity in the 75–125 μm passband represents 5×10^8 $L_\odot$ at the distance of the galactic center.

b) an extended ($>1°$) region with a strong concentration towards the center, radiating in the 1–5 μm region. This is probably a large concentration of unresolved stars in the nucleus of our galaxy. The observed energy distribution agrees with an integrated stellar energy distribution altered by 27 magnitudes of visual extinction (BECKLIN and NEUGEBAUER, 1968, 1969).

c) an extended 30 seconds of arc area of predominant 10 μm and 20 μm radiation centered on Sgr A; at 10 μm this source has been shown to contain at least 4 concentrations.

d) a pointlike source within the boundaries of the 30 seconds of arc source; this may be a late-type supergiant, luminosity $3\times10^{5}\,L_{\odot}$.

RIEKE and LOW (1971) estimate the total flux from the galactic center (source *c*) to be $5\times10^{-24}\,\mathrm{W\cdot m^{-2}\cdot Hz^{-1}}$ at 10 μm and $23\times10^{-24}\,\mathrm{W\cdot m^{-2}\cdot Hz^{-1}}$ at 20 μm. These numbers may be underestimates as pointed out in the introduction to this paper where the beamswitching technique was discussed; HOUCK *et al.* (1971) from rocket observations found evidence for an extended source in the galactic center of a few degrees along the galactic plane at flux density levels of $50\times10^{-24}\,\mathrm{W\cdot m^{-2}\cdot Hz^{-1}}$ both at 13 μm and 20 μm. Their raw data suggest that they have observed the 30 seconds of arc extended source together with radiation at these wavelengths from the center of the extended 100 μm source.

There is a tendency in the literature to describe the observations of the galactic center region as if the radiation would originate from different sources which can be physically distinguished. Probably this is due to the large range of spatial resolutions and wavelengths at which the observations have been made or presented. We have attempted to construct a coarse radiation model on the basis of observations scaled to a common aperture. For this we have chosen a circular diaphragm of 22 seconds of arc, centered on Sgr A. The results are in Table 2. Numbers for 1.06 and 1.18 μm are from SPINRAD *et al.* (1971) who actually used a 22 seconds of arc diaphragm. Numbers for 1.65, 2.2, 3.5 and 4.8 μm are from BECKLIN and NEUGEBAUER (1969). Numbers for 10.2 and 20 μm have been averaged from BECKLIN and NEUGEBAUER (1969) and RIEKE and LOW (1971); the last two papers give information on the aperture dependence of the flux density. The value for 100 μm corresponds to the peak brightness in the map of HOFFMANN *et al.* (1971 b). The values for 8 and 13 μm have been read from a 8–14 μm spectrum of the galactic center (see Fig. 9) reported by WOOLF (1973); the profile has been drawn in Fig. 10. An attempt has been made to understand the observations for $\lambda\leqq2.2$ μm as a continuum of late-type giants, modified by interstellar reddening. The assumption on the stellar population is based on the observed near infrared energy distribution at the surface of the nucleus of M31 (SANDAGE *et al.* 1969). For the interstellar extinction curve we have adopted VAN DE HULST's (1949) monochromatic curve No. 15 as modified by the broadband response functions, large extinction and a low temperature source. Such computations have been made by LEE (1968) and we have adopted his effective extinction law for 8.7 mag visual extinction (the largest extinction in LEE's tables) and a black body source of 3500 K. The "monochromatic" 1.06 μm and 1.18 μm extinction values have been interpolated. Using the energy distribution of α Boo (K2 III, typical for late-type giants) and the extinction law defined above, while requiring the modified curve to fit the 1.06 μm

Table 2. Galactic center data scaled to a 22 seconds of arc circular diaphragm

λ (μm)	1.06	1.18	1.65	2.2	3.5	4.8	8	10.2	13	20	100
$-\log F_\nu$ (W · m^{-2} · Hz^{-1})	28.33	27.80	26.03	25.35	24.86	24.35	23.63	23.52	22.92	22.92	23.66
$-\log \lambda F(\lambda)$ (W · m^{-2})	13.88	13.39	11.77	11.22	10.93	10.55	10.06	10.05	9.56	9.74	11.18
A_λ (A_v=28 mag)	10.6	8.9	4.7	2.9	1.3	0.7	—	—	—	—	—
A_λ (A_v=200 mag)	—	—	—	21	9.6	5.0	2.4	1.8	1.1	0.6	0

and 2.2 μm points a unique solution is obtained: 28 mag of visual extinction. Adopting $M_v=0$ we need 7×10^3 late-type giants corresponding to $1.2 \times 10^6 L_\odot$ at $M_{\text{bol}} = -1$ mag at 10 kpc within our 22 seconds of arc diaphragm. These results agree with earlier calculations of BECKLIN and NEUGEBAUER (1968, 1969) and SPINRAD *et al.* (1971), which is not surprising because the same observations have been used in a somewhat different procedure.

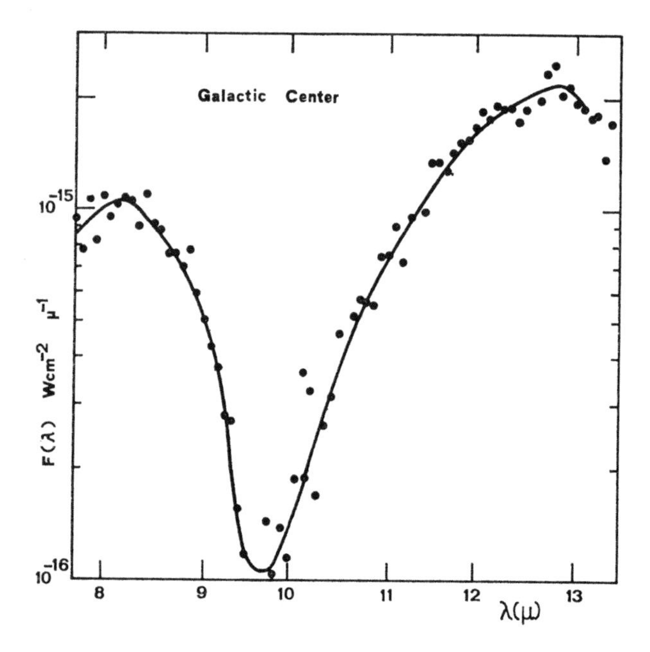

Fig. 9. 8–14 μm spectrum of the galactic center. (WOOLF, 1973)

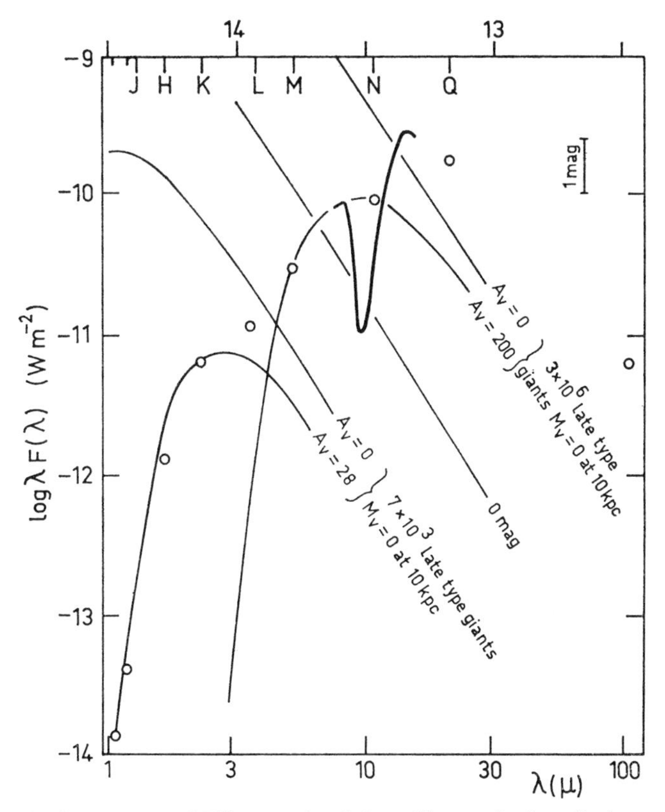

Fig. 10. The galactic center. Log $\lambda F(\lambda)$ versus λ scaled to a 22 seconds of arc diaphragm. Open circles represent data from BECKLIN and NEUGEBAUER (1969), RIEKE and LOW (1971) and HOFFMANN *et al.* (1971b); the drawn profile at 10 μm has been adapted from WOOLF (1973)

The contribution of the continuum discussed above to the observed flux density at 4.8 μm is not more than 0.1 mag (Fig. 10). Therefore, the observations at $\lambda > 5$ μm can be analyzed independently of those at $\lambda < 2.2$ μm, assuming that our analysis so far has been correct. We proceed to discuss the hypothesis that the far infrared radiation is caused by thermal radiation from a dust cloud which is heated by a dense cluster with radiation characteristics which are determined by late-type giant stars. We assume that the intervening dust cloud is sufficiently far away from the cluster to render thermal radiation of the dust at 8 μm and below negligible. The shape of the 9.7 μm feature (Fig. 9) suggests that at 8 μm the cluster continuum is observed. Using the technique described above we find a unique solution by fitting the 4.8 μm and 8 μm points; the solution requires 200 mag visual extinction and 3×10^6 late-type giants (corresponding to $5 \times 10^8 L_\odot$ at $M_{bol} = -1$ mag) at 10 kpc within the 22 seconds of arc diaphragm. The total luminosity may be a factor 2 higher, adopting the aperture dependence of the 10 and 20 μm radiation. This would bring us at a luminosity of $1 \times 10^9 L_\odot$.

The large extinction (200 visual magnitudes) and luminosity ($1 \times 10^9 L_\odot$) in this radiation model is consistent with the following independent evidence:

a) Observations by WOOLF and GILLETT, reported by WOOLF (1973), show the absorption feature at 9.7 μm which has been reproduced in Fig. 9. The depth of the absorption feature can transformed to total visual extinction as has been explained by WOOLF (1973). The central depression is 3.4 mag relative to a continuum that has been interpolated between shoulders of 1.1×10^{-15} W·cm^{-2}·μm^{-1} at 8 μm and 2.1×10^{-15} W·cm^{-2}·μm^{-1} at 13 μm. Adopting the arguments on the nature of Becklin's star (section IIId) we would predict $50 \times 3.4 = 170$ mag of visual extinction, or somewhat more if the depression has been underestimated.

b) As mentioned above, the observed 75–125 μm luminosity in a $2° \times 4°$ field is equivalent to $5 \times 10^8\,L_\odot$. Adopting the ratio of total infrared flux to 75–125 μ flux to be 2–4 (HII regions, see section IVa) or 4 (for the central 4 minutes of arc of the galactic center as follows from LOW and AUMANN, 1970, and HOFFMANN *et al.*, 1971b) we arrive at a luminosity of $1 \times 10^9\,L_\odot$ or $2 \times 10^9\,L_\odot$, resp., consistent with the implicit assumption that the central stellar luminosity of $1 \times 10^9\,L_\odot$ would be the primary energy source.

The deviation of the 20 μm observation in Fig. 10 from the radiation model is relatively large but, on an absolute scale, not very significant. Allowing a generous 10^{13} Hz bandwidth the excess radiation at 20 μm corresponds to $10^6\,L_\odot$, which might qualitatively be explained as thermal radiation from the dust within the 22 seconds of arc diaphragm. Note that at 100 μm the stellar contribution to the observed radiation within this aperture would be 10%.

Part of the 20 μm and 13 μm excess radiation can be accomodated by adopting a model with a continuous concentration of luminosity and extinction towards the center. This refinement seems premature and would be justified only if more and better observations were available.

The schematic radiation model of the galactic center as described above takes only the infrared measurements into account. This is not unreasonable as the energy emitted in other regions of the spectrum is relatively small. Moreover, the detailed maps in the infrared and the radio region lack in correspondence (cf. DOWNES and MARTIN, 1971; EKERS and LYNDEN-BELL, 1972; RIEKE and LOW, 1971). It is worthwhile to note that the required total mass, $10^7\,M_\odot$ presents no immediate problems; however the mass distribution and the contribution of low-luminosity stars needs further investigation.

V. Extragalactic Sources

Since the general review paper by NEUGEBAUER *et al.* (1971) only few new observations have been published. We restrict ourselves to mentioning some new references.

YOUNG *et al.* (1972) report photometry of Markarian 231, a Seyfert galaxy which was not included in the papers by KLEINMANN and LOW (1970a, b) where many observations of galaxies can be found. An intriguing feature is the variability of some Seyfert galaxies. This would set upper limits to the size of the radiating region. WICKRAMASINGHE (1971) has discussed the observations in relation to some models of the primary energy source; he claims that sufficiently small regions of dust could produce the observed energy if cosmic ray particles supply the primary energy.

BECKLIN *et al.* (1971) observed the nucleus of Cen A from 1–10 μm. The energy distribution, from the radio to the X-ray region, is similar to what has been observed in the Crab-nebula and in the Seyfert galaxy NGC 1275.

VAN DER KRUIT (1971) has observed a number of Seyfert and normal galaxies with the Westerbork Synthesis Radio Telescope at 1415 MHz. He finds a linear relation between the power at 1415 MHz and at 10 μm.

M82, measured and mapped by KLEINMANN and LOW (1970a, b) has been observed by KRONBERG *et al.* (1972); their data suggest that there is no detailed correspondence between the optical (infrared) and radio sources.

STRITTMATTER *et al.* (1972) announce the identification of 4 objects of the BL Lac type. BL Lac has been observed at 3.5 and 11 μm by STEIN *et al.* (1971). The four sources of STRITTMATTER are OJ 287, ON 231, ON 325 and PKS 1514–24. OJ 287 has been observed at 3.5 and 11 μm; the infrared colour index is similar to that of BL Lac; as BL Lac, OJ 287 appears to be variable on a timescale of months.

VI. The Future

The explosive development of infrared astronomy is likely to continue for a while. The 0.1 μm resolution spectroscopy of GILLETT and others is going to reveal the existence and abundance of hitherto unidentified solids in interstellar space, circumstellar shells, dense clouds and planetary nebulae. The development of high resolution spectroscopy in the far infrared is taking shape and may allow the detection of several new forbidden lines. Higher spatial resolution in the far infrared is to be expected from better stabilized balloonplatforms or aircraftplatforms, equipped with larger telescopes. These instruments will allow us to study HII regions and other extended sources at better spectral and spatial resolution. Comparable developments in groundbased infrared astronomy necessitate the construction of large dedicated reflectors or spatial interferometers.

Apart from the high resolution work in the near infrared, notably by J. and P. CONNES, the present level of productive activity in infrared astronomy in Europe is very modest. Nearly all the observational work reviewed in this paper has been done in the U.S. by American astronomers. This situation is not likely to change in the near future.

Though this review is not complete in a broad sense an attempt has been made to include the latest references on the selected topics, as far as they were not mentioned in earlier reviews quoted in this paper. The survey of the literature to which we had access was finished on August 1, 1972. The helpful suggestions of M. DE VRIES are grateful acknowledged. I thank OLTHOF and DE VRIES for a critical reading of the manuscript.

References

ACKERMANN, G.: Astron. Astrophys. **8**, 315 (1970).
AUMANN, H. H., LOW, F. J.: Astrophys. J. **159**, L 159 (1970).
BECKLIN, E. E., NEUGEBAUER, G.: Astrophys. J. **151**, 145 (1968).

BECKLIN, E. E., NEUGEBAUER, G.: Astrophys. J. **157**, L 31 (1969).
BECKLIN, E. E., FROGEL, J. A., HYLAND, A. R., KRISTIAN, J., NEUGEBAUER, G.: Astrophys. J. **158**, L 133 (1969).
BECKLIN, E. E., FROGEL, J. A., KLEINMANN, D. E., NEUGEBAUER, G., NEY, E. P., STRECKER, D. W.: Astrophys. J. **170**, L 15 (1971).
BREGER, M., DYCK, H. M.: Astrophys. J. **175**, 127 (1972).
BORGMAN, J., KOORNNEEF, J., SLINGERLAND, J. H.: Astron. Astrophys. **4**, 248 (1970).
COHEN, M.: Astrophys. J. **173**, L 61 (1972).
COHEN, M., WOOLF, N. J.: Astrophys. J. **169**, 543 (1971).
DAVIES, R. D., MASHEDER, M. R. W., BOOTH, R. S.: Nature **237**, 21 (1972).
DOWNES, D., MARTIN, A. H. M.: Nature **233**, 112 (1971).
EKERS, R. D., LYNDEN-BELL, D.: Astrophys. Letters **9**, 189 (1972).
FORREST, W. J., GILLETT, F. C., STEIN, W. A.: Astrophys. J. **170**, L 29 (1971).
GEHRZ, R. D.: Thesis, University of Minnesota (1972) (reported by WOOLF, 1972).
GILLETT, F. C., HYLAND, A. R., STEIN, W. A.: Astrophys. J. **162**, L 21 (1970).
GILLETT, F. C., LOW, F. J., STEIN, W. A.: Astrophys. J. **149**, L 97 (1967).
GILLETT, F. C., MERRILL, K. M., STEIN, W. A.: Astrophys. J. **172**, 367 (1972).
GILLETT, F. C., STEIN, W. A.: Astrophys. J. **159**, 817 (1970).
GOLDBERG, L.: Astrophys. Letters **2**, 101 (1968).
GRASDALEN, G. L., GAUSTAD, J. E.: Astron. J. **76**, 231 (1971).
HACKWELL, J. A.: Thesis, University College of London (1971) (reported by WOOLF, 1972).
HARPER, D. A., LOW, F. J.: Astrophys. J. **165**, L 9 (1971).
HARRIS, D. H.: In J. M. GREENBERG and H. C. VAN DE HULST (eds.), Interstellar Dust and Related Topics. IAU Symp. No. 52, p. 31. Dordrecht-Holland: D. Reidel 1973.
HERBIG, G. H.: Astrophys. J. **162**, 557 (1970).
HERBIG, G. H.: Astrophys. J. **169**, 537 (1971).
HERBIG, G. H., ZAPPALA, R. R.: Astrophys. J. **162**, L 15 (1970).
HOFFMANN, W. F., FREDERICK, C. L.: Astrophys. J. **155**, L 9 (1969).
HOFFMANN, W. F., FREDERICK, C. L., EMERY, R. J.: Astrophys. J. **170**, L 89 (1971a).
HOFFMANN, W. F., FREDERICK, C. L., EMERY, R. J.: Astrophys. J. **164**, L 23 (1971b).
HOUCK, J. R., SOIFER, B. T., PIPHER, J. L., HARWIT, M.: Astrophys. J. **169**, L 31 (1971).
HULST, H. C. VAN DE: Rech. Astr. Obs. Utrecht **11**, part 2 (1949).
HYLAND, A. R., BECKLIN, E. E., FROGEL, J. A., NEUGEBAUER, G.: Astron. Astrophys. **16**, 204 (1972).
JOHNSON, H. L.: Astrophys. J. **141**, 923 (1965).
JOHNSON, H. L.: Ann. Rev. Astron. Astrophys. **4**, 193 (1966).
JOHNSON, H. L.: Astrophys. J. **150**, L 39 (1967).
JOHNSON, H. L.: In B. M. MIDDLEHURST and L. H. ALLER (eds.), Nebulae and Interstellar Matter, p. 167. Chicago: Chicago University Press 1968.
JOYCE, R. R., GEZARI, D. Y., SIMON, M.: Astrophys. J. **171**, L 67 (1972).
KLEINMANN, D. E., LOW, F. J.: Astrophys. J. **159**, L 165 (1970a).
KLEINMANN, D. E., LOW, F. J.: Astrophys. J. **161**, L 203 (1970b).
KRONBERG, P. P., PRITCHET, C. J., BERGH, S. VAN DEN: Astrophys. J. **173**, L 47 (1972).
KRISHNA SWAMY, K. S., O'DELL, C. R.: Astrophys. J. **151**, L 61 (1968).
KRUIT, P. C. VAN DER: Astron. Astrophys. **15**, 110 (1971).
LEE, O. J., CORE, G. D., BARTLETT, T. J.: Ann. Dearborn Obs. **5**, part. 6 (1947).
LEE, T. A.: Astrophys. J. **152**, 913 (1968).
LOCKWOOD, G. W.: Astrophys. J. **160**, L 47 (1970).
LOW, F. J.: In L. PEREK (ed.), Highlights of Astronomy, p. 136. Dordrecht-Holland: D. Reidel 1968.
LOW, F. J.: Semi-ann. Techn. Rep. Contract No. F 19628-70-C-0046, Project No. 5130 (ARPA) (1970).
LOW, F. J.: In B. T. LYNDS (ed.), Dark Nebulae, Globules and Protostars, p. 115. Arizona: University of Arizona Press (1971).
LOW, F. J., AUMANN, H. H.: Astrophys. J. **162**, L 79 (1970).
MILLER, J. S.: Astrophys. J. **161**, L 95 (1970).
NEUGEBAUER, G., BECKLIN, E. E., HYLAND, A. R.: Ann. Rev. Astron. Astrophys. **9**, 67 (1971).

NEUGEBAUER, G., LEIGHTON, R. B.: NASA SP-3047 (1969) (Govt. Printing Office).
PENSTON, M. V., ALLEN, D. A., HYLAND, A. R.: Astrophys. J. **170**, L 33 (1971).
PECKER, J. C.: In Les spectres des astres dans l'infrarouge et les microondes. Liège Symp. 1971, p. 243. Liège (1972)
PRICE, S. D.: Astron, J. **73**, 431 (1968).
RANK, D. M., HOLTZ, J. Z., GEBALLE, T. R., TOWNES, C. H.: Astrophys. J. **161**, L 185 (1970).
RIEKE, G. H., LOW, F. J.: Nature **233**, 53 (1971).
SANDAGE, A. R., BECKLIN, E. E., NEUGEBAUER, G.: Astrophys. J. **157**, 55 (1969).
SCHILD, R., NEUGEBAUER, G., WESTPHAL, J. A.: Astron. J. **76**, 237 (1971).
SCHRAML, J., MEZGER, P. G.: Astrophys. J. **156**, 269 (1969).
SPINRAD, H., LIEBERT, J., SMITH, H. E., SCHWEIZER, F., KUHI, L. V.: Astrophys. J. **165**, 17 (1971).
SPINRAD, H., WING, R. F.: Ann. Rev. Astron. Astrophys. **7**, 249 (1969).
STEIN, W. A., GILLETT, F. C.: Nature **233**, 72 (1971).
STEIN, W. A., GILLETT, F. C., KNACKE, R. F.: Nature **231**, 254 (1971).
STRITTMATTER, P. A., SERKOWSKI, K., CARSWELL, R., STEIN, W. A., MERRILL, K. M., BURBIDGE, E. M.: Astrophys. J. **175**, L 7 (1972).
STROM, S. E., STROM, K. M., BROOKE, A. L., BREGMAN, J., YOST, J.: Astrophys. J. **171**, 267 (1972a).
STROM, K. M., STROM, S. E., BREGER, M., BROOKE, A. L., YOST, J., GRASDALEN, G., CARRASCO, L.: Astrophys. J. **173**, L 65 (1972b).
THADDEUS, P., WILSON, R. W., KUTNER, M., PENZIAS, A. A., JEFFERTS, K. B.: Astrophys. J. **168**, L 59 (1971).
TOOMBS, R. I., BECKLIN, E. E., FROGEL, J. A., LAW, S. K., PORTER, F. C., WESTPHAL, J. A.: Astrophys. J. **173**, L 71 (1972).
VOELCKER, K., ELSÄSSER, H.: In J. M. GREENBERG and H. C. VAN DE HULST (eds.), Interstellar Dust and Related Topics. IAU Symp. No. 52, p. 529. Dordrecht-Holland: D. Reidel 1973.
VRIES, M. DE: Private communication (1972).
WEBBINK, R. F., JEFFERS, W. Q.: Space Sc. Rev. **10**, 191 (1969).
WICKRAMASINGHE, N. C.: Nature **230**, 166 (1971).
WOOLF, N. J.: Astrophys. J. **157**, L 37 (1969).
WOOLF, N. J.: In J. M. GREENBERG and H. C. VAN DE HULST (eds.), Interstellar Dust and Related Topics. IAU Symp. No. 52, p. 485. Dordrecht-Holland: D. Reidel 1973.
WOOLF, N. J., STEIN, W. A., STRITTMATTER, P. A.: Astron. Astrophys. **9**, 252 (1970).
YOUNG, E. T., KNACKE, R. F., JOYCE, R. R.: Nature **238**, 263 (1972).

Discussion

SMYTH, M. J.:
Do we yet know enough about sky noise at 10 microns, and its spatial dependence, to be certain that the small apertures you mention are essential?

BORGMAN, J.:
I have not yet seen the results of WESTPHAL's sky noise survey.

OORT, J.:
The suggestion that the far-infrared radiation from the galactic center is due to some 10^6 normal K giants is quite fascinating. One might ask whether the total mass which would be expected to accompany these giants would be compatible with what we have from the observations of the rotation of the nuclear disk. These measures indicate that the total mass within 100 parsec from the center is about 0.9×10^9 $M_{\odot}$. This is certainly sufficient to accomodate the 3×10^6 K giants if the accompanying population is like that in a globular cluster.

BORGMAN, J.:
Apparently we are is that respect in a satisfactory situation. Meanwhile we should be aware that our radiation model has been adapted to the energy distribution that is seen e.g. at the surface of the nuclear bulge in M31. Surely we must e.g. accomodate a reasonable number of early-type stars which excite the HII regions. However, they need not necessarily be within the 30 seconds of arc source and their mass should be relatively small.

Interpretation of Far Infrared Observations

(Invited Lecture)

By S. R. Pottasch
Kapteyn Astronomical Institute, Groningen, The Netherlands

With 9 Figures

Abstract

Possible interpretations of the far infrared (20 μm to 900 μm) observations are discussed. Attention is especially directed to compact HII regions, which may be a place of present star formation. The region of the galactic center and the nuclei of other galaxies are also considered. Emission by dust grains is considered in detail, as is line emission in this spectral region. The possibility of synchrotron emission is also discussed.

I. Introduction

In recent years observations of celestial objects in the far infrared have reached the stage that interpretation has a reasonably solid observational basis, and the interpretation can in turn provide a fruitful guideline for future observations.

In this paper, only those objects known to radiate at 20 μm and 100 μm will be discussed. This limitation is placed to allow a deeper discussion of the remaining objects.

II. Summary of the Measurements

a) The Survey of Hoffmann, Frederick *and* Emery *(1971b)*

These authors, using a balloon-borne telescope and a detector sensitive between 75 μm and 135 μm, surveyed about 750 square degrees of the sky, including 110 degrees along the galactic equator. Their diaphragm was 12′, which, as we shall see, is entirely insufficient to distinguish any spatial structure, which sometimes may be as small as 10″ or even smaller. Thus if Hoffmann *et al.* describe the source as extended, we cannot distinguish between a truly extended object and several smaller objects close together in the sky.

Of the 72 objects which these authors observed, 56 lie in or close to the galactic plane. Of these 56 sources, it is possible to identify 41 with HII regions (and as HII regions we include those which can be so identified either because of observable $H\alpha$ emission, or a thermal radio spectrum, or the measurements of radio recombination lines). Six sources are identifiable with radio sources in the galactic center and most if not all of these are also HII regions. Four sources have the same position as known dark nebulae and may be associated with them. The remaining four sources cannot be identified with objects known in either optical or radio frequencies. Of the sources measured outside the galactic plane, eight were measured in the Orion region. Here three seem associated with HII regions, one perhaps with a dark nebula, and four are unidentified. Of the remaining nine sources, one may be associated with a dark nebula, two are radio sources listed in the 4C catalog, and the remaining six are not at present identified.

b) The Measurements of LOW *and Collaborators*

These authors (KLEINMANN and LOW, 1967, 1970a, b; LOW and AUMANN, 1970; HARPER and LOW, 1971) have measured selected HII regions, the galactic center, planetary nebulae, and galaxies in the infrared. They have shown that the HII regions (including the galactic center) are the strongest sources with flux increasing toward 100 μm. Selected galaxies also are observable in the far infrared, and the intrinsic luminosity of these galaxies is substantially greater than HII regions. We shall discuss these measurements and several other relevant observations in more detail presently.

c) Objects to be Discussed

It is clear that many HII regions are strong far-infrared emitters. This was not predicted before the measurements were made and in this sense it is surprising. In section III we will, therefore, discuss the HII regions in detail. In fact this will be the main task of this review, since the small amount known about infrared radiation from galaxies can best be discussed by comparison to the HII regions.

III. HII Regions and Compact HII Regions

a) The Size of the Emitting Regions

For several of the infrared sources detailed maps have been made in the infrared. To my knowledge W3, W51, and NGC 7538 have been mapped at several wavelengths between 2 μm and 20 μm (BECKLIN *et al.*, 1972). In all three cases a similar result has been obtained, which is illustrated in Fig. 1, which shows a map of the emitting regions in W3. The upper diagram shows the infrared source at 2.2 μm, the center diagram shows the region at 20 μm, while the lowest diagram shows the map obtained at 6 cm in the radio region by WYNN-WILLIAMS. All maps were made with a resolving power of from 5″–10″ arc. The agreement is striking, and together with similar results from the other two nebulae, lead us to the following tentative conclusions:

(1) the infrared radiation originates in a very small region, probably with a diameter of between 0.1 and 1 parsec;

(2) the sizes and intensity distribution do not change substantially with wavelength;

(3) there is a definite relation between the infrared radiation and the radio radiation (both seem to come from approximately the same region in space).

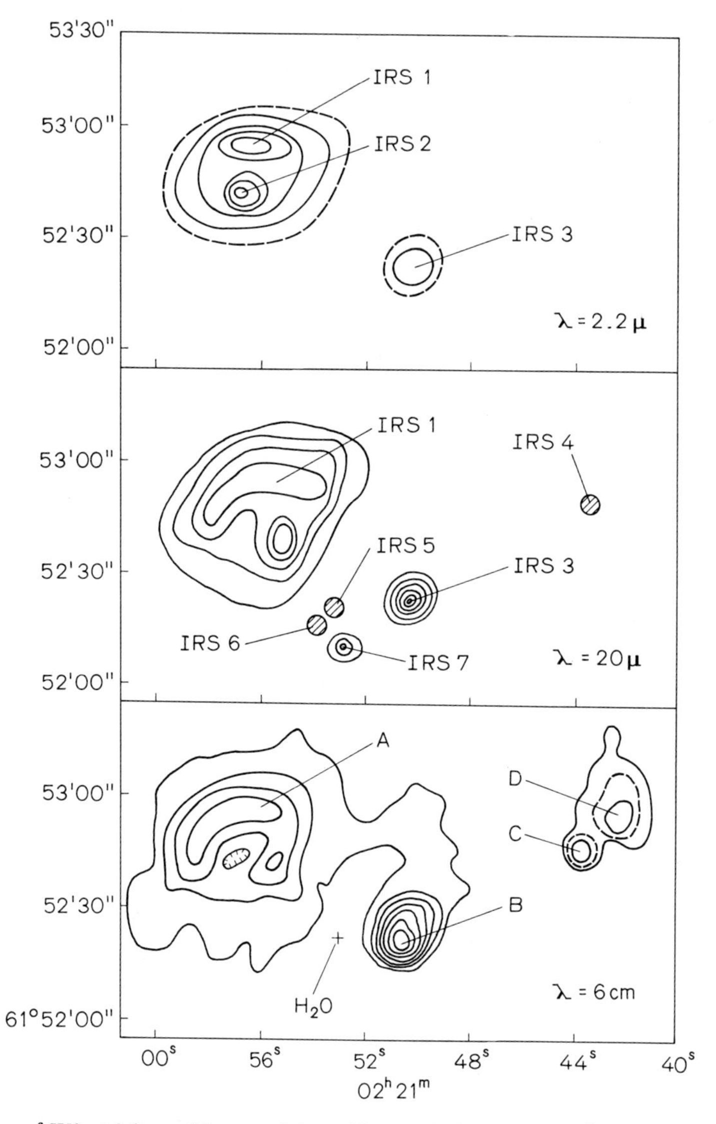

Fig. 1. Map of W3 at 2.2 μm, 20 μm and 6 cm. The resolution is about 5″ arc

It is interesting to look at the optical picture of W3 which is in the region of the well-known nebula IC 1795, shown in Fig. 2. In the region of the infrared objects, deliniated by the small rectangle, there is nothing to see optically, except to note that there is a substantial amount of interstellar absorption in the region, but all discernable features show a much larger size than the infrared objects.

In the case of W3 the 100 μm measurements are due to FURNISS *et al.* (1972). As in W51, where 100 μm measurements are also available, there is no indication of the size of the 100 μm source, other than that it is less than 10′ arc. HARPER

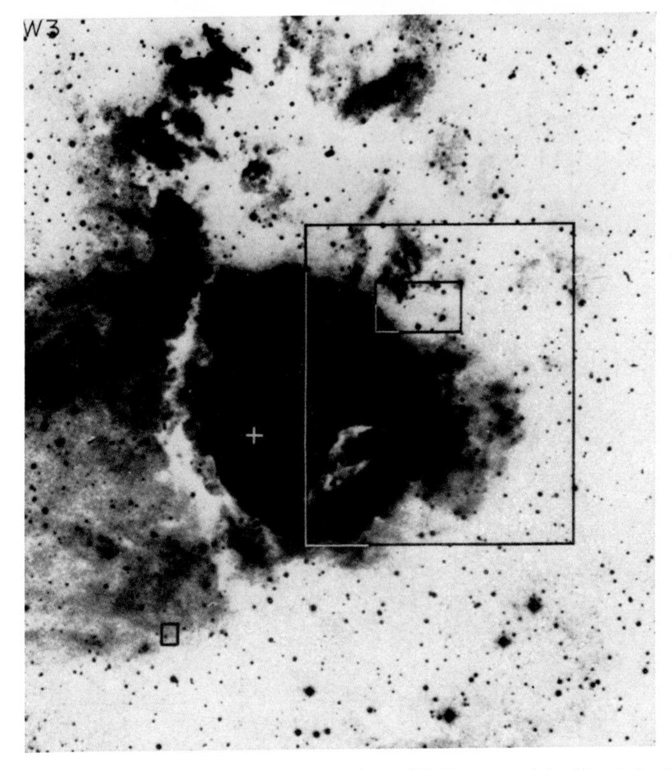

Fig. 2. Optical ($H\alpha$) picture of IC 1795. The region of W3 mapped in Fig. 1 is shown by the smaller rectangle

and LOW (1971) observe that when they increase the diaphragm size from 4'.8 to 8'.4 they obtain a 30% increase in flux. They use this observation to deduce that the diameter is indeed several minutes of arc. This argument is not at all strong, since MARTIN (1972) has shown that there are at least eight radio sources within a region of 10' in size, all of them small. Increasing the size of the diaphragm could have caused one of the less intense regions to come within the field. It should, therefore, be kept in mind that the 100 μm flux may be the sum of several nearby regions.

b) The Relationship between the Infrared Flux and the Radio Emission

Since it appears that both of these emissions come from essentially the same region, we now investigate in more detail their relation. The radio continuum emission is due to thermal free-free emission. This can be established in two ways. First, the spectrum is that expected from free-free emission, optically thin at the shorter wavelengths, optically thick at the meter wavelengths. Secondly, the electron density found on the assumption that it is free-free emission can be used to predict the hydrogen radio recombination line fluxes. These predicted fluxes are observed even for the compact HII region shown in Fig. 1. Thus we know the average

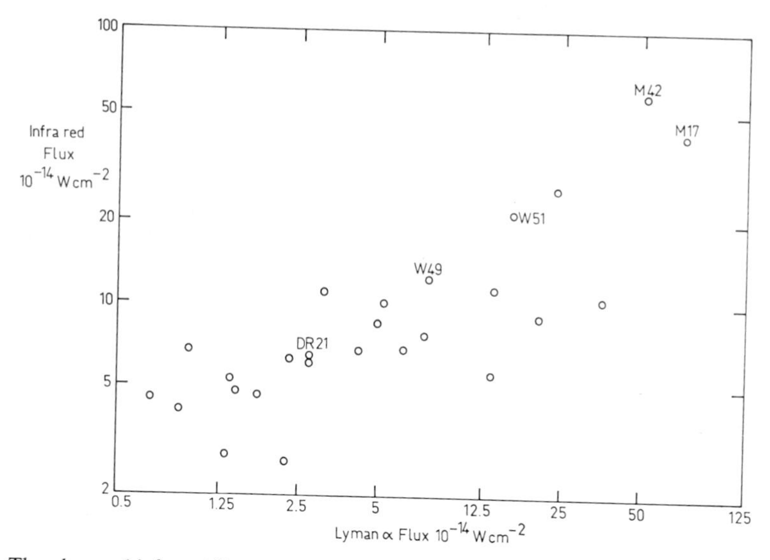

Fig. 3. The observed infrared flux (HOFFMANN *et al.*, 1971 b) is plotted against the observed radio emission, which is given in units of the Lyman α flux which an ionized region produces at the same time as it produces the observed thermal radio emission

values of the electron density n_e ($\sim 10^4$ cm^{-3}) and electron temperature T_e (~ 7000 K) in these regions.

The radio radiation not only originates from about the same region as the infrared radiation, but also its intensity is related to that of the infrared radiation. This was first pointed out by HARPER and LOW (1971) and is illustrated by the plot shown in Fig. 3. The ordinate in this figure is the infrared flux as measured by HOFFMANN *et al.* (1971 b) in the 80–130 μm band. The abscissa is labelled the Lyman α flux produced in the nebulae. It is essentially proportional to the radio flux. This is because each recombination of an electron and proton produces a single Lyman α quantum and the recombination rate is proportional to a slowly varying function of T_e and to n_e^2. The radio and Lyman α flux have exactly the same n_e^2 dependence and almost the same T_e dependence, so that the Lyman α flux differs only by a constant from the radio radiation.

Thus

$$\frac{I(\text{radio cont})}{I(\text{Lyman }\alpha)} = 2.3 \times 10^{-16}\, \nu^{-0.1}\, T_e^{0.15}\ \text{Hz}^{-1},$$

where ν is measured in GHz.

The values of the radio flux used were taken from ALTENHOFF *et al.* (1970), both because they have observed most of the galactic plane which HOFFMANN *et al.* measured, and because their half-power beam width, 11′ arc, is very similar to that used by HOFFMANN *et al.* We may notice in Fig. 3, that there is a definite correlation between the infrared radiation and the radio radiation (or the Lyman α flux) produced in the nebula. The scatter, however, is probably somewhat greater than the observational error, but that might be expected since the infrared radiation is only measured in a limited band and does not refer to the total radiation. The sources in the galactic center are not plotted on this diagram: they tend to fall above the plotted points by a factor of 2 to 3.

It is interesting to note on the diagram that the energy contained in the infrared flux between 80 μm and 135 μm is somewhat greater than what is being produced in the Lyman α radiation. It suggests that the Lyman α radiation is, by some means, converted into infrared radiation, and that the source which produces the ionized region is sufficiently energetic to account for the infrared radiation. This is probably true even if the total IR radiation is several times what is measured in the 80–135 μm band, since the source of energy is probably a factor of at least 3–10 more greater than that required to maintain the ionization.

We may conclude that the similarity of spatial position and dimension of the radio and infrared objects, and their similarity in energy content point to a physical association of the ionized nebula producing the radio emission with the source of infrared radiation. We may, for example, think of a hot star embedded in a dense medium, which ionizes this medium (at least in part) and an important part of its energy is converted into infrared radiation. We return to this picture in Section IV.

c) Broadband Spectral Measurements of Some HII Regions

Measurements are presently available for some HII regions, from 1.65 μm to 900 μm. Because the position and the size are sometimes not accurately known, assembling a complete spectrum from the individual observations is somewhat uncertain. Also the wide spectral band used makes it difficult to determine the effective wavelengths measured.

Fig. 4 shows the spectra of three compact HII regions from 2 μm to 100 μm. The measurements are taken from Hoffmann *et al.* (1971b), Becklin *et al.* (1972), and Neugebauer and Garmire (1970). The spectra increase in energy Hz^{-1} by about six orders of magnitude from 2 to 100 μm and in that sense are very similar. On the other hand, the spectra cross each other, i.e. although W51–IRS2 is about an order of magnitude brighter than K3–50 at 100 μm, it is definitely less intense at 2 μm. This difference in the spectra in the wavelength range 2–20 μm can be seen in the different spectra observed by Becklin *et al.* for the several components of W3. An example of this can be seen in Fig. 5.

The radio emission ($\sim$10 cm) for each of the objects is shown in Fig. 4 as well. The dashed line in the figure shows an extrapolation of the free-free radio emission toward shorter wavelengths. (In addition to the free-free emission, free-bound and two-quantum transitions also make an important contribution to the radiation in the 1–10 μm range.) It can be easily seen in the figure that below about 5–10 μm less radiation is observed than would be expected simply from thermal radiation which we know must be emitted. The effect is substantial, amounting to more than a factor 100 at the shortest observed wavelength. We shall discuss this effect in more detail in section IV; it is clear that substantial absorption is occurring in all these objects at wavelengths shorter than 5 μm.

In Fig. 6 the spectrum of the infrared nebula in Orion is shown, described at 22 μm by Kleinmann and Low (1967). The points between 70 μm and 200 μm

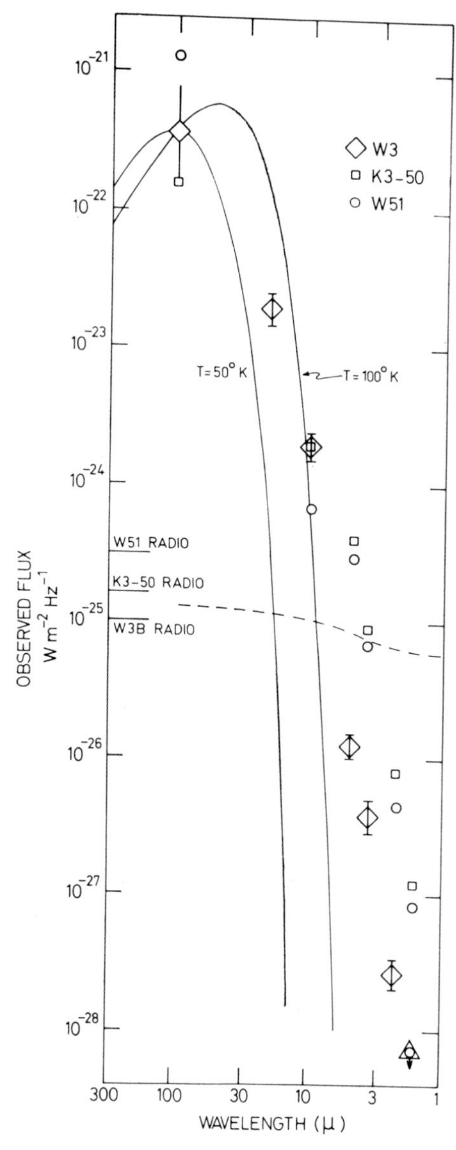

Fig. 4. The infrared spectra of several compact HII regions. The radio emission is also marked on the figure, as well as the thermal emission expected in the infrared for W3B, on the basis of the radio emission

are taken from Erickson *et al.* (1972); the intensities agree very well with the 100 μm point taken from Hoffmann *et al.* (1971 b). The point at 900 μm has been measured by Park *et al.* (1970); these authors measured over a broad band so that it is difficult to obtain an effective wavelength. A group at Meudon (Bensammar, Biraud, Epchtein, Gay and Sèvre, unpublished) have measured a somewhat

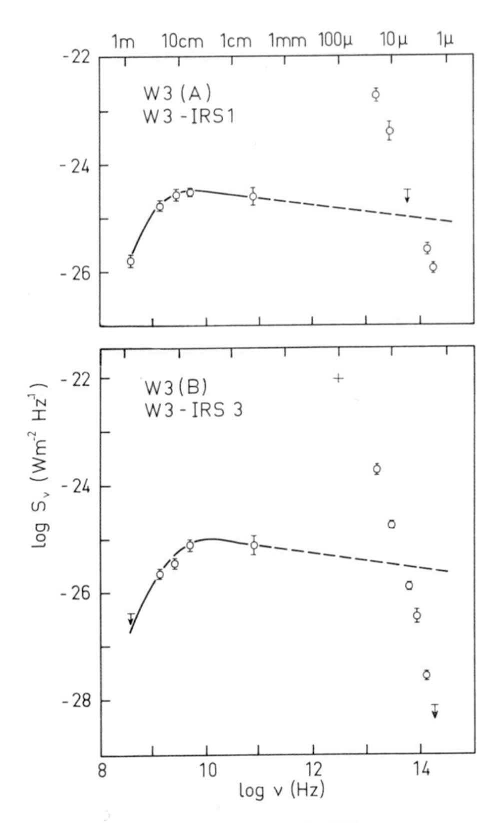

Fig. 5. The infrared spectra of two compact sources in W3

lower flux in this wavelength region. The size of this object is not well defined. NEY and ALLEN (1969) find a size of about 40″. KLEINMANN and LOW place a lower limit of 30″ diameter. HOFFMANN *et al.* had a diaphragm of 11′ diameter, ERICKSON *et al.* measured with a square of 4′ on each side, and PARK *et al.* indicate that their observations are consistent with a diameter of 1′ arc. It is worthwhile noting that this nebula appears coincident with a nebula measured in the formaldehyde line (THADDEUS *et al.*, 1971) which has a half-power size of about 1′×3′. The continuum radio emission appears to be displaced from the KLEINMANN-LOW nebula by 1′, but the fact that infrared intensity and radio intensity fit well on the correlation in Fig. 3, indicate that there is a physical connection.

In this case we see that the maximum energy occurs at between 60 to 100 µm and declines rather slowly on the longwave side. At 900 µm the flux is only one order of magnitude less intense than at its maximum value. Only in the case of M17 has the decline in flux come from the maximum on the longwave side been followed. JOYCE *et al.* (1972) measure a peak flux of $11\times10^{-22}\cdot W\cdot m^{-2}\cdot Hz^{-1}$ at 345 µm. This may be compared with the value given by HOFFMANN *et al.* at 100 µm of $23\times10^{-22}\cdot W\cdot m^{-2}\cdot Hz^{-1}$. This wavelength dependence is similar in the two cases (Orion and M17).

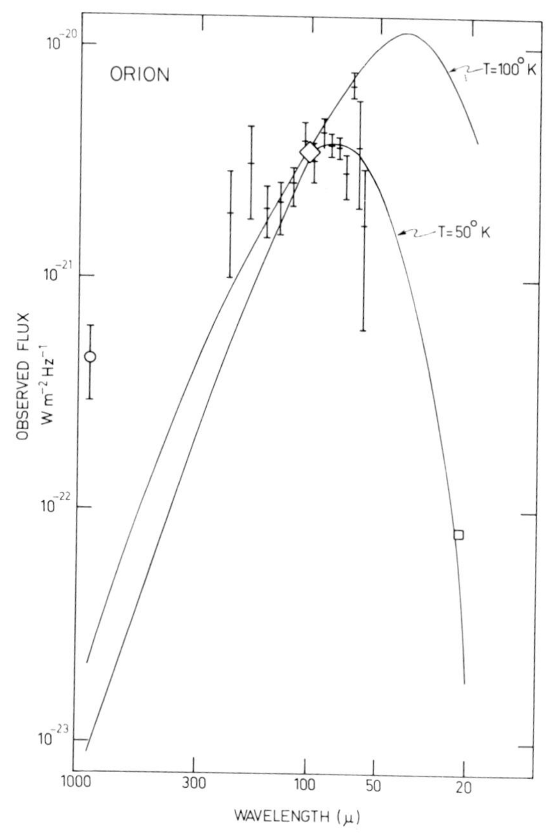

Fig. 6. The infrared spectrum of the nebula in Orion. The solid curves are $B_\nu(T)/\lambda$ on an arbitrary scale such that they pass through the 100 μm point

The measurements summarized in Fig. 6 do not definitely show any features in the spectrum, but are suggestive that better spectral resolution may yield interesting results. For example the strong peak near 60 μm might be caused by the 63 μm line of O.

From these spectra we may integrate as a function of wavelength to obtain a total infrared luminosity. Some typical values are:

	Distance	L
Orion	0.5 kpc	5×10^{38} erg·s^{-1}
W51a	7 kpc	3×10^{40} erg·s^{-1}
W3B	3 kpc	6×10^{38} erg·s^{-1}
K3–50	8 kpc	6×10^{39} erg·s^{-1}

The distances are uncertain, and this may change the result. For comparison, the luminosity of a blackbody of $T=40000$ K and $R=10\,R_\odot$ is 8.7×10^{38} erg·s^{-1}; that of $T=60000$ K and $R=20\,R_\odot$ is 1.8×10^{40} erg·s^{-1}.

IV. Possible Interpretations of the Infrared Measurements in HII Regions

a) Dust Emission

Emission by dust grains has been suggested as the source of the IR radiation. Let us consider this in detail. The emission by a grain, assumed spherical for simplicity, will be

$$F_\nu = Q_{abs}\,\pi B_\nu(T)\,\mathrm{erg\cdot s^{-1}\cdot cm^{-2}\cdot Hz^{-1}}, \tag{1}$$

where $B(T)$ ($\mathrm{erg\cdot s^{-1}\cdot Hz^{-1}\cdot sterad^{-1}}$) is the Planck function of temperature T and Q_{abs} is a factor which is a property of the material radiating, and of its size and the wavelength emitted. For spherical particles the value of Q may be calculated using Mie's theory:

$$x = \frac{2\pi a}{\lambda},$$

where a is the particle radius and λ the wavelength. For values of x of the order of unity or greater, Q_{abs} is also of the order of unity. However, when $\lambda \gg a$, the value of Q_{abs} decreases. Physically this is because it becomes increasingly difficult for an antenna to emit radiation of a wavelength much larger than its own dimension. For this case VAN DE HULST (1957) gives the following approximate expression

$$Q_{abs} = \tfrac{8}{3}\,x\,m'', \tag{2}$$

where m'' is the imaginary part of the index of refraction of the material.

Probably this last expression is that which we wish to use if the dust grains in these compact HII regions are assumed to be anything like the dust in the general interstellar medium, whose size is of the order $a = 0.2\ \mu\mathrm{m}$. The value of m'' is a function of the material and the wavelength.

When the optical depth through the emitting dust is less than unity (most of the radiation emitted by the individual grains comes through the nebula), the radiation received at the Earth is

$$S_\nu = \frac{2}{3}\pi a^2 n F_\nu\, 2R\frac{R^2}{r^2}, \tag{3}$$

where n is the density (number per $\mathrm{cm^3}$) of the grains, R is the radius of the (spherical) nebula, and r is the distance to the nebula. When the dust just becomes optically thick, the radiation received is

$$S_\nu = F_\nu \frac{R^2}{r^2}. \tag{4}$$

It may be expected that eventually the optical depth must become high enough so that the individual properties of the grains are no longer important, and the radiation will simply be that of a black-body

$$S_\nu = \pi B_\nu \frac{R^2}{r^2}. \tag{5}$$

This change-over from equations (4) to (5) will occur only at very high optical depths. The expression for the optical depth is

$$\tau_\nu = Q_{\mathrm{abs}} \pi a^2 n \Delta l . \tag{6}$$

b) Application to Several Nebulae

We have already noted that the observed emission at wavelengths less than 10 μm is less than what would be expected from an extrapolation of the free-free emission. This requires the following optical depths for W3B, assuming pure absorption

λ	τ
1.65	>6.9
2.2	5.6
3.5	3.0
4.8	1.9

It can be seen that the optical depth is roughly proportional to $1/\lambda$. Let us assume that the absorption is caused by dust grains. This dust may be foreground or it may be part of the nebula. We choose the latter possibility for the following reasons:

(1) the explanation of the far infrared radiation will be sought in emission from dust grains; since the grains are thus present in or near the nebula, they can be thought to produce the absorption.

(2) the absorption can be calculated for each of the W3 infrared objects and is found to vary strongly for objects 10–20″ arc apart in space; this is not so likely for foreground absorption.

In order to make the discussion more definite we assume that the grains have a radius of $a = 0.2$ μm and that the value of Q_{abs} ($\lambda = 1.65$ μm)$= 0.2$. These values seem to be used in the literature to describe "dirty ice" in the interstellar medium.

An extrapolation of the values of τ to the further infrared can be made on the basis of the $1/\lambda$ relation and are shown in Table 1. The 20 μm and 100 μm radiation can then be predicted from equation (3), as a function of the temperature of the grains. It is compared to the observed radiation in Table 1. Agreement can be obtained at about $T = 85$ K for the 20 μm observation, but the 100 μm observation is then about an order of magnitude higher than what can be predicted. We can perform this same analysis for the sources K3–50 and W51a, with a similar result: agreement is obtained at 20 μm for reasonable temperatures ($T = 110°$ and 80° respectively) but the observed 100 μm emission is more than an order of magnitude higher than the prediction.

This result essentially does not depend on the value of a or of Q_{abs} assumed. It does depend on the fact that $Q_{\mathrm{abs}} \sim 1/\lambda$. But Q_{abs} would have to be a constant or increase with increasing wavelength to explain the observations.

This same effect is noted also for the last nebula listed in Table 1, the Orion nebula. Here we have assumed the nebula to be optically thick at all wavelengths,

Table 1. Predicted dust radiation between 20–100 μm

Nebula		λ	τ_v	Predicted dust radiation			Observed radiation
				$T=150°$	$T=100°$	$T=70°$	
W3B	$2R=0.25$ pc	20 μm	0.55	7.7×10^{-22}	6.1×10^{-23}	3.3×10^{-24}	1.9×10^{-23}
	$n_e=1.2\times10^4$	100 μm	0.11	9.7×10^{-23}	4.9×10^{-23}	2.3×10^{-23}	3.7×10^{-22}
K3–50	$2R=0.23$ pc	20 μm	0.33	8.0×10^{-23}	6.3×10^{-24}	3.4×10^{-25}	$1\ \times10^{-23}$
	$n_e=3.9\times10^4$	100 μm	0.065	1.0×10^{-23}	5.1×10^{-24}	2.4×10^{-24}	1.6×10^{-22}
W51a	$2R=0.74$ pc	20 μm	0.45	1.1×10^{-21}	8.5×10^{-23}	4.6×10^{-24}	$1\ \times10^{-23}$
	$n_e=1.0\times10^4$	100 μm	0.091	1.4×10^{-22}	6.8×10^{-23}	3.2×10^{-23}	1.3×10^{-21}
M42	$2R=0.15$ pc	20 μm	optically	1.8×10^{-21}	1.4×10^{-22}	7.7×10^{-24}	$8\ \times10^{-23}$
		100 μm	thick to	2.3×10^{-22}	1.2×10^{-22}	5.4×10^{-23}	3.5×10^{-21}
		900 μm	900 μm	5.1×10^{-25}	3.7×10^{-25}	2.6×10^{-25}	4.5×10^{-22}

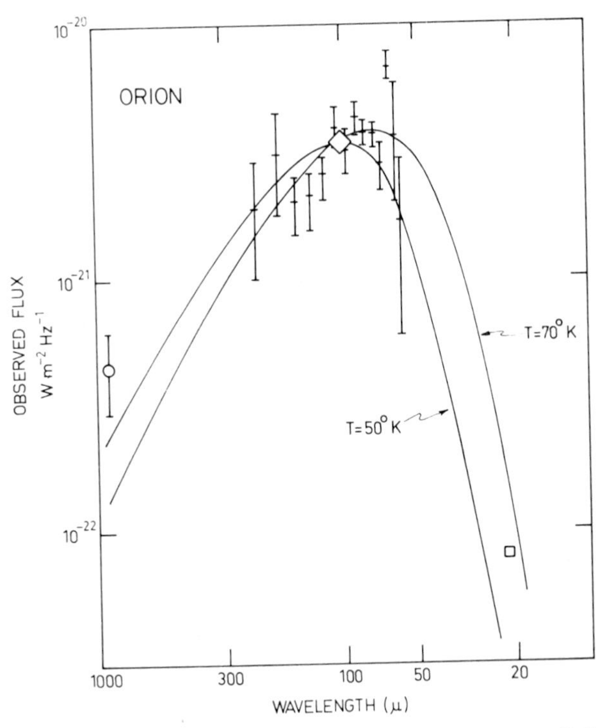

Fig. 7. The infrared spectrum of the nebula in Orion. The solid curves are $B_v(T)$ on an arbitrary scale such that they pass through the 100 μm point

and have used Eq. (4) to compute the expected flux. If the nebula is not optically thick, we will predict less radiation and our problems would increase. For the Orion nebula we see the same problem as for the other nebula, and we see that the problem becomes even more serious at 900 μm where the prediction falls short of the observations by about a factor 1000. Another way of seeing this is by comparing Figs. 6 and 7. The solid curve in Fig. 6 is proportional to $\frac{1}{\lambda}B_v$ while in Fig. 7 it is proportional to B_ν. The second figure gives a better fit although the present theory predicts a $\frac{1}{\lambda} B_\nu$ dependence. There are several possible ways to remove this difficulty.

(1) Q_{abs} is constant or increases slightly with increasing λ. This is not the general behaviour of most materials at values of $\frac{2\pi a}{\lambda} \ll 1$. We could think of much larger particles than 0.2 μm, however, or of collective effects of the grains at longer wavelengths.

(2) The size of the emitting region increases with increasing λ. Since no measurements of the size of these objects have been made above 25 μm we cannot rule this out. If the temperature were decreasing outward in a uniform grain distribution, it could produce such an effect. In the measurements between 2–20 μm there is no indication of such an effect, however.

Further we may compute grain densities and masses. It is perhaps rather too early to do this since the agreement between theory and observation is so poor, but an order of magnitude may still be useful. The total dust mass assumed to be emitting is given by:

$$M_{dust} = \rho n \frac{4\pi a^3}{3} \frac{4\pi R^3}{3} \, gm, \tag{7}$$

where ρ is the specific density of the grains and the other quantities have been defined before. The luminosity of the observed nebula, L_ν, in the optically thin case, is given by (Eqs. 1 and 3)

$$L_\nu = Q_{abs} 4\pi a^2 \pi B_\nu n \frac{4\pi}{3} R^3. \tag{8}$$

Solving Eq. 8 for R^3 and substituting in Eq. 7, we have:

$$M_{dust} = \frac{\rho a L_\nu}{3 Q_{abs} \pi B_\nu}. \tag{9}$$

The luminosity L_ν is derived from the measure at 100 μm and the known distance. We use $\rho = 2\,\mathrm{gm \cdot cm^{-3}}$ (SPITZER, 1968), $a = 2 \times 10^{-5}$ cm, and Q_{abs} ($\lambda = 100$ μm) $= 4 \times 10^{-3}$. The value of the Planck function B_ν is specified by the temperature of the grain. We use the value $T = 100°$; the value of B_ν at 100 μm is not very sensitive to the precise value of T chosen. The values of dust mass found are given in Table 2, for several nebulae selected because it is known that the radio emission originates in one or several compact regions. For comparison the total mass of hydrogen in the compact regions is also given. From abundance considerations (if one assumes the dust consists primarily of elements other than hydrogen and helium) we would expect that $M_{HII} > 100\, M_{dust}$. This is clearly not the case. We may ask what quantities in Eq. 9 may be changed to substantially lower M_{dust}. Increasing the temperature to 500 K is out of the question because it would produce an entirely different spectrum. It is unlikely that ρ or a could be decreased by an order of magnitude or more. The remaining parameter, Q_{abs}, is more uncertain, but from present

Table 2. Dust masses in infrared nebulae

Nebula	$\frac{L_\nu}{4\pi r^2}$ (100 μ) erg·cm^{-2}·s^{-1}·Hz^{-1}	Distance r pc	M_{dust} M$_\odot$	M_{HII} M$_\odot$
Orion	3.5×10^{-18}	500	0.35	5.7
K3–50	1.6×10^{-19}	8000	4.1	90.0
W51a	1.3×10^{-18}	7000	25.6	340.0
DR 21	4.2×10^{-19}	1500	0.38	3.0
W3	6×10^{-19}	3000	2.2	17.0
W49	7.6×10^{-19}	14000	61.0	170.0

knowledge of the general behaviour of emissivity when $\lambda \gg 2\pi a$, we might expect that it is smaller rather than larger.

If the calculation of the dust mass is correct in the order of magnitude, it poses severe problems for this interpretation. Possible ways out of this dilemma are:

(1) the grains consist primarily of hydrogen;

(2) the 100 μm emission comes from a much more extended region than the radio emission or the 20 μm emission.

Regarding the first possibility, there is little evidence that the interstellar dust grains contain primarily hydrogen. As to the second possibility, we have already discussed it above as an explanation for the observed spectral variations. It is clearly important to obtain accurate pictures of these sources at 100 μm.

We may conclude that while it is not unreasonable to consider emission by dust grains as the best available explanation of the observed emission, there are still serious difficulties which must be explained before this hypothesis can be verified.

c) Line Emission

The spectrum of HII regions in the spectral region from 3000–10000 Å is dominated by emission lines. These lines are mostly "forbidden" transitions, which are formed by collision of an electron with an ion in the ground state; this collision raises the ion to an excited state from which only a "forbidden" transition can occur. The energy emitted in the visible has always been considered as the principal cooling mechanism in diffuse nebulae.

These lines should also be observed, in similar or greater strength, in the infrared. They have been discussed by PETROSIAN *et al.* (1969) and POTTASCH (1968) in the context of HII regions and the interstellar medium. Following these authors we can predict the strengths of these lines for the compact HII regions. We have also made use of the calculations of OLTHOF (1972). Because collisional deexcitations may be important we specify the electron density $n_e = 10^4$ cm^{-3} used. The results are not sensitive to T_e, for which the value 7500 K was used.

Consider the measurements made in the spectral band between 17–26 μm. Six important lines are expected in this range from an ionized region; these are listed in Table 3. Assuming the abundances and the fractional ionization of each element

Table 3. Expected line emission between 17–26 μm

λ (μm)	Ion	Abundance	Fraction in Ion	$F/EM\,\theta^2$ ($W \cdot cm^{-2}$)
17.95	Fe^{+}	10^{-4}	1/2	4.0×10^{-18}
18.7	S^{++}	5×10^{-5}	1/3	0.9×10^{-18}
22.93	Fe^{++}	10^{-4}	1/2	7.5×10^{-18}
24.2	Ne^{+4}	5×10^{-4}	1/10	0.9×10^{-18}
25.8	O^{+3}	6×10^{-4}	1/4	2.2×10^{-18}
25.99	Fe^{+}	10^{-4}	1/2	9.4×10^{-19}

Table 4. Expected line emission between 17–26 μm

Nebula	EM	θ^2	Total expected line flux ($W \cdot cm^{-2}$)	Observed flux ($W \cdot cm^{-2}$)
W3B	3.3×10^{7}	2.2×10^{-5}	1.8×10^{-14}	8×10^{-16}
Orion A	3.5×10^{6}	5.6×10^{-3}	5.0×10^{-13}	4×10^{-14}
W51a	7.3×10^{7}	3.1×10^{-5}	2.5×10^{-14}	10^{-15}
M17	5×10^{6}	10^{-3}	1.3×10^{-13}	10^{-13}

also given in the table, the fluxes listed in the last column per unit emission measure (EM) and per square degree, are found. A check on this calculation can be made by computing the expected intensity of the S^{++} line at λ 9532 for which we would predict a flux of $3 \times 10^{-14}\,W \cdot cm^{-2}$ from the Orion nebula. The measured flux for this (strong) line is about $10^{-14}\,W \cdot cm^{-2}$ from a region near the trapezium in a region about $4' \times 4'$. The difference may in part be due to extinction, and in part due to the abundance or ionization assumption. The agreement is reasonably good, and supports the reliability of these calculations.

In the fourth column of Table 4, the total expected line flux is shown, and compared to the observed flux in this spectral band. It is noteworthy that the predicted line radiation is more than an order of magnitude greater than the observed flux; more so because the observed flux is usually considered to be continuous emission. Some of this difference is due to the fact that water vapour absorption attenuates some of these lines. Probably the Fe^{+} line at 25.99 μm does not penetrate the atmosphere and perhaps the O^{+3} line at 25.8 μm is equally absorbed. It is possible that some of the other lines are partly absorbed, but it would be surprising if they were completely absorbed. Thus, it is possible to explain the observed 20 μm emission as wholy or partly line emission. This is also true of the 10 μm observations where especially the lines of Ne^{+} at 12.8 μm and S^{+3} at 10.5 μm may be strong. These lines should be looked for: if they are not present this could indicate that the abundances used are much too high, or that substantial absorption is present.

Although line radiation is expected throughout the infrared region, the lines will become somewhat weaker toward the longer wavelengths. This is because the spontaneous transition rate, the Einstein A value, decreases, and thus collisional de-excitations become relatively more important, at the expense of the radiative transition. This is especially noticeable at higher densities. Even if the predicted intensities did not decrease toward longer wavelength, the observed intensity has increased by one or two orders of magnitude, which makes it very unlikely that line emission is the most important source of emission at 100 μm. We can illustrate this by considering the expected intensity of the O^{++} line at 88.1 μm. If $n_e = 10^4$ cm^{-3} and the abundance $O/H = 6 \times 10^{-4}$, then we would predict a flux

$$F/EM\,\theta^2 = 3.5 \times 10^{-20}\,\mathrm{W \cdot cm^{-2}}.$$

For the nebula K3–50 this would mean a total flux in the line of 3.5×10^{-17} W·cm^{-2}. Even ten lines of a similar strength would not approach the value of 2.6×10^{-14} W·cm^{-2} observed by HOFFMANN *et al.* (1971 b). Some lines may be stronger, e.g. one would expect the O^{++} line at 51.8 μm to be almost an order of magnitude stronger. Thus, one would not expect to see spectral lines from an ionized region if one has only a spectral resolution of 10 μm. This is in agreement with the observations of ERICKSON *et al.* (1972) of the Orion nebula which show no indication of either of these lines with the above spectral resolution. If the resolution were 1 μm then the peak intensity of the line would be expected to come about 10% above the continuum, and with a resolution of 0.1 μm the lines should be easily visible. ERICKSON *et al.* indicate that they may see a line in the region of 65 μm (see Fig. 6). A line due to O is expected at 63 μm, but it is expected to be strong in the atmosphere of the Earth, so that it would be necessary to check these observations. If it is really present in the Orion nebula, it probably does not come from the ionized region which emits the radio continuum, but from a nearby neutral region. It is clear that further observations are necessary.

d) Summary of the Present Situation

There are two reasons for believing that the origin of the high luminosity of the compact HII regions is a hot O star. First of all, an O star is capable of emitting the same total energy as observed. Secondly, the O star can account for the ionized nebula which is observed in the radio continuum to be approximately coincident with the infrared emitting region (except in the case of Orion, where the situation is more complicated). The O star is never observed in the visual (Orion may again be an exception) and it is postulated that strong absorption by dust is the reason.

It is attractive to further postulate that this same dust is the mechanism for transforming the predominantly ultraviolet energy of the O star into the observed infrared radiation. The observed coincidence of the infrared and ionized nebula would require that the dust be distributed in or near the ionized material. For this to be true the optical depth of the dust for absorption of Lyman continuum radiation must not be much greater than 3 in the nebula, otherwise the ionization would not occur. It may be necessary that it be greater than 1, however, since the Lyman α radiation produced in the nebula probably falls short by a factor of 3–10 of having enough energy to produce the observed infrared radiation; therefore some of the Lyman continuum energy must be directly absorbed by the dust. This requires a material which is more strongly absorbing in the visual part of the spectrum than in the ultraviolet. This is not a usual property for the most materials, but there is evidence that the general interstellar dust may have a similar property (WITT and LILLIE, 1972).

We have attempted to use the simplest assumptions concerning the dust radiation, i.e. the emissivity of the dust grains, to reproduce the observed infrared radiation. A dust temperature of 80–130 K seems indicated, but it is impossible in detail to reproduce the observed spectrum. This does not necessarily mean that the dust hypothesis must be discarded, it may mean that the model is too simplified, or that we do not know enough about the properties of the emitting grains. It is, however, wise to be cautious about the acceptance of the dust hypothesis.

The possibility of line emission in the infrared has also been discussed. It may contribute substantially in the region of the spectrum near 20 μm, and may be observable throughout the spectrum.

V. The Galactic Center

The information available at present leaves doubts as to whether the galactic center is similar to compact HII regions or not.

At 100 μm at least six sources have been recognized in the vicinity of the galactic center. The total energy in these six sources (in a 12′ diaphragm) is about 7×10^{-21} $W \cdot m^{-2} \cdot Hz^{-1}$ (HOFFMANN *et al.*, 1971 b). Probably most of the sources are HII regions, although they are not seen in the visual part of the spectrum. The evidence that they are HII regions is the thermal nature of the radio spectrum and the presence of radio recombination lines in hydrogen. If we plot these points on the diagram shown in Fig. 3, they are somewhat above the other points, indicating an excess of infrared radiation for a given radio emission of about a factor of 2 or 3. While this is not substantial, it is noteworthy.

In addition to these six sources, there is an extended emission at 100 μm which covers a region of about $3\overset{\circ}{.}2 \times 2^{\circ}$ around the galactic center. The flux emited by this region is 7.6×10^{-20} $W \cdot m^{-2} \cdot Hz^{-1}$ (HOFFMANN *et al.*, 1971 a; HOUCK *et al.*, 1971) or about a factor of 10 higher than that emitted by the six sources. The question is open as to whether this is simply the sum of many smaller compact HII regions or the extension of the six observed sources (in which case they are much larger than the compact regions) or a general flux emitted by a continuous medium.

This general emission has an energy of about 3×10^{42} $erg \cdot s^{-1}$, or perhaps somewhat more. This may be compared to the average compact HII region, which emits about 3×10^{38} $erg \cdot s^{-1}$. Thus, if we wish to involve hot stars as the source of energy, at least 1000 of them would be required. This is not impossible, since we are considering a region of radius 250 pc.

In the 5–25 μm region there is a small extended region near the galactic center which emits a considerable flux in this wavelength region. Its size is not precisely determined, but probably most of the radiation is confined to within a diameter of 25″ (LOW and AUMANN, 1970, report that when the beam diameter is increased from 25″ to 120″ less than a 20% increase in flux is observed). Its spectrum is known between 8 and 13 μm (WOOLF and GILLETT, preprint) and it shows a strong absorption band between 9 and 11 μm. The broader band measurements between

12–25 μm show a roughly constant flux of 2×10^{-23} W·m^{-2}·Hz^{-1} (Low *et al.*, 1969).

An extended (3° × 2°) source also exists at these wavelength (HOUCK *et al.*, 1971). It also has a roughly constant flux between 12–23 μm of 5×20^{-23} W·m^{-2}·Hz^{-1}. It is remarkable to note that there is less than a factor of 3 difference between these completely different sized regions. It means that the ratio of the 100 μm/20 μm flux is much different in the small source (where it must be less than 100) and in the "extended" region (where it is at least 1000). For comparison, this ratio is about 50 in the Orion nebula. It is probably necessary to await further observational evidence before trying to say anything more profound than that the two mechanisms discussed for the compact HII regions, are probably the most likely possibilities to explain the infrared emission.

We may compute the mass of dust required to produce the observed 100 μm flux in the same way as described earlier for the HII regions, with the following results:

	M_{dust}
Sgr B2	160 $M_{\odot}$
Entire extended region	3000 $M_{\odot}$

We may compare the value of dust required for the extended 2° × 3° region with the value of ionized hydrogen in this region $M_{\text{HII}} \leqq 7\times10^5\, M_{\odot}$ (MEZGER, 1974). The ratio of dust to ionized hydrogen is somewhat lower than for the compact HII regions.

VI. Galaxies

Only two galaxies (NGC 1068 and M 82) have been measured in the infrared at wavelengths longer than 50 μm. These two, and nine others, have been measured at 10 and 20 μm, and at shorter wavelengths (KLEINMANN and LOW, 1970a, b; NEUGEBAUER *et al.*, 1971). Six of the eleven are Seyfert galaxies and one a QSO; the others have characteristics which indicate that they are active. The two galaxies observed at wavelengths longer than 50 μm are also the brightest at 10 μm.

a) Comparison of the Spectra of Galaxies with HII Regions

Figs. 8 and 9 show the measurements at different wavelengths of NGC 1068 and M 82, as well as the compact HII regions K3–50 and M 17. The spectra show a remarkable similarity; they are certainly not more different from an HII region than the HII regions are from each other. The only possible exception is the 345 μm point (JOYCE *et al.*, 1972) measured for M 82: it is as high as the upper limit at 100 μm measured by HOFFMANN *et al.* (1971b). If this is confirmed, it means that the spectrum is flat between 100–300 μm. Furthermore, the higher resolution spectrum of NGC 1068 between 6–13 μm shows no feature, corresponding to the strong absorption at 9.7 μm noted in the spectrum of the small core at the Galactic Center.

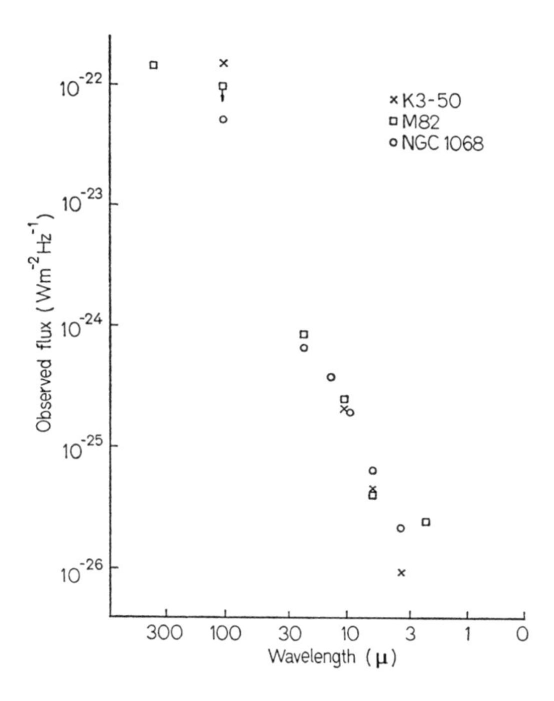

Fig. 8. Comparison of the infrared spectra of two galaxies with a compact HII region in our galaxy

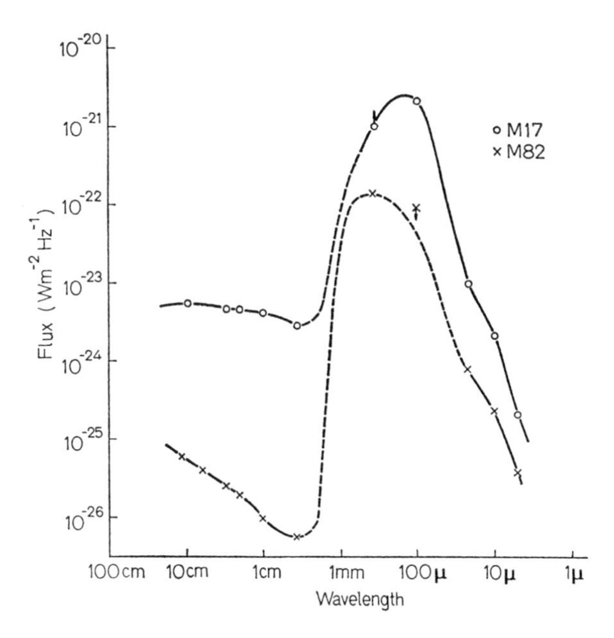

Fig. 9. Spectral measurements of the HII region M17 and the galaxy M82 extending from the near infrared to the radio region

b) Variability

Some of these sources may be variable in the millimeter wavelengths. Reports as to their variability between 2–20 μm are also found in the literature. We will not discuss them in detail; at this moment they are not firmly established. If they are present, the time scale may be of the order of days.

c) Total Luminosity

For NGC 1068 and M 82 we can calculate the total luminosity with reasonable accuracy, since the spectrum has been measured through 100 μm. The results are:

	Distance	L
NGC 1068	1.3×10^7 pc	1.6×10^{46} erg·s^{-1}
M 82	4.3×10^6 pc	1.8×10^{45} erg·s^{-1}

This is 10^3 times as luminous as the galactic center region, and in the case of NGC 1068, would require an equivalent of 10^6 O stars to produce this luminosity.

d) Dimensions

Attempts have been made to measure the dimensions between 1–10 μm (Kleinmann and Low, 1970a, b; Neugebauer *et al.*, 1971). At 10 μm M 82 and NGC 4151 appear to have a resolvable size, 400 pc and 100 pc respectively. NGC 1068 is not resolved, indicating that its size is less than 300 pc. However, at 1.65 and 2.2 μm, the emitting region in NGC 1068 is larger (at least 1300 pc). This is in constrast to the compact HII regions which appear to have the same dimension between 1.65 and 20 μm. At longer wavelengths it is only known that the size of M 82 at 345 μm is less than 2000 pc (Joyce *et al.*, 1972).

e) Possible explanations

(i) *Radiation by Dust.* This is suggested by the similarity of the spectra with compact HII regions. The temperature of the dust must, therefore, also be of the order of 100 K. If we assume that the dust has the same properties as interstellar dust and that it is optically thin, we may compute the required dust mass and the radius R of the region emitting at 100 μm. These are:

	M_{dust}	R
NGC 1068	$5 \times 10^6\ M_\odot$	>180 pc
M 82	$10^6\ M_\odot$	>100 pc

The value of R was calculated assuming that the optical depth was unity at 100 μm. If it is substantially greater this will not reduce R but will increase M_{dust}. If the optical depth at 100 μm is less than unity, R must be greater.

These parameters are at first glance not so unreasonable. The only observation which conflicts with this interpretation is the short term variability, which would restrict the size of the emitting region to less than 0.01 pc. Thus, if the variability proves to be real, this hypothesis may be eliminated.

(ii) *Line Emission*. The O^{++} line at λ 5007 was measured by OSTERBROCK and PARKER (1965) in a region $10'' \times 20''$ at the center of NGC 1068. These authors measured a flux of 2.0×10^{-18} W·cm^{-2} for the strength of this line. On the basis of the strength of this line we may predict the strength of the O^{++} 88.1 μm line. The ratio of the strength of these two lines depends on T_e and n_e. If we use the values $T_e = 5000$ K and $n_e = 10^3$ cm^{-3}, the infrared line is about an order of magnitude stronger than the visible line. Further a correction must be made for the very considerable reddening (see e.g. WAMPLER, 1968). If this causes an order of magnitude increase of the actual visible line intensity, we would predict a flux in the 88.1 μm line of 2×10^{-16} W·cm^{-2}, which is 2% of the total measured flux in the 75–135 μm band (10^{-14} W·cm^{-2}).

In the 22 μm band we would expect that the six lines previously discussed, produce at least as much emission as the 88.1 line. But only about 3×10^{-16} W·cm^{-2} has been measured in this band. Thus, we would expect line emission to be an important contribution to the emission in the infrared.

This situation is reminiscent of the situation in compact HII regions, although the arguments used in reaching this conclusion are somewhat different. High spectral resolution is clearly necessary before we can disentangle these measurements.

(iii) *Synchrotron Emission*. The possibility that synchrotron emission is the source of the infrared emission is suggested by the fact that it is probably responsible for most of the non-thermal radio emission. However, as LEQUEUX (1970) has noted, the low-frequency part of the synchrotron spectrum of any distribution of relativistic electrons cannot produce a flux which is steeper than $\nu^{1/3}$, if no low-frequency absorption mechanism exists. Extrapolating following this law from the 100 μm or 300 μm fluxes toward longer wavelengths clearly produces a much too high flux in the radio frequencies. Thus, synchrotron emission can be eliminated, unless there is a substantial low-frequency absorption. In the galactic center it is easy to show that this absorption does not exist: the HII regions are seen clearly with a thermal spectrum in the radio frequencies.

For these galaxies we have no direct information concerning low-frequency absorption. BURBRIDGE and STEIN (1970) have suggested that synchrotron self-absorption becomes important if high-energy particles are accelerated in magnetic fields in a small volume (the small volume follows naturally if one believes variability on a short time scale). In the frequency range where the self-absorption is important (the optical depth is high) the flux is proportional to $\nu^{5/2}$. Thus, extrapolating the point of JOYCE *et al.* to longer wavelengths, following this law we would expect a flux of 5×10^{-25} W·m^{-2}·Hz^{-1} at 3.5 mm; this is a lower limit because it assumes that the spectrum immediately turns down sharply after 345 μm. The observed radio flux is 6×10^{-27} W·m^{-2}·Hz^{-1} (KELLERMANN and PAULINY-TOTH, 1971) at 3.5 mm. This is a factor 100 too small to be explained as a $\nu^{5/2}$ extrapolation of the spectrum observed at 345 μm. This is clearly illustrated in Fig. 9. The radio measurements also indicate a size of about $20'' \times 10''$ for M 82 (very similar to the 10 μm measurements) which also argues against the small diameter synchrotron self-absorption source. This same argument can be applied to NGC 1068. KELLERMANN and PAULINY-TOTH measure an upper limit of 3×10^{-27} $W \cdot m^{-2} \cdot Hz^{-1}$ for this source at 3.5 mm. Extrapolating with $\nu^{5/2}$ spectrum gives a flux of 2×10^{-23} $W \cdot m^{-2} \cdot Hz^{-1}$ at 100 μm, which is about a factor of 5 less than observed.

(iv) *Conclusion.* None of these suggestions seems very attractive. The dust model would be much more plausible if the variations prove not to be real. Even so, the large dust masses required are somewhat difficult to explain. Further, although the similarity of the broad-band spectra in the infrared between the compact HII regions and the active galaxies indicate a similar mechanism for the infrared emission, the strong dissimilarity of their spectra in the radio frequency region (see Fig. 9) indicates that the source of energy is entirely different. For example, 10^6 O stars would produce much ionized hydrogen, which in turn would produce more free-free radio emission than observed.

Line spectra might be an important contributor, and should be carefully investigated observationally but it is unlikely that it will explain the entire spectrum. The synchrotron mechanism appears extremely unlikely because of the rapid fall of the flux toward millimeter wavelengths.

References

ALTENHOFF, W. J., DOWNES, D., GOAD, L., MAXWELL, A., RINEHART, R.: Astron. Astrophys., Suppl. **1**, 319 (1970).

BECKLIN, E. E., WYNN-WILLIAMS, C. G., NEUGEBAUER, G.: Monthly Notices Roy. Astron. Soc.: **160**, 1 (1972).

BURBRIDGE, G. R., STEIN, W. A.: Astrophys. J. **160**, 573 (1970).

ERICKSON, E. F., SWIFT, C. D., WITTEBORN, F. C., MORD, A. J., AUGASON, G. C., CAROFF, L. J., KUNZ, L. W., GIVER, L. P.: Astrophys. J. **183**, 535 (1973).

FURNISS, I., JENNINGS, R. E., MOORWOOD, A. F. M.: Astrophys. J. **176**, L 105 (1972).

HARPER, D. A., LOW, F. J.: Astrophys. J. **165**, L 9 (1971).

HOFFMANN, W. F., FREDERICK, C. L., EMERY, R. J.: Astrophys. J. **164**, L 23 (1971 a).

HOFFMANN, W. F., FREDERICK, C. L., EMERY, R. J.: Astrophys. J. **170**, L 89 (1971 b).

HOUCK, J. R., SOIFER, B. T., PIPHER, J. L., HARWIT, M.: Astrophys. J. **169**, L 31 (1971).

HULST, H. C. VAN DE: In Light Scattering by Small Particles. New York: John Wiley 1957.

JOYCE, R. R., GEZARI, D. Y., SIMON, M.: Astrophys. J. **171**, L 67 (1972).

KELLERMANN, K. I., PAULINY-TOTH, I. I. K.: Astrophys. Letters **8**, 153 (1971).

KLEINMANN, D. E., LOW, F. J.: Astrophys. J. **149**, L 1 (1967).

KLEINMANN, D. E., LOW, F. J.: Astrophys. J. **159**, L 165 (1970a).

KLEINMANN, D. E., LOW, F. J.: Astrophys. J. **161**, L 203 (1970b).

LEQUEUX, J.: Astrophys. J. **159**, 459 (1970).

LOW, F. J., AUMANN, H. H.: Astrophys. J. **162**, L 79 (1970).

LOW, F. J., KLEINMANN, D. E., FORBES, F. F., AUMANN, H. H.: Astrophys. J. **157**, L 97 (1969).

MARTIN, A. H. M.: Monthly Notices Roy. Astron. Soc. **157**, 31 (1972).

MERGER, P. G.: In L. N. MAVRIDIS (ed.) Stars and the Milky Way System. Berlin-Heidelberg-New York: Springer-Verlag 1974.

NEUGEBAUER, G., GARMIRE, G.: Astrophys. J. **161**, L 91 (1970).

NEUGEBAUER, G., GARMIRE, G., RIEKE, G. H., LOW, F. J.: Astrophys. J. **166**, L 45 (1971).

NEY, E. P., ALLEN, D. A.: Astrophys. J. **155**, L 193 (1969).

OLTHOF, H.: Technical Rept., University of Groningen (1972) (unpublished).

OSTERBROCK, D. E., PARKER, R. A. R.: Astrophys. J. **141**, 892 (1965).

PARK, W. M., VICKERS, D. G., CLEGG, P. E.: Astron. Astrophys. **5**, 325 (1970).

PETROSIAN, V., BAHCALL, J. N., SALPETER, E. E.: Astrophys. J. **155**, L 57 (1969).
POTTASCH, S. R.: Bull. Astron. Inst. Neth. **19**, 469 (1968).
SPITZER, L. JR.: Diffuse Matter in Space. New York-London-Sydney-Toronto: Interscience Published 1968.
THADDEUS, P., WILSON, R. W., KUTNER, M., PENZIAS, A. A., JEFFERTS, K. B.: Astrophys. J. **168**, L 59 (1971).
WAMPLER, E. J.: Astrophys. J. **154**, L 53 (1968).
WITT, A. N., LILLIE, C. F.: Astron. Astrophys. **25**, 397 (1973).
WYNN-WILLIAMS, C. G.: Monthly Notices Roy. Astron. Soc. **151**, 397 (1971).

Discussion

BORGMAN, J.:
In your diagram of infrared flux versus Lyman α flux I noticed a deviation from the expected 45° line in the area of large fluxes. Can that be understood in your models?

POTTASCH, S. R.:
The larger fluxes measured are not necessarily an indication of large luminosities, thus, have no physical meaning. However, scatter of the points on this diagram could have a physical meaning, in addition to the observational uncertainty; e.g. the spectra of the IR emitter varies from one source to the other or the temperature of the star causing the ionization could vary from one source to the other.

It may be that this intrinsic scatter, coupled with a selection effect caused by the fact we can observe much more sensitively in the radio region than in the infrared, cause the slope to be smaller than 45°.

MEZGER, P. G.:
In comparing IR and radio observations one should be aware of a few pit falls: 1. Radio maps are compared with IR flux measurements that pertain to a "point source"; this leads to an underestimate of the IR flux densities. 2. Masses of HII derived for the inner "compact" core of the HII region are compared to the total amount of dust needed to explain thermal radiation of dust. The low-density extended HII region in which the compact core is embedded usually contains a total mass of HII an order of magnitude or two higher. 3. One certainly should take into account a decrease of grain temperature with increasing distance from the source of grain heating. Taking these facts and the still preliminary quality IR measurements into account, I feel that a comparison of radio and presently available IR observations does not contradict an interpretation of IR radiation from HII region as being thermal radiation from dust which is heated by stellar radiation.

Infrared Techniques

(Invited Lecture)

By M. J. Smyth
University of Edinburgh, Department of Astronomy, Royal Observatory, Edinburgh, Scotland, United Kingdom

Abstract

This article reviews for the non-specialist the current state and limitations of infrared techniques for astronomical observations in the range 1 μm to 1 mm. A brief historical survey is followed by sections on the special problems of infrared measurements, detection systems, infrared photometers, spectrometers, and polarimeters, infrared telescopes, infrared observatory sites, and the role of airborne and space observations.

I. Introduction

The striking developments in infrared astronomy during this past decade have been possible only through the exploitation and refinement of new techniques. It is the purpose of this review to draw the attention of the non-specialist to the technical problems involved, the ways in which they have been met, and possible lines of development. Relevant technical articles are widely scattered throughout the literature, and rather than attempt any sort of complete bibliography, I have given only typical references in each field, including reviews. For present purposes the infrared is taken to mean the range from 1 μm to 1 mm, bounded at the one end by conventional photographic and photoelectric techniques, and at the other by radio techniques.

II. Development of Infrared Techniques

The first exploitation of new infrared techniques has frequently occurred in astronomical applications. The existence of radiation beyond the visible spectrum was indeed discovered by an astronomer, Sir William Herschel (1800), although to him it was an unwanted phenomenon which caused discomfort when observing the Sun through dark eyepiece filters.

The thermopile was developed by Nobili and Melloni about thirty years later, and was used by Piazzi Smyth (1858) for what was probably the first true measurement of infrared radiation from a celestial source other than the Sun, when he compared the radiation of a candle with that of the Moon observed from a very dry site at an altitude of 2717 m on Tenerife. The thermopile remained the principal infrared detector for over a century, thus confining observations to rather bright sources even when large telescopes were used. The bolometer developed by LANGLEY, also a thermal (non-selective) detector, revealed the existence of those relatively transparent "windows" in the atmospheric absorption (see Fig. 9 of article by BORGMAN), whose limits define the scope of ground-based infrared astronomy. It is possible that EDISON measured radiation from α Boo in 1878 (EDDY, 1972), but systematic infrared measurements of stars did not begin until 1914 (COBLENTZ, 1922; PETTIT and NICHOLSON, 1928).

As in many fields, the technological developments of the Second World War led to rapid advances in infrared astronomy. Lead sulphide photoconductive detectors, a thousand times more sensitive than bolometers within a narrower region of spectral response, were exploited by KUIPER *et al.* (1947) to obtain spectra of planets and of a few bright stars, and by FELLGETT (1951 b) for stellar photometry.

The use of detectors cooled by liquid nitrogen or liquid helium led to the further extensions of sensitivity and wavelength range that have revolutionised infrared astronomy during the past decade. JOHNSON (1962) established photometric bands at 3.4 μm and 5 μm by the use of indium antimonide detectors, while the use of doped germanium photoconductors (WILDEY and MURRAY, 1964) and the germanium bolometer (LOW and JOHNSON, 1964) extended stellar photometry to 10 μm and 20 μm, which is the normal limit for ground-based observation. Extension to longer wavelengths in most cases demanded the use of balloons, pioneered by the Johns Hopkins group in 1956 (STRONG, 1965), of aircraft (KUIPER and CRUICKSHANK, 1968; LOW *et al.*, 1969), and of rockets (SHIVANANDAN *et al.*, 1968).

An important advance in spectrometry, by no means confined to the astronomical field, was made by FELLGETT (1951 a) when he developed the multiplex technique, in the form of the Fourier transform spectrometer, in order to improve the efficiency of recording the spectra of astronomical sources. Indeed the multiplex advantage in high-resolution spectroscopy, as developed by the CONNES, was such as to yield infrared spectra of the brighter planets (CONNES *et al.*, 1969) superior in resolution to the best spectra of the Sun obtained by conventional scanning procedures. MERTZ (1965) developed the rapid-scan Fourier spectrometer, a compact moderate resolution instrument suitable for use in aircraft (JOHNSON *et al.*, 1968).

III. Problems of the Infrared

The first and obvious property of the infrared is its range in wavelength of $1:10^3$; evidently a variety of techniques is required to cover the range efficiently. A glance at relevant numerical quantities for typical photometric regions will indicate some of the problems:

	Wavelength	Wavenumber	Photon energy	Tp
(B)	0.44 μm	22700 cm^{-1}	2.8 eV	6600 K
(K)	2.2 μm	4500 cm^{-1}	0.6 eV	1300 K
(N)	10.2 μm	1000 cm^{-1}	0.1 eV	300 K
	1 mm	10 cm^{-1}	0.001 eV	3 K

Here Tp is the temperature of a black body whose emission per unit wavelength peaks at the wavelength concerned. B is an indicator for stars of moderate temperature, K for cool stars, N for cool circumstellar shells, and 1 mm for cosmic background. Furthermore, the room-temperature black-body peak at 10 μm implies that measurements in this region will be dominated by thermal emission from the telescope and its surroundings. The listed photon energies imply that photoemissive cathodes are not applicable much beyond 1 μm, but photoconductors (internal photoelectric effect) will operate to 10 μm and beyond. The wavenumber values show that a given path-difference leading to a resolution of, say, 1 cm^{-1}, will correspond to moderately high resolving power at 2 μm but very low resolving power at 1 mm. Finally, the number of photons per watt is higher in the infrared, so that photon noise on the signal may be expected to be less important than in the visible.

Now by way of illustration consider the brightest non-solar system sources in the sky, at the wavelengths corresponding to the B, K, and N photometric bands, observed with the 'popular' telescope aperture of 1.5 m, and with no allowance for losses or extinction:

(a) brightest visible star is α CMa, $B=-1.46$;
flux $\simeq 4\times10^{-8}$ W $\simeq 10^{11}$ photons s^{-1};
sky background (10″ aperture) $\simeq 10^{-15}$ W.

(b) brightest 2 μm star is α Ori, $K=-4.0$;
flux $\simeq 10^{-8}$ W $\simeq 10^{11}$ photons s^{-1};
thermal background $\simeq 7\times10^{-12}$ W.

(c) brightest 10 μm source is η Car, $N=-8$;
flux $\simeq 7\times10^{-9}$ W $\simeq 4\times10^{11}$ photons s^{-1};
thermal background $\simeq 10^{-6}$ W.

Thermal background is calculated notionally for an f/1 beam 'seeing' areas of 0.1 emissivity at 300 K. The photon rates are rather similar, so that photon noise on the signal is not more important than in visible photometry. What is significant, of course, is that background becomes comparable with signal not until $B=17.5$, but already at $K=+4$; while the thermal background is two orders of magnitude greater than the signal from the brightest 10 μm source.

Present detection limits with a 1.5 m telescope are about $K=+11$ (flux $\simeq 10^{-14}$ W in above example) and $N=+6$ (flux $\simeq 2\times10^{-14}$ W in above example). RIEKE and LOW (1972) report measurements of galaxies at a level of about 0.2 flux-units which again corresponds to $N\simeq+6$.

An approximate calculation for the thermal background value cited shows that, since it corresponds to 4×10^{13} photons s^{-1}, the photon noise in a one-second integration is of order 6×10^{6} photons or 10^{-13} W. Thus it is photon noise on the thermal background that sets limitations on detection in the neighbourhood of 10 μm. A BLIP detector (background-limited photodetector) is one whose internal noise is less than that imposed by the thermal background in the particular mode of operation.

IV. Detectors

For the general properties of IR detectors and their associated cryogenic, optical, and electronic systems, reference may be made to a number of standard text-books, such as WOLFE (1965), HOUGHTON and SMITH (1966), SMITH *et al.* (1968), and HUDSON (1969). LÉNA (1972) has reviewed the state of the art in infrared detection systems, taking into account fundamental limitations and the background limitations in different classes of ground-based and airborne astronomy. LOW and RIEKE (1972) have extensively reviewed techniques of IR astronomy with special reference to the germanium bolometer.

Photo-emissive detectors (photomultipliers) are only marginally infrared detectors, the 'classical' S1 photocathode having low quantum efficiency and response only to 1.1 μm. Compound semiconductive cathodes now have 10 percent quantum efficiency at wavelengths just short of 1.1 μm (SPICER and BELL, 1972), and the threshold is gradually being pushed towards 2 μm (WILLIAMS, 1969).

Among a number of available intrinsic photoconductors, liquid-N_2 cooled PbS detectors (HUMPHREY, 1965) are useful to 4 μm, and the noise equivalent power (N.E.P.) of a $\frac{1}{4}$ mm detector exposed to 2π steradians of 295 K background is of order 10^{-13} W·Hz$^{-\frac{1}{2}}$.

In the neighbourhood of 2 μm, when BLIP operation is not achieved, optical immersion can be used to concentrate radiation on a detector of small surface area and low N.E.P. Liquid-N_2 cooled InSb is useful beyond 5 μm but a little inferior to PbS at the shorter wavelengths.

A doped germanium extrinsic photoconductor, Ge:Hg cooled to 35 K with liquid H_2, was used for the first stellar measurements at 10 μm (WILDEY and MURRAY, 1964), and QUIST (1968) has shown that Ge:Cu achieves BLIP detection for backgrounds as low as 10^{-11} W in the 8–12 μm region, with a corresponding N.E.P. of 10^{-15} W·Hz$^{-\frac{1}{2}}$.

Thermal (non-selective) detectors, which were the only detectors for the first 150 years of infrared astronomy, are fundamentally limited by the statistics of energy exchange with their surroundings, the noise equivalent power for a thermal detector of area A cm^2 in an enclosure at T K being

$$(\text{N.E.P.})_{\min} = 3.54 \times 10^{-17}\, T^{\frac{5}{2}} A^{\frac{1}{2}}\ \text{W} \cdot \text{Hz}^{-\frac{1}{2}}. \tag{1}$$

The corresponding N.E.P. for a 1 mm^2 room-temperature detector is thus 5.5×10^{-12} W·Hz$^{-\frac{1}{2}}$, which is much inferior to a cooled photoconductor. Fundamental limits for cryogenic temperatures are:

77 K: 1.8×10^{-13} W·Hz$^{-\frac{1}{2}}$,
4 K: 1.1×10^{-16} W·Hz$^{-\frac{1}{2}}$,
0.3 K: 1.7×10^{-19} W·Hz$^{-\frac{1}{2}}$.

A great part of the broadband infrared photometry of the past decade is due to the liquid-He cooled germanium bolometer introduced by LOW (1961), and now available commercially from at least two sources.

Its properties and applications are comprehensively described by LOW and RIEKE (1972). Under low-background conditions (10^{-8} W), laboratory N.E.P. values as low as 10^{-14} W·Hz$^{-\frac{1}{2}}$ have been achieved for a $\frac{1}{4}$ mm detector. NOLT and MARTIN (1971) describe an absorption pumped ^{3}He-cooled bolometer operating at 0.3 K, but have not yet achieved low enough backgrounds to exploit the potential reduction in fundamental N.E.P.

The choice between thermal detector and photoconductor at 10 μm depends on thermal background conditions. For high background, e.g. broadband photometry 8–13 μm, the thermal detector is superior on account of its 100 percent quantum efficiency. Photoconductors, with quantum efficiencies of the order of 10 percent, are superior in narrow-band photometry where the thermal background is correspondingly reduced. The crossover point is a matter of discussion, but WOOLF (1972a) finds that 4K photoconductors are substantially better than bolometers at 1K for photometry with a bandwidth $\frac{\delta\lambda}{\lambda}$ of order 1 percent.

Bolometers are effectively used in broadband photometry at wavelengths in the region of 100 μm, e.g. the balloon surveys of HOFFMANN *et al.* (1971a, b). In the submillimetre region, however, CLEGG and HUIZINGA (1972) claim the InSb hot electron bolometer (Rollin detector) to be markedly superior in the range 300 μm–3 mm.

Of possible new detection systems, LOW and RIEKE (1972) discuss temperature-sensitive inductances operating at extremely low temperatures. Heterodyne detection is discussed by TEICH (1968) and LÉNA (1972), but appears to be advantageous only for very narrow band spectroscopy against low background (e.g. for detection of narrow interstellar lines from a manned space station), and for long base-line interferometry. The Josephson junction is also discussed by LÉNA (1972). Upconversion—the conversion of infrared photons into visible photons to be detected by conventional means—is interesting but inefficient at present (MILTON, 1972), and in any case does not remove the background limitation.

It is evident that the whole infrared range 1 μm–1 mm is now covered by efficient detectors, and that background radiation is in most cases the limiting factor. Further major advances will require telescopes designed to minimize thermal background, especially in airborne and space applications discussed later.

V. Infrared Instruments

This section deals with the photometers, spectrometers, and polarimeters that form the interface between the infrared detector and the telescope.

As regards photometers, apart from the practical problems of mounting the detector in a suitable cryostat—and of obtaining reliable supplies of liquid helium at remote mountain observatories!—it is clear from what has already been said that elimination of background must be the major distinguishing feature of infrared photometry. Since in any case semiconductor detectors yield optimum signal/noise at moderate a.c. frequencies, it is convenient to remove the background contribution by chopping between a sky area containing the star and an adjacent empty sky area. The fact that the background may be 10^6 times greater than the star signal implies that, in order to eliminate a large unbalance component or "offset", the two beams must be very accurately matched as regards radiating surfaces "seen" by the detector in the two positions. The best means of sky chopping is controversial: a rotating two- or three-level mirror can generate large beam displacements but is liable to misadjustment and edge effects; an oscillating single mirror is now generally preferred. LOW and RIEKE (1972) and WOOLF (1972), whose groups between them represent a major fraction of current infrared photometry, agree however that small stable offsets are most readily achieved by wobbling the secondary mirror of a cassegrain telescope. This evidently brings the telescope itself within the scope of photometric design, and its implications are considered in the next section. A final point concerning photometers is that, although fixed focal-plane apertures have been normally used, it will be important in future to have variable apertures in order to study the spatial structure of extended infrared sources.

Intermediate between broadband photometers and high-resolution spectrometers are spectrometers with 1–2 percent resolving power utilizing commercial circular variable interference filters, which can cover the range 3–14 μm on two semicircles. The reduction in thermal background due to the narrow bandwidth can be enhanced by enclosing the filter within the cryostat, and permits use of detectors with low N.E.P. These photometers, introduced by GILLETT *et al.* (1968), have produced many spectra of the "dust-bump" attributed to silicates in circumstellar shells. Another beautiful example is the absorption spectrum at 1 percent resolving power, of the galactic centre, made by WOOLF and GILLETT and illustrated by WOOLF (1973).

High resolution over a limited wavelength range has been obtained with a scanning Fabry-Perot interferometer, using a circular variable interference filter to isolate single orders. RANK *et al.* (1970) have detected the SIV emission of a planetary nebula at 950 cm^{-1} with a resolution of 0.3 cm^{-1}; and GEBALLE *et al.* (1972) have studied the fundamental CO band near 2140 cm^{-1} in several cool stars with a resolution of 0.18 cm^{-1}.

Scanning of the spectrum formed by a diffraction grating has played an important role in infrared spectroscopy, especially in the production of solar spectral atlases and in a series of papers on stellar spectra by the Caltech group following MCCAMMON *et al.* (1967). It was however the multiplex property of Fourier transform spectrometry that permitted extensive spectral ranges to be analysed at high resolution. There is an extensive literature on the subject, e.g. the Aspen Conference on Fourier Spectroscopy (VANASSE *et al.*, 1971); P. and J. CONNES solved the technical problems of precise mirror motion, of compensation for image scintillation by means of internal modulation, and of computing very large Fourier transforms; as already mentioned, their spectra of Venus and Mars at 0.08 cm^{-1} resolution are superior to the best solar atlases obtained by scanning. Their spectrometer has been used also to obtain stellar spectra, and a similar instrument has been constructed at J. P. L. by BEER *et al.* (1971) and successfully operated over the wavelength range 1.2–5 μm. The compact rapid-scan Fourier spectrometer developed by MERTZ (1965) has, in improved commercial versions, been used by JOHNSON *et al.* (1968), JOHNSON and MÉNDEZ (1970), and SMYTH *et al.* (1971), to obtain stellar spectra at resolutions down to 2 cm^{-1}, and is capable of 0.5 cm^{-1} or better. GAMMON *et al.* (1972) have used a rapid-scan interferometer to obtain stellar spectra in the 8–13 μm region, where background cancellation is a major problem.

The electronically tunable acousto-optic filter (Isomet Corporation) is an interesting recent development, with resolving power over 10^3 in the 1–3.5 μm region.

Little work on infrared polarimetry has been reported. CAPPS and DYCK (1972) have used wire-grid polarizers at 10 μm to detect possible polarization at the 1 percent level in cool stars.

VI. Infrared Telescopes

The fact that infrared wavelengths are longer than visible wavelengths leads immediately to the idea of using optics with relaxed tolerances; but the concept of the "light-bucket" must not be carried too far. The high throughput of interferometric spectrometers aids the problem of high resolution coupled with large stellar images, but the problems of concentrating the flux upon a detector probably less than 1 mm in diameter, and of thermal background, remain. FELLGETT (1969) uses a compact diagram to summarize the relationships between telescope aperture, spatial resolution, and mechanical flexure, for various wavelengths, spectral resolutions, and detector sizes.

In order to minimize sky background radiation, image sizes should not much exceed the seeing limit of a few seconds of arc (indeed it is suspected that infrared seeing is superior to that in the visible). Moreover, diffraction plays a significant role at the longer wavelengths. BORGMAN (1972) has shown that an aperture that is to include 92 percent of the energy in the diffraction will subtend approximately $5\lambda/D$ radians on the sky, where D is telescope aperture; for a telescope using a 6 seconds of arc aperture for photometry at 20 μm, a telescope aperture of 3.4 m is required. Thus, reflectors made up from an array of small mirrors are ruled out, so far as 10 μm and 20 μm photometry are concerned, unless they can be phase-matched, a difficult technical problem.

Nevertheless a number of relatively inexpensive infrared telescopes has been constructed, some using Kanigen-coated aluminium alloy mirrors (JOHNSON, 1968). The British 152-cm infrared telescope recently installed on Tenerife utilizes a rather thin glass mirror. CONNES is constructing at Meudon a 36-mirror composite flux-collector for spectrometry with an equivalent aperture of 4 m. A proposal for a composite of six telescopes feeding one detection system is described by WEYMANN and CARLTON (1972).

There has been some discussion of the optimum size for infrared photometric telescopes, but STEIN and WOOLF (1969) make it clear that 1.5 m is not the limit.

Reduction of thermal background arising in the telescope is an important problem discussed in detail by LOW and RIEKE (1972). They advocate the use of a Cassegrain focal ratio as large as f/45; this implies a small secondary mirror that can be wobbled for sky-chopping and can act as an aperture stop of low emissivity, and also a small central hole in the primary. A penalty is increased tube length and dome size. Ultra-high vacuum-deposited silver is suggested as a mirror coating of low emissivity.

Some degree of automatic control, including programmed offsets and the ability to "nod" in declination between adjacent sky regions, is an important feature of infrared telescope systems.

BECKMAN and SHAW (1972) have described a telescope designed for efficient energy collection in the submillimetre region.

VII. Infrared Observatories

Following the example of PIAZZI SMYTH (1858), modern infrared observatories have been placed at sites over 2000 m altitude in order to reduce absorption and emission by atmospheric H_2O. Examples are Mauna Kea, Hawaii (4215 m), Jungfraujoch, Switzerland (3500 m), Mt. Lemmon, Arizona (2800 m), and Izaña, Tenerife (2385 m).

KUIPER (1970a, b) has extensively surveyed the qualities of numerous sites for infrared astronomy. In general, precipitable atmospheric H_2O should not exceed 1–2 mm. Some of the sites considered would give substantially lower values, but at the cost of severe physiological and logistic problems. Mauna Kea appears to be the highest site at which these problems have not become excessive.

"Sky noise" is a problem sometimes met at all sites and not very thoroughly understood. It arises from spatial inhomogeneities in the atmospheric emission that are not completely cancelled by sky-chopping. Low and RIEKE (1972) state that "sky noise increases rapidly with field of view and is usually dominant in good systems with fields larger than 15 arcsec on a 60-inch telescope". WESTPHAL (1972) is conducting a survey of 10 μm sky noise, indicating well defined differences between a number of sites.

During dry weather at high altitude sites, atmospheric windows at 350 μm and 460 μm become partially transparent and useful ground-based observations are possible. HARPER *et al.* (1972) report observations from Catalina (2560 m) at 350 μm.

VIII. Airborne and Space Observations

To make observations other than in the atmospheric windows, and to attain drastic background reduction at all wavelengths, it is necessary to carry the telescope above the major part of the Earth's atmosphere. Many aspects of airborne and space infrared observation were discussed at the ESLAB/ESRIN Symposium at Noordwijk (MANNO and RING, 1972) and need be mentioned only briefly here.

Aircraft provide stable platforms for telescopes up to 32 cm aperture at heights up to 12.5 km, when residual H_2O is of order 1 μm (precipitable) and emissivity is of order 0.5 at 200 K. JOHNSON *et al.* (1968) have used lunar spectra obtained from NASA Convair 990 aircraft in order to correct for H_2O absorption in ground-based stellar spectra. NASA plans a 91-cm airborne infrared telescope to be operational in 1973.

At balloon altitudes of 30 km, atmospheric emissivity is as low as 0.01, but is still non-negligible. The 100 μm surveys of HOFFMANN *et al.* (1971a, b) have already been mentioned. Instrumental thermal background at present exceeds that from the atmosphere, so that a cryogenically cooled telescope will substantially increase the sensitivity.

Rocket observations above 100 km, e.g. SHIVANANDAN *et al.* (1968), are necessary when the lowest possible thermal background levels must be attained, as in studies of cosmic background radiation. Telescopes up to 1 m aperture are now being flown.

Difficulties of long-term cryogenic cooling of detectors have so far inhibited satellite infrared experiments. Passive coolers radiating into space appear to be limited to temperatures above 40 K or so, while active refrigerators have heavy power demands. Several systems are however under study. Use of the manned space shuttle may remove the difficulty.

IX. Conclusion

Evidently a wide variety of new infrared techniques has become available during the past decade, and has led to the development of an exciting and active branch of astronomy. Inevitably, as techniques become more refined, or move into space, they also become more expensive, and careful consideration must be given to the whole system of detector, photometer, telescope, and site, in relation to the astronomical ends that are to be pursued.

References

Beckman, W. A., Shaw, J. A.: Infrared Physics **12**, 219 (1972).
Beer, R., Norton, R. H., Seaman, C. H.: Rev. Sci. Instr. **42**, 393 (1971).
Borgman, J.: Private communication (1972).
Capps, R. W., Dyck, H. M.: Astrophys. J. **175**, 693 (1972).
Clegg, P. E., Huizinga, J. S.: In V. Manno and J. Ring (eds.), Infrared Detection Techniques for Space Research. Fifth ESLAB/ESRIN Symp., p. 132. Dordrecht-Holland: D. Reidel 1972.
Coblentz, W. W.: Astrophys. J. **55**, 20 (1922).
Connes, J., Connes, P., Maillard, J.-P.: Atlas of Near Infrared Spectra of Venus, Mars, Jupiter and Saturn. Paris: C.N.R.S. 1969.
Eddy, J. A.: J. Hist. Astron. **3**, 165 (1972).
Fellgett, P. B.: PhD Thesis, University of Cambridge (1951a).
Fellgett, P. B.: Monthly Notices Roy. Astron. Soc. **111**, 537 (1951b).
Fellgett, P. B.: Phil. Trans. Roy. Soc. London Series A, **264**, 309 (1969).
Gammon, R. H., Gaustad, J. E., Treffers, R. R.: Astrophys. J. **175**, 687 (1972).
Geballe, T. R., Wollman, E. R., Rank, D. M.: Astrophys. J. **177**, L 27 (1972).
Gillett, F. C., Low, F. J., Stein, W. A.: Astrophys. J. **154**, 677 (1968).
Harper, D. A. Jr., Low, F. J., Rieke, G. H., Armstrong, K. R.: Astrophys. J. **177**, L 21 (1972).
Herschel, Sir W.: Phil. Trans. Roy. Soc. London Pt. II **90**, 255 (1800).
Hoffmann, W. F., Frederick, C. L., Emery, R. J.: Astrophys. J. **164**, L 23 (1971a).
Hoffmann, W. F., Frederick, C. L., Emery, R. J.: Astrophys. J. **170**, L 89 (1971b).
Houghton, J., Smith, S. D.: Infra-Red Physics. Oxford: Clarendon Press 1966.
Hudson, R. D.: Infrared System Engineering. New York: Wiley-Interscience 1969.
Humphrey, J. N.: Applied Optics **4**, 665 (1965).
Johnson, H. L.: Astrophys. J. **135**, 69 (1962).
Johnson, H. L.: In A. Beer (ed.), Vistas in Astronomy, vol. 10, p. 149 New York: Pergamon Press 1968.
Johnson, H. L., Coleman, I., Mitchell, R. I., Steinmetz, D. L.: Commun. Lunar Planet. Lab. **7**, 83 (1968).
Johnson, H. L., Méndez, M. E.: Astron. J. **75**, 785 (1970).
Kuiper, G. P.: Commun. Lunar Planet. Lab. **8**, 121 (1970a).
Kuiper, G. P.: Commun. Lunar Planet. Lab. **8**, 337 (1970b).
Kuiper, G. P., Cruikshank,, D. P.: Commun. Lunar Planet. Lab. **7**, 179 (1968).
Kuiper, G. P., Wilson, W., Cashman, R. J.: Astrophys. J. **106**, 243 (1947).
Léna, P.: In V. Manno and J. Ring (eds.), Infrared Detection Techniques for Space Research. Fifth ESLAB/ESRIN Symp., p. 103. Dordrecht-Holland: D. Reidel 1972.

Low, F. J.: J. Opt. Soc. Am. **51**, 1300 (1961).
Low, F. J., Johnson, H. L.: Astrophys. J. **139**, 1130 (1964).
Low, F. J., Kleinmann, D. E., Forbes, F. F., Aumann, H. H.: Astrophys. J. **157**, L 97 (1969).
Low, F. J., Rieke, G. H.: To be published (1972).
Manno, V., Ring, J. (eds).: Infrared Detection Techniques for Space Research, Dordrecht-Holland: D. Reidel 1972.
McCammon, D., Münch, G., Neugebauer, G.: Astrophys. J. **147**, 575 (1967).
Mertz, L.: Astron. J. **70**, 548 (1965).
Milton, A. F.: Applied Optics **11**, 2311 (1972).
Nolt, I. G., Martin, T. Z.: Rev. Sci. Instr. **42**, 1031 (1971).
Pettit, E., Nicholson, S. B.: Astrophys. J. **68**, 279 (1928).
Piazzi Smyth, C.: Report on the Teneriffe Astronomical Experiment of 1856, London 1858.
Quist, T. M.: Proc. Inst. Elec. Electron. Engrs. **56**, 1212 (1968).
Rank, D. M., Holtz, J. Z., Geballe, T. R., Townes, C. H.: Astrophys. J. **161**, L 185 (1970).
Rieke, G. H., Low, F. J.: Astrophys. J. **176**, L 95 (1972).
Shivanandan, K., Houck, J. R., Harwit, M. O.: Phys. Rev. Letters **21**, 1460 (1968).
Smith, R. A., Jones, F. E., Chasmar, R. P.: The Detection and Measurement of Infra-Red Radiation. Oxford: Clarendon Press 1968.
Smyth, M. J., Cork, G. M. W., Harris, J., Wallace, T.: Nature Phys. Sci. **231**, 104 (1971).
Spicer, W. E., Bell, R. L.: Publ. Astron. Soc. Pacific **84**, 110 (1972).
Stein, W. A., Woolf, N. J.: Applied Optics **10**, 655 (1969).
Strong, J.: Sci. Am. **212**, 28 (1965).
Teich, M. C.: Proc. Inst. Elec. Electron. Engrs. **56**, 37 (1968).
Vanasse, G. A., Stair, A. T., Baker, D. J.: Aspen International Conference on Fourier Spectrometry, 1970, AFCRL-71-0019 (1971.)
Westphal, J. A.: Private communication (1972).
Weymann, R. J., Carlton, N. P.: Sky Telesc. **44**, 159 (1972).
Wildey, R. L., Murray, B. C.: Astrophys. J. **139**, 435 (1964).
Williams, B. F.: Appl. Phys. Letters **14**, 273 (1969).
Wolfe, W. L. (ed.): Handbook of Military Infrared Technology. Washington: Office of Naval Research 1965.
Woolf, N. J.: Private communication (1972).
Woolf, N. J.: In J. M. Greenberg and H. C. van de Hulst (eds.), Interstellar Dust and Related Topics, I.A.U. Symp. No. 52, p. 485, Dordrecht-Holland: D. Reidel 1973.

Discussion

Borgman, J.:

1. The argument of Dr. Smyth on the obsolence of multiplex techniques when photon-noise limited detectors become available might have to be reconsidered in case of sky-noise limited operation.

2. The Josephson junction holds great promise for high resolution spectroscopy.

3. I have confidence that the 9.7 μm absorption feature has been correctly measured in spite of the 9.6 μm ozone absorption which is relatively stable.

4. There is promise, at least in the near infrared, that two-dimensional image coding will offer opportunity for image recording with reasonable resolution e.g. Hadamard transforms and derivations of the relevant coding matrices are convenient to handle.

SMYTH, M. J.:
1. I suspect that sky noise would act like scintillation noise against the multiplex advantage and would be compensated in the same way by internal modulation or rapid scanning.

2.–4. You will gather that in the severely condensed version of this paper presented at the meeting it was not possible to mention everything.

MEZGER, P. G.:
Could you elaborate on your statement that improved IR detectors will eliminate the advantages of Fourier spectroscopy?

SMYTH, M. J.:
An ideal detection system would be limited only by the photon noise on the signal, and in these circumstances the multiplex advantage which was FELLGETT'S motive for introducing Fourier transform spectrometry would disappear. The other advantages, of high throughput and clean line-profile, of course remain, and in any case we are quite far from realising the ideal detection system.

Infrared Photometry of Galactic HII Regions

By H. Olthof, J. J. Wijnbergen, Th. J. Helmerhorst, and R. J. van Duinen
Department of Space Research, Kapteyn Astronomical Institute, University of Groningen, The Netherlands

With 3 Figures

Abstract

Far-infrared photometric observations of two galactic HII regions are presented together with observations of Jupiter for calibration.

The results indicated higher values than observed by Hoffmann *et al.* (1971) which might be due to a larger extend of the sources.

I. Introduction

The recent results of infrared photometric studies have shown a strong far infrared excess in several sources. Recent measurement of Hoffmann *et al.* (1971) have shown that most of their detected sources coincide with galactic HII regions. Our studies of HII regions are concentrating on the determination of the spectral distribution in certain well-defined bands in the far infrared. Here results are presented from a photometric study in two wavelength bands around 100 µm. These observations may elucidate the analysis of the relevant radiation mechanisms.

II. Instrumentation

The photometers consist of multiple reststrahlen reflection filters at pumped liquid helium temperature (Wijnbergen *et al.*, 1972), followed by Gallium doped Germanium bolometers (Infrared Laboratories Inc.). Two filter-detector combinations are located at the cold plate of a helium dewar. Each photometer is fed by a 20-cm parabolic reflective telescope, observing off-axis. Both telescopes are mounted vertically in a balloon gondola looking outward via a rocking flat mirror. A 26 degrees rocking motion of the flat mirror around a horizontal axis driven by a step motor allows the sky to be scanned in elevation while the gondola is oriented in a southern direction by means of an active system that uses the local horizontal magnetic vector as reference. The siderial motion leads to series of drift scans at different elevations. A command system can change the offset of the magnetic sensor in both directions, thus allowing repeated observations of the same celestial area during a single flight. The aperture stop used restricts the full field of view on the sky to 30′ arc.

The most recent flight of the gondola has taken place in the night of June 6, 1972. It was launched in the south of France in cooperation with the French space research organisation C.N.E.S. The performance of the gondola was satisfactory except for occurrence of an oscillatory motion both in azimuth and elevation. The character and amplitude of this motion varied during the flight. The present report contains results of this one flight only.

III. Observations

During the flight we have measured in two well defined wavelength bands 71–95 μm and 84–130 μm. Fig. 1 shows the transmission of the second filter system which is also typical for the first band. The scanned area on the sky, given in galactic coordinates, is shown in Fig. 2. It roughly corresponds to the right ascension-declination range of

α: 17^h15^m–19^h30^m,
δ: $-37°$– $-10°$.

Several sources as NGC 6334, 6357, galactic center, M 8, W 30, W 31, W 33 and M 17 in the galactic plane between longitudes 350° and 15° have been detected in both photometers.

Together with observations of Jupiter we have laboratory measurements to calibrate the detection system. A temperature of 140 K gives the best fit through our observations of Jupiter based on the laboratory calibration (Fig. 3) which is in

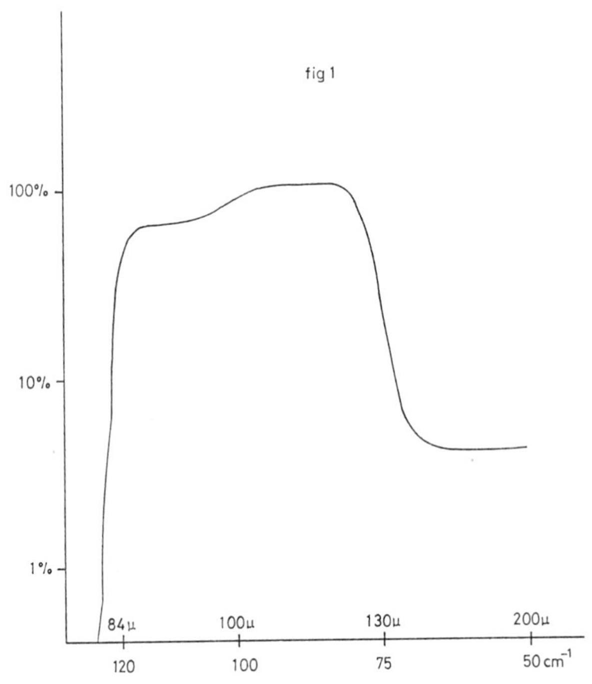

Fig. 1. Relative reflection characteristic of the CsBr cristal which has been used as filter for the 84–130 μm photometer

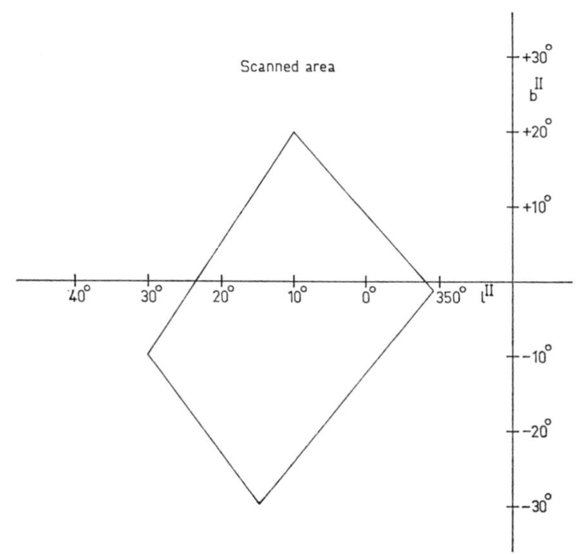

Fig. 2. Scanned area in galactic coordinates

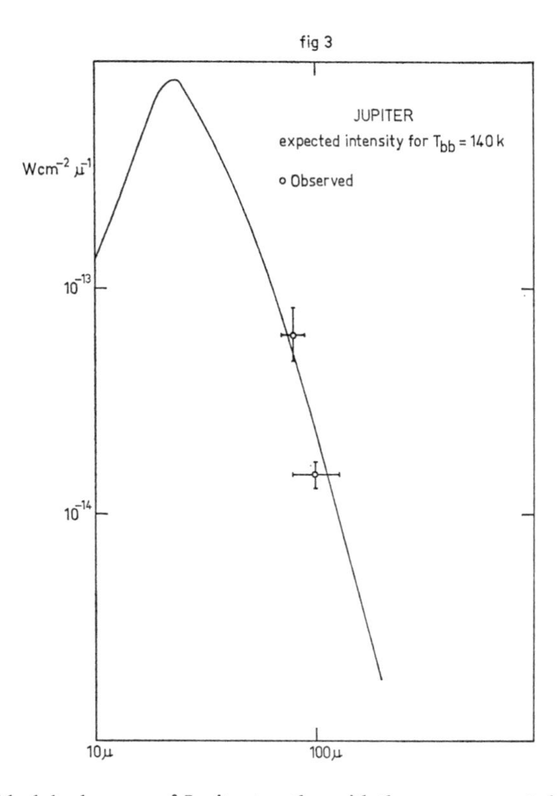

Fig. 3. Expected black body curve of Jupiter together with the measurements based on laboratory calibration

Table 1. Ratio between source and Jupiter

	71–95 μm	84–130 μm
NGC 6334	0.5 ± 0.2	0.6 ± 0.1
NGC 6357	0.3 ± 0.1	0.4 ± 0.1

Table 2. Observed fluxes in $W \cdot m^{-2} \cdot Hz^{-1}$

	71–95 μm	84–130 μm
NGC 6334	$5 \pm 2 \times 10^{-21}$	$3.8 \pm 0.6 \times 10^{-21}$
NGC 6357	$3 \pm 1 \times 10^{-21}$	$2.6 \pm 0.6 \times 10^{-21}$

Table 3. Comparison with HOFFMANN *et al.* (1971), SOIFER *et al.* (1972)

	Flux in 84–130 μm ($\times 10^{22}\ W \cdot m^{-2} \cdot Hz^{-1}$)		
	This work	HOFFMANN *et al.*	SOIFER *et al.*
NGC 6334	38	16	
NGC 6357	26	6	24

good agreement with the value of 134 K determined by AUMANN *et al.* (1969). Although we have measurements on many sources, this paper will concentrate on the results of NGC 6334 and NGC 6357. The flux ratio of these sources with respect to Jupiter are shown in Table 1 while Table 2 gives the results in $W \cdot m^{-2} \cdot Hz^{-1}$ under the assumption that Jupiter radiates as a black body of 140 K.

IV. Discussion

Comparison of these measurements with earlier measurements of these sources by HOFFMANN *et al.* (1971) and SOIFER *et al.* (1972) is shown in Table 3. HOFFMANN *et al.* have observed these sources with a balloon gondola using a differential chopping system while SOIFER *et al.* have observed NGC 6357 with a rocket telescope using full field chopping, the same technique we have used. Table 3 shows that the fluxes we measured for these two sources are substantially larger than the numbers given by HOFFMANN *et al.*, whereas we find good agreement with the flux given by SOIFER *et al.* for NGC 6357. The discrepancy might be ascribed to the extension of this HII region, causing an underestimate of its total flux with a differential chopping technique over an angle less than the source diameter. The scan of SOIFER *et al.* through NGC 6357 shows a source with a half width of the order of 0°.7. In the table of HOFFMANN *et al.* this source is said to be a point source which means that the size of the source should be less than their beamwidth of 0°.2 degrees. This suggests that the actual source sizes have been underestimated by HOFFMANN *et al.* A similar argument may explain the discrepancy between our results and the results of HOFFMANN *et al.* on NGC 6334 for which the latter finds indication of an extension over 0°.3 which might be an underestimate of the actual source size. If it is true that the far infrared radiation from these sources can be ascribed mainly to thermal radiation from dust grains, these two color measurements indicate that the maximum emission occurs somewhere below 80 μm. These measurements indicate that there is almost no doubt that the flux density in the 71–95 μm band is higher than in the 84–130 μm band.

Observations in wavelength bands around 40 μm and 60 μm are necessary to determine the peak in this thermal radiation model.

Acknowledgement

We want to thank G. DE GROOT, P. B. VAN DER WAL and T. DUURSMA for assistance during the construction and launching phase of the gondola together with T. WIERSTRA, A. KONING and N. ZUIDEMA for assistance during the reduction phase of the data. The stimulating discussions with Drs. S. R. POTTASCH, J. BORGMAN and C. D. ANDRIESSE are greatfully acknowledged.

References

AUMANN, H. H., GILLESPIE, C. M. JR., LOW, F. J.: Astrophys. J. **157**, L 6 (1969).
HOFFMANN, W. F., FREDERICK, C. L., EMERY, R. J.: Astrophys. J. **170**, L 89 (1971).
SOIFER, B. T., PIPHER, J. L., HOUCK, J. R.: Preprint (1972).
WIJNBERGEN, J. J., MOOLENAAR, W. H., GROOT, G. DE: In V. MANNO and J. RING (eds.), On Infrared Detection Techniques for Space Research. Fifth ESLAB/ESRIN Symp. Dordrecht-Holland: D. Reidel 1972.

Discussion

COURTÉS, G.:
What is your pointing accuracy?

OLTHOF, H.:
The pointing accuracy is of the order of 0°.5. Exact positions are very difficult to obtain with a balloonborne telescope which is only magnetically oriented.

BORGMAN, J.:
Are these the only sources observed by your group?

OLTHOF, H.:
No, at this moment the results on these sources are ready. We are working at this moment on the analysis of the signals from the galactic center, M 8, W 30, W 31, W 33 and M 17.

8–13 μm Spectrum of the Galactic Centre

By D. AITKEN and B. JONES
Department of Physics and Astronomy, University College London, United Kingdom

Abstract

We have observed the galactic centre in July 1972 over the wavelength range 8–13.5 μm at a number of resolutions ($\Delta\lambda = 0.05$ μm, $\Delta\lambda = 0.09$ μm and $\Delta\lambda = 0.2$ μm) using the 74-inch Radcliffe telescope. The data is only partially analysed and the results presented are preliminary.

The spectrum can be interpreted in terms of interstellar silicate absorption with an optical depth ($\tau \simeq 1.3$ at 10.5 μm corresponding to $\sim 6 \times 10^{-4}$ g·cm^{-2} of silicate material in the line of sight, assuming the galactic centre emission to follow a power law spectrum in this region. If, however, the emission is taken as appropriate to an optically thin thermal source consisting of similar material to the interstellar matter this estimate must be increased to about 1.5×10^{-3} g·cm^{-2}.

The detailed results are being prepared for publication.

Discussion

BORGMAN, J.:

1) The feature between 12 and 13 μm in your spectrum, tentatively identified as a neon line, might also be atributed to interstellar ice absorption at 12.5 μm.

2) Though you analyse your numbers while adopting 30 magnitudes visual extinction there is as far as I can see no evidence in your observations which contradict my suggestion of a large, e.g. 200 magnitudes, local extinction in the region.

AITKEN, D.:

1) The wavelength scale is good to better than ± 0.1 μm so a discrepancy of 0.3 μm is very unlikely. Also the broad absorption feature of interstellar ice we would be very difficult to separate from the range of possible silicate mixtures which could fit the spectrum.

2) May be the extinction to the galactic centre is more typical of interstellar matter than for the Orion cloud in front of BECKLIN's object? However, if the extinctions are compared at 10.5 μm (instead of at 9.7 μm where ozone corrections may not be reliable) then WOOLF's data are similar to ours and similar values are obtained for BECKLIN's object. This reduces the discrepancy between the two regions and an extinction ~ 80 visual magnitudes is implied for the galactic centre.

High Speed Evaluation of Photographic Plates

(Invited Lecture)

By V. C. Reddish
Royal Observatory Edinburgh, United Kingdom

With 1 Figure

Abstract

Various automatic and semi-automatic machines for measuring photographs are reviewed briefly in relation to astronomical requirements. The automatic machine GALAXY at the Royal Observatory Edinburgh is described in more detail and it is explained how it finds stars at a rate of 10^4/hour and measures their positions to ± 1 μm and magnitudes to $\pm 0\overset{m}{.}016$ at 10^3 stars/hour. Users of the machine are listed. The GALAXY MK 2 in course of construction will make measurements of stars and galaxies with an accuracy of about 3 μm and $0\overset{m}{.}1$ at rates up to 10^7 images/hour, or 0.5 μm and $0\overset{m}{.}016$ at 10^3/hour. It will also measure the shapes, sizes, densities and orientations of images at similar speeds.

I. Introduction

Many fields of research utilise photography and require the extraction of large amounts of data from photographs; for examples, medicine and biology, X-ray crystalography, high energy particle physics and astronomy.

The needs of astronomers generally differ from those in other fields in two respects – in the high precision required of the data, especially positional information, and in the large sizes of the photographs employed.

Nevertheless there may be some astronomical problems which can effectively make use of relatively low cost machines (say £ 10000) such as that developed by the Medical Research Council at Hammersmith Hospital, London, which counts the number of objects above a given density level and gives their average size; measurements of clumpiness in galactic nebulae, or counts of galaxies in clusters, are problems which come to mind.

Machines which give detailed information about each image, such as its position, size and density, are much more expensive and often utilise real time computer analysis of the measurements. I can do no more than mention some of them. They include the Image Reading Instrument System by Saab-Scania in Sweden; the Cellscan-Glopr system for medical research by Perkin-Elmer in the U.S.A.; the Microdensitometer Data Acquisition systems by Photometric Data Systems, U.S.A; the coordinate measuring machines at the Lick and U.S. Naval Observatories; the Control Data Corporation machine developed with W. J. Luyten at Minnesota; Sweepnick at the Cavendish Nuclear Physics Laboratory in Cambridge; the range of image analysing systems developed by Image Analysing Computers

Ltd in the U.K. and used in both industry and the laboratory, and the GALAXY machines at the Royal Observatories in Edinburgh and now also at Herstmonceux.

I do not intend to say anything about machines for measuring photographs of spectra because I presume that photoelectric detectors will soon be universally used for spectrophotometry.

Since I have some direct knowledge of the GALAXY machines I shall talk about them at greater length, particularly to describe the facilities which will be incorporated in the MK2 version due to come into operation next year.

II. Galaxy MK1

The GALAXY machine has four main features; a precise mechanical carriage to hold the photograph, a system to measure the carriage position, two CRT scanners, one to search for images on the plate and the other to measure them in more detail, each with photocells to measure the light transmitted through the plate, and an electronic system to control the operations of the machine and to convert the signals received into meaningful and desirable data.

(i) The Carriage
This can hold photographic plates up to 358 millimetres (14 inches) square. The axes are orthogonal to 2″ arc and linear to 0.3 μm.

(ii) Carriage Position Measurement
Moiré fringe systems measure the carriage position in increments of 1 μm.

(iii) Scanners
a) Search. The search mode utilises a CRT with ÷4 optics giving a choice of spot sizes 8, 16 or 32 μm at the plate. A linear raster scan searches the plate at a rate of 900 mm^2/h, most of the time being occupied in mechanically stepping on the carriage from one line scan to the next.
b) Measurement. This mode employs a CRT with a choice of optical systems, currently ÷240, ÷75 and ÷30 demagnifications. Using the ÷240 system, the spot size is 1 μm at the plate and it scans 1024 concentric circles $^1/_8$ μm apart at 30 KHz, thus closely representing a spiral scan, one complete scan taking 30 milliseconds.

(iv) Control and Data System
a) Search. Each image is generally scanned several times, but there is logic to cause only the position of the centre of the image to be output. This mode gives a finding list of image positions, accurate to about 30 μm, typically at a rate of 10^4 images/h. The positions are output on paper tape; this data can be used directly for star count statistics, or the tape can be regarded as an intermediate storage facility.

b) Measurement. A list of finding positions—from the search mode of operation or from any other source, is fed into the machine on paper tape. The machine reads the first pair of coordinates, moves the carriage to bring that image under the measurement scanner, and then transfers control to the servo systems associated with the spiral scan.

As the image is scanned, variation in transmitted light with azimuth gives centering information—the carriage moves until the image is centred on the spiral scan and the position of the carriage is measured by the Moiré fringe system in increments of 1 μm. Simultaneously, transmission through the image is measured as a function of radius. This measured image profile is differenced from one of a progressive set of 1024 ideal profiles stored by program in the machine; the difference is weighted and integrated and the ideal profile is changed until the integral is zero, that is until the measured and ideal profiles match. The location of the matching profile in the store then labels the profile parameters. The weighting function is related to the gradient of the ideal profile, and is also applied to the measurement of of position. The ideal profiles and weighting functions are determined by parameters chosen to suit the particular photographic plate being measured. Their object is to cause the machine to make the best use of the information contained within each star image in measuring its position and strength. The measurement of position to 1 μm and of profile to one part in 1024 are coded and output on punched paper tape. The whole process takes 4 seconds per star. At this speed of 900 images/h, measurements are made with r.m.s. errors of repeatability of ± 0.5 μm in position and $\pm 0\overset{m}{.}016$ in magnitude.

This technique of profile matching has several advantageous consequences. Because it makes better use of the information in the image, the external variance of magnitude measurements is only half that obtained using an iris photometer. By appropriate choice of profiles, the shape of the measurement versus magnitude calibration curve can be varied, and in particular it can be made linear over a range of several magnitudes. The measurements do not, as in the case of an iris photometer, become asymptotic to the background density at the faint end, because GALAXY progresses smoothly from the faintest star images to grain noise clumps; thus errors of measurement remain quite low to the faintest recorded images.

The accuracy of positional measurements benefit similarly. Using measures of a Schmidt plate obtained with all three sets of optics ($\div 30$ for largest images, $m_{pg} \sim 5$, to $\div 240$ for smallest images, $m_{pg} \sim 14$) no positional dependence on magnitude or colour is found over ranges of 9 magnitudes and 1.5 magnitudes respectively.

For very bright ($m_{pg} < 4$) and very faint images ($m_{pg} > 14$) the standard deviation of a single measurement increases but no systematic error in position is introduced.

(v) Use

GALAXY MK 1 was first switched on as a complete machine in March 1969 and by June of that year its performance was regularly exceeding specification. Since January 1970 it has been in operation 24 hours/day, 7 days/week, with interruptions only for regular routine maintenance. There have been no serious faults or breakdowns. In these $2\frac{1}{2}$ years, five million images have been measured on 500 plates for astronomers from 16 Observatories. These include, in addition to several British Observatories, the Copenhagen, Kitt Peak, Laurentian, Leiden, and Lund Observatories, the Max Planck Institute, NASA, and the Padua, Palomar and Steward Observatories. Requests for measurements on the machine have greatly exceeded both our expectations and the time available to meet them, and the variety of astronomical programs for which it has been used has also been surprising.

A number of minor modifications to the machine have been made since 1969 including the provision of a carriage large enough to hold 14″ (358 mm) square plates, and a digital surface photometry mode. The latter has been used especially for surface photometry of nebulae and faint galaxies and quasars.

From the technical point of view I think that the success of GALAXY has been due to two factors. Firstly, a careful choice of digital and analogue systems. In some instances, for example averaging around a circular scan, analogue systems are faster than digital ones. Secondly, and related to the example just given, if the features of a pattern are known, for example if it is expected to be circular or elliptical and more dense in the centre, it is much quicker to compare the measurements to an ideal representation of the pattern than to carry out a pattern analysis procedure on the measurements.

With the prospect of large numbers of survey plates expected to come from the U.K. 48-inch Schmidt Telescope to be erected in Australia next year, we thought it desirable to increase the speed of operation, and to make the machine capable of measuring galaxies, as well as stars. Following a design study, detailed design and manufacture of a new control system began in July 1971 and is due to be completed and in operation in July 1973. The MK2 machine like the MK1 is a joint enterprise between Faul-Coradi Scotland Ltd and the Royal Observatory Edinburgh.

III. Galaxy MK2

There are three major changes in comparison with the MK1: greatly increased speed of the raster scan by driving the carriage continuously instead of stepping it; the addition of measuring facilities to the raster scan, turning the search mode into a coarse measurement mode; and the generalisation of the circular scans to ellipses in the fine measurement mode to enable faint galaxies as well as stars to be measured rapidly. At this half way stage in design and manufacture, the expectation is that the GALAXY MK2 will measure with the following speeds and accuracies:

(i) Coarse Measurement

Position: ± 2 μm to ± 4 μm depending on image size,
Size: $\pm 3\%$ or better depending on image size,
Density: ± 0.02,
Magnitude: $\pm 0\overset{m}{.}1$,
Shape: a/b to 20%,
Orientation: quadrant,
Speed: $2\ \mathrm{mm}^2 \cdot \mathrm{s}^{-1}$ (14-inch square plate/day),
Output: up to 10^7 images/hour on magnetic tape.

Facilities will include automatic setting of the threshold level, and a digital surface photometry mode which will give surface densities with an accuracy of 0.02 at a speed of $1\ \mathrm{mm}^2 \cdot \mathrm{s}^{-1}$ and a resolution of 8 μm.

(ii) Fine Measurement

Position: $\pm 0.5\ \mu m$,
Size: $\pm 0.2\ \mu m$,
Density: ± 0.02,
Magnitude: $\pm 0\overset{m}{.}016$,
Ellipticity: $(a \pm 0.2\ \mu m)/(b \pm 0.2\ \mu m)$,
Orientation: $\pm 3°$,
Speed: stars; 900/h,
galaxies; 300/h.

Non-elliptical images will be tagged. There will also be a facility for measuring the positions and strengths of absorption or emission lines in objective prism spectra with similar accuracies and speeds.

In the case of survey plates coming from the 48-inch Schmidt telescope in Australia, we envisage the operation of the machine as follows.

Firstly, the plate will be scanned in the coarse measurement mode. On the basis of the data available to us from simulation experiments on the MK 1, carried out by N. M. Pratt who is developing all the machine control software, it is probable that off-line computer analysis of the coarse measurements will enable us to distinguish between stars and nebulae or galaxies. The magnitude measures of the stars will probably be as good as a single survey plate can give, $0\overset{m}{.}1$, and will be available at this stage for analysis. The positions of the galaxies will be read back into the machine to be scanned in the fine measurement mode, giving precise geometrical parameters, magnitudes and positions for good elliptical galaxies and noting those images with a degree of non-ellipticity. These latter will be scanned in the iso-densitometry mode to give detailed maps of them.

In this way we hope to provide the kinds of quantitative information that astronomers generally appear to require from the survey plates. I should say that the southern survey plates will have intensity calibrations on them and these will be measured at the same time, together with fiducial marks imposed for position reference.

As the uniformity and sizes of photoelectric cathodes increase, the time will come when direct electrical read-out from the telescope will replace our present survey techniques; but I think that it will be some years before we can relinquish the need for high speed extraction of data from photographic plates. And when photoelectric devices do replace them for survey purposes, all the computer based analytical techniques required to operate on the output will be the same as those that have been developed for the GALAXY machine.

Discussion

Turlo, Z.:
Would you like to say what GALAXY machine will do when it encounters on the photographic plate overlapping pictures, diffraction patterns, scratches and other defects of the photographic emulsion or simply the edge of the plate. It appears to me that such trivial things can create not so trivial practical problems when one starts to think in terms of software needed in order to exploit fully those impressive potential possibilities offered by the MK I and MK II GALAXY machines.

WERLINGER, A. P.:
I gather that the MK II will be commercially available in the near future. What other observatories are planning to install the GALAXY MK II? What is the approximate cost and delivery time for the MK II?

REDDISH, V. C.:
I imagine that the Royal Greenwhich Observatory will have one and that others will follow in about a year and a half. I think that the price of the machine will be about £ 150.000.

CARBIERI, C.:
What is your opinion about machines using large CRT and flying spot techniques? Could they be much faster than GALAXY?

REDDISH, V. C.:
GALAXY uses flying spot scanners - micro-spot CRT tubes. We correct for brightness variations by sampling the spot brightness and comparing the transmitted light to that. Positional information is calibrated by simultaneously scanning a 64 micron grating.

BORGMAN, J.:
The GALAXY should in principle be a good instrument for detecting variations e.g. variable stars, large proper motions etc. In order to do this it would be necessary to store the complete fine resolution information on magnetic tape. Could you give us some indication of the cost on the tape that would be involved?

REDDISH, V. C.:
Between £ 30.000 and £ 60.000. We have not yet decided whether we should scan all the survey plates systematically or only those for which specific requests are made by individual astronomers carrying out research programes.

RIJSBERGEN, R. VAN:
Can you tell something about the profiles that are stored in the memory of the machine, are they standard?

REDDISH, V. C.:
They have the following form:

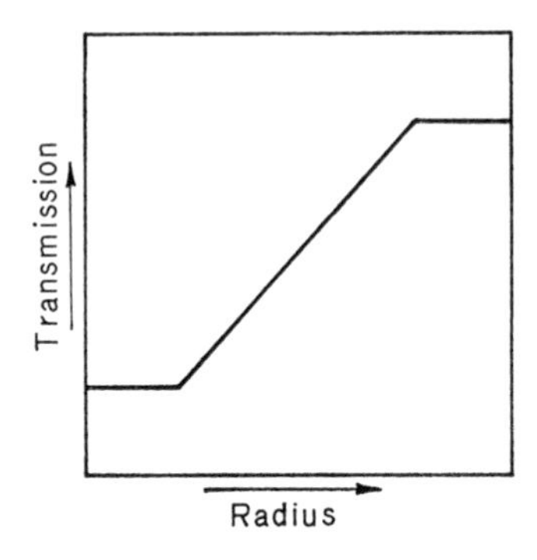

The gradient and radius is determined from sample scans of a number of stars on the plate and the whole set of 1024 is set up by computer interpolation.

Analysis of Electronographic Images

By M. R. S. Hawkins
The Observatories, Cambridge University, United Kingdom

With 3 Figures

Abstract

Techniques are described for correcting, filtering and plotting electronographic images as isophote plots. The method is then illustrated by the analysis of an electronograph of NGC 4278 and 4283.

I. Introduction

The advent of image tubes has opened up the way for significant advances in certain important areas of astronomy. The potential gain in information of an electronograph over a photograph has been estimated as forty times, and yet advantage is not normally taken of the image tubes special properties, while the very real difficulties encountered in obtaining these highly accurate results are largely ignored.

Of considerable importance in contemporary extra-galactic astronomy is the study of extended objects very much fainter than the night sky itself, for example haloes surrounding galaxies, quasar envelopes, galactic bridges, etc. In this field discrimination between night sky and object is vital, and for this purpose electronographic image tubes offer pronounced advantages over all other image devices.

Electronography involves the conversion of signals from a photon-carrier to an electron-carrier. It should be emphasized that there is no question of "information enhancement". The gain is in the response of nuclear emulsion or other electron detectors to electrons compared with the response of a photographic plate to photons. Electronographic methods have three particular advantages over most other devices:

a) linearity of response (i.e. photographic density is proportional to incident photon flux);

b) no reciprocity failure;

c) very little image spread;

d) high signal-to-noise ratio;

and it is the purpose of this paper to describe how these characteristics may be used to realize the limiting performance of the electronographic image tube to study faint fields with particular reference to galactic haloes. The method is illustrated by the analysis of NGC 4278.

II. System Parameters

The system to be considered here consists of:

(i) a spectracon at the prime focus of a telescope (or on a bench for certain calibrations). This produces an electronographic image on a piece of nuclear emulsion.

(ii) The film is measured as a rectangular sampled array of transmitted intensities by a microphotometer with digitized 10 -bit output.

(iii) The array of intensities is converted by computer to an array of densities which are then available for Fourier analysis and other corrections.

The final output is normally in the form of a computer compiled isophote plot of the area of interest on the film, although occasionally other formats are more appropriate.

The design and construction of the spectracon is described in detail in McGEE *et al.* (1966) and summarized with a number of improvements in McGEE *et al.* (1969). We shall only discuss those aspects of design which are important in image analysis. The performance characteristics of interest are as follows (see McGEE *et al.* 1969):

a) Photocathode Sensitivity. This may be summarized as follows:

S 1: 1–2% quantum efficiency,
S 11: 5–8% quantum efficiency,
S 20: 10–15% quantum efficiency,

as compared with ~1% under the best conditions for photography. This great gain must be offset against non-uniformity of the photocathode which is of two kinds. Firstly, there is a continuous variation in sensitivity over the whole surface usually between about 5% and 10%. Secondly, most photocathodes contain "dead spots" i.e. discrete spots of very low sensitivity. In principle, a picture free from both these kinds of variation may be constructed, and the techniques involved are discussed below.

b) Image Stability and Geometry. For about half an hour after switching on the tube the image may drift slightly, but after this the image will be stable providing operating conditions remain constant. In addition, recent improvements have rendered the image geometry sufficiently accurate for the most precise astronomical measurements.

c) Background and Resolution. For L4 emulsion the resolution obtained is about 120 lp/mm which on most telescopes means that the limit on resolution is set by atmospheric seeing.

The background emission from the photocathode has presented one of the most serious difficulties in the development of image tubes, however, mumetal screening has now made tubes virtually independent of their surroundings, and providing the tube is kept cool the background should give an increase in density of only 0.0004/hour for L4 emulsion.

There is, however, another form of dark current. This always appears and disappears suddenly and usually forms a distinctive pattern on the emulsion. Whenever it appears it will ruin the exposure. It is probably caused by a static build up on the walls of the tube and may usually be cured by switching the H.T. off and on.

From these remarks we observe that providing the system is working properly, the only distortions due to the image tube which must be removed are

a) non-uniformity of photocathode,

b) dead spots on photocathode.

III. Noise

Under this heading we shall consider random fluctuations superimposed upon the input signal at various stages of the system, and typically proportional to the square root of the signal. This should be distinguished from the sky background.

a) Granularity

This covers a number of effects which are more or less indistinguishable. Firstly, there will be a fluctuation in the statistical distribution of photons reaching the photocathode from a given object; then, there will be scattering in the mica window, and the projected track length in the emulsion. Also variation in grain size and clumping of grains will contribute an unpredictable element to the signal. There is also a further effect which may be distinguished from the preceding ones. This arises from minute particles being deposited on the film during the development process. These two effects may be distinguished by comparing the noise power spectra of a uniformly exposed and an unexposed piece of film, and it is found that actual exposure of the film to electrons does not make a significant additional contribution above the noise level of development and measurement. It should be remarked here that there is nothing that the observer can do to decrease the non-development noise, but that at any rate the spatial wavelength involved will be small. Developer noise, however, may have a sizable contribution in all spatial wavelengths, e.g. a continuous gradient of fog over the whole exposure, and scrupulous care in development is necessary to reduce the noise level at the longer (and indeed all) wavelengths.

b) Microphotometer Noise

The microphotometer contributes to noise in two ways – variation in the lamp, and amplifier drift and flutter. These two effects may be dealt with together. We observe that providing the time scale of the variation is small compared with the time between samples then the noise contribution to the array will mainly be in the small spatial wavelengths, and so it is of the utmost importance that the equipment be set up to eliminate all long-term drifts as these cannot be removed mathematically without distorting the signal.

IV. Image Recovery

In sections II and III we have seen that the only serious forms of image degradation are:

(i) variation in photocathode sensitivity,

(ii) dead spots on photocathode,

(iii) granularity noise,

(iv) microphotometer noise.

In order to make the corrections we require two exposures, A and B, of the object of interest, on different parts of the photocathode, and an exposure S of a uniform source of illumination (often the dawn sky). We first divide each reading of A and B by the corresponding reading of S, i.e. the reading taken from the same point on the photocathode. The linearity of response of the image tube ensures that the variation in photocathode sensitivity will now have been corrected.

To remove the dead spots we simply replace readings of A spoilt by dead spots with suitably normalized readings of B of the same area of sky. Again, the linearity of response makes this possible. However, it should be said at this point that this is not a particularly satisfactory procedure since among other things it requires two separate exposures, and the only really satisfactory solution lies in the production of photocathodes without dead spots. It is to be hoped that this will soon become possible.

The array that we now have has been corrected for variation in sensitivity and dead spots on the photocathode. It remains to filter the noise, but first we shall discuss the aperture size of the microphotometer.

We first observe that an aperture of side d will heavily damp all spatial wavelengths smaller than d (i.e. the smallest detail seen in the signal will be greater than d). So the recorded signal will be a bandlimited signal and the highest spatial wavelength will be $f=1/d$. Now let S be the sampling interval. Then the Shannon sampling theorem states that for the signal function to be exactly represented by discrete samples $S \leqq \frac{1}{2f} = \frac{d}{2}$. In fact we shall assume $S = \frac{1}{2}d$, where d is chosen smaller than the largest detail it is hoped to observe but otherwise as large as possible in order to reduce the number of data points required and suppress high frequency noise.

It might be thought that having chosen the aperture size as above all spatial wavelengths smaller than this would have been removed and no further filtering would be necessary. However, this is not the case, for although emulsion noise will have been damped, the microphotometer noise will not—indeed it will only just have been introduced. This must, therefore, be removed mathematically, together with any further filtering of the emulsion noise which may be necessary.

The choice of an appropriate frequency filter is a matter of some difficulty. Various attempts have been made to derive an "optimum" filter (e.g. BRAULT and WHITE 1971) where the filter

$$F(S) = \frac{P_s(S)}{P_s(S) + P_N(S)}$$

is derived, P_s and P_N being the signal and noise power spectra respectively. However, this sort of filter can only be called "optimum" in a rather limited sense since both the signal and noise power spectra must be specified in a smoothed form which is basically undetermined. In fact a suitable function (e.g. a Voigt function or Gaussian) is generally specified for the signal power spectrum, while the noise power spectrum is normally taken as flat. It is true that by using this method the errors will be of the second order, due to the variational approach used to evaluate the filter, but even so it would seem that in view of the requirement to specify the noise and signal power spectra little is gained over a more pragmatic approach.

In general, two cases arise. Firstly, the signal may be a profile of an object whose spectral shape is known, e.g. a star. In this case an appropriate filter may be chosen from the known shape of the spectrum. A more difficult situation arises when the signal represents a complicated pattern, e.g. a spiral galaxy. In this case there is no way of knowing what the signal power spectrum should look like and so some assumptions must be made as to the nature of the noise contribution. It is usually safe to say that most of the high frequency contribution is noise and so a filter which attenuates steadily more of the higher frequencies is required. A Gaussian has been found to give good results (see Figs. 2 and 3) but other filters may be tried.

V. Observations of NGC 4278/4283

To demonstrate the use of the techniques described above we shall consider the analysis of an electronograph of the two galaxies NGC 4278 and NGC 4283. The following details are relevant:

	NGC 4278	NGC 4283
Type	E1	E1
Magnitude	11.6	13.4
Angular diameter	1'.90	0'.87
Radial velocity	672 km·s^{-1}	1078 km·s^{-1}
Angular separation	3'.7	

These details are taken from De Vaucouleur's Reference Catalogue of Bright Galaxies.

It will be observed that according to this data the apparent distance between the outer limits of these galaxies is 1'.4, however, their apparent proximity suggested that it might be interesting to examine the space between the galaxies for any evidence of a bridge between them. The difference in the redshifts would clearly make such a bridge of some astronomical interest and it was felt that should such a bridge exist then with its high signal to noise ratio an image tube was the ideal instrument to use.

A one hour exposure was obtained on the Cambridge 36-inch telescope in April 1972 under conditions which though fairly good for the site, would be considered

Fig. 1. Isophote plot of high density regions of NGC 4278 and NGC 4283

most unfavourable for most observations. A sky exposure was obtained at dawn on the same night and in May 1972 a second exposure of the galaxies.

The films, L4 nuclear emulsion, were developed as recommended by the manufactures. The exposures were then measured on a microphotometer as arrays of 90×212 points with a square aperture of side 100 μm and a sampling interval of 50 μm. The arrays of transmissions were then converted by computer to densities and the resulting raw signal of the galaxies plotted in Fig. 2. The signal was then corrected and filtered as described in section IV using a Gaussian filter of standard deviation corresponding to spatial wavelength of 80 μm (this effectively cuts out

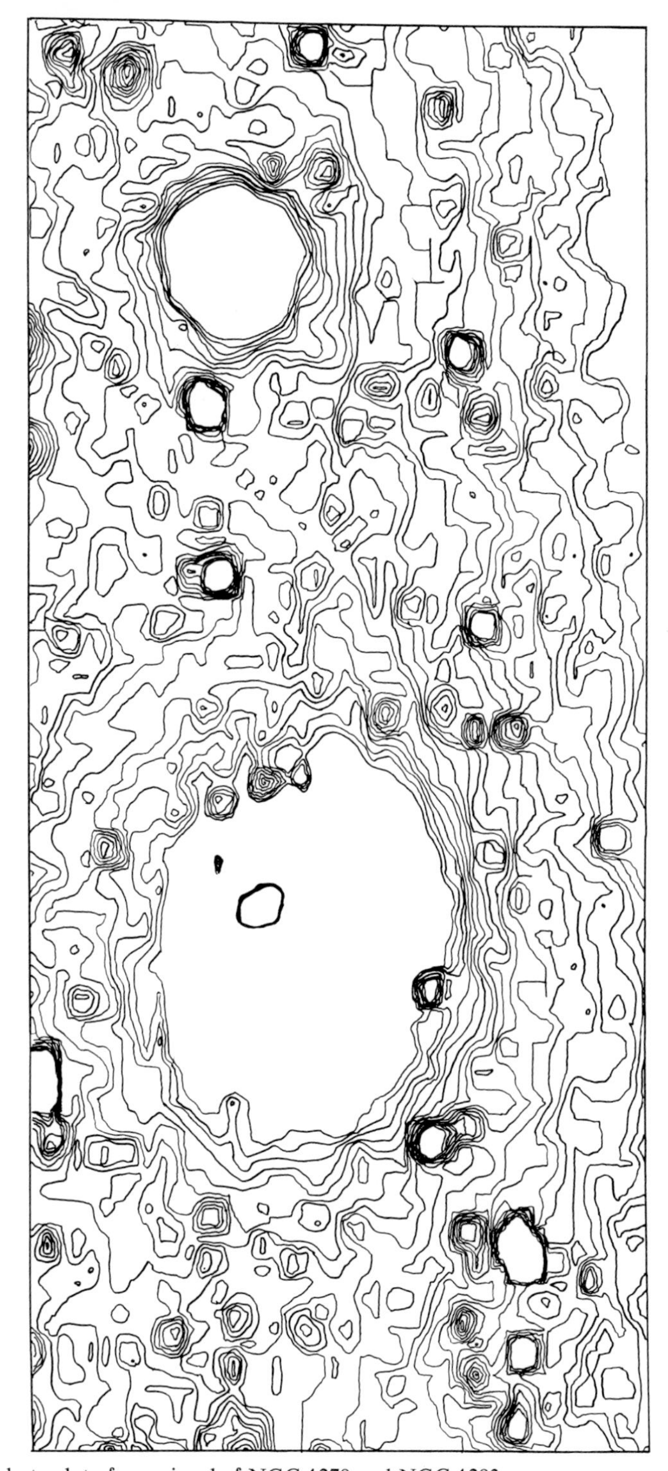

Fig. 2. Isophote plot of raw signal of NGC 4278 and NGC 4283

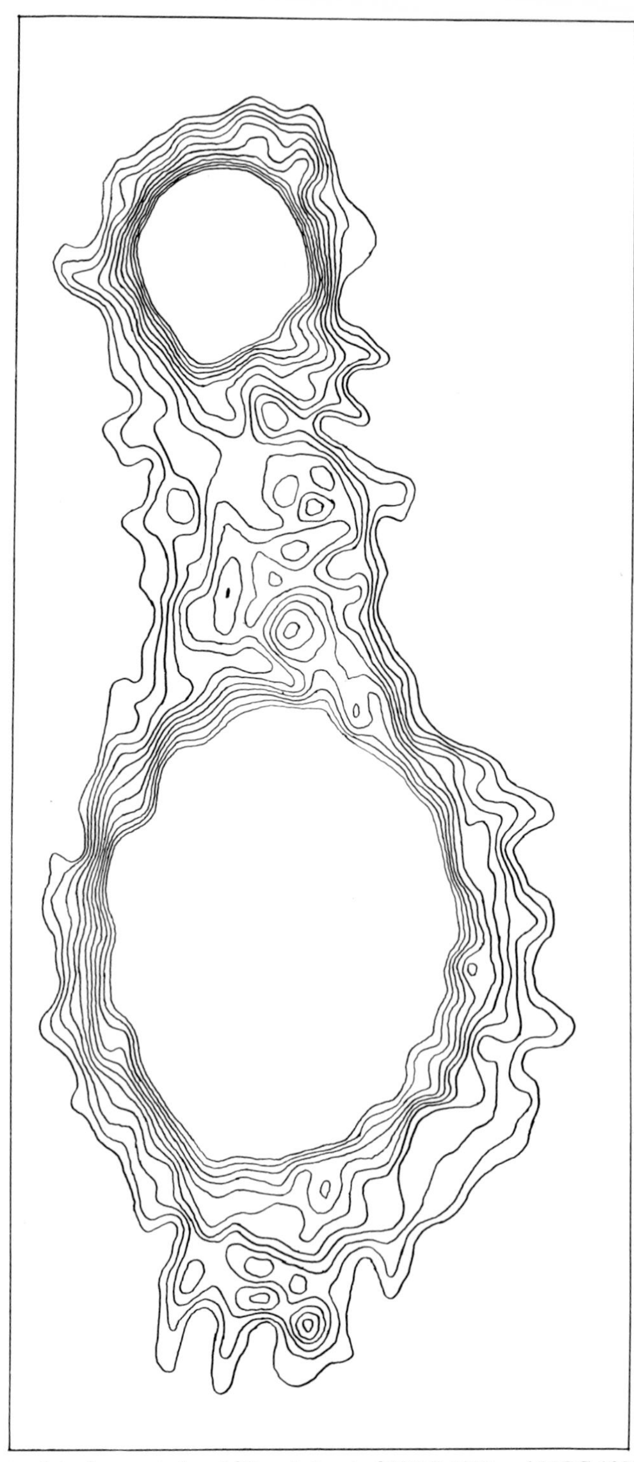

Fig. 3. Isophote plot of corrected and filtered signal of NGC 4278 and NGC 4283

all spatial wavelengths less than 160 μm). The resulting corrected signal was then plotted as Fig. 3.

Examination of Figs. 2 and 3 shows that while Fig. 2 gives very little information as to the emission from between the galaxies, in Fig. 3 a clear picture has emerged. The contours are spaced at density intervals of 0.005 and the lowest shows a density of 1.26. The mean sky background is at a density of 1.24 and so the lowest plotted contour is thus 1.6% above the night sky.

To appreciate the significance of this figure it is constructive to compare these results with those obtained by photographic methods. The percentage above the night sky of the lowest reliable contour is a figure which depends upon the signal to noise ratio and in the case of photographs with their low signal to noise ratio about 20% is the best that can be hoped for. Furthermore, this only applies to the linear part of the response curve. In the case of very faint signals reciprocity failure may mean that the plate will not respond at all in any reasonable exposure time.

Fig. 1 shows an isophote plot which roughly corresponds to the best which could be obtainable by photographic means. It is in fact taken from an uncorrected electronograph; it is interesting to notice that when an electronograph is plotted in this range the effect of dead spots (which can just be seen) and other distortions is insignificant. If we now consider Fig. 2 we see that the galaxies in Fig. 1 will fit into the gaps in Fig. 2, and we have in effect plotted the outer contours of Fig. 1 at 1/20th the spacing. Finally, consider Fig. 3. The improvement over Fig. 2 is obvious and makes clear the value of using proper correction and filtering techniques.

VI. Conclusions

It is believed that the figure of 1.6% above the night sky is a considerable improvement over any other existing technique with comparable output. Furthermore, there would appear to be considerable scope for improvement on this figure, by for example advances in the manufacture of photocathodes.

It is also worth remarking that the exposure was taken under conditions far worse than those to be found on many modern sites. It seems possible that a number of important questions may be answerable by the combination of modern observing facilities and these electronographic techniques.

References

McGee, J. D., Khogali, A., Ganson, A., Baum, W. A.: In J. D. McGee, D. McMullan and E. Kahan (eds.), Advances in Electronics and Electron Physics, vol. 22A, p. 11. London: Academic Press 1966.

McGee, J. D., McMullan, D., Bacik, H., Oliver, M.: In J. D. McGee, D. McMullan, E. Kahan and B. L. Morgan (eds.), Advances in Electronics and Electron Physics, vol. 28A, p. 61. London: Academic Press 1969.

Brault, J. W., White, O. R.: Astron. Astrophys. **13**, 169 (1971).

Discussion

SMYTH, M. J.:
Can you tell us briefly how the isodensitometric tracings were made and how the spatial frequency filtering was performed?

HAWKINS, M. R. S.:
The film was measured on the microphotometer at Cambridge. The signal was filtered using a Gaussian filter. "Optimum filter" techniques were considered but not enough information available for the expected spectral shape and so it was felt that a Gaussian provided a good compromise, besides tying in with radio techniques.

WEHINGER, A. P.:
Your optical measure of the electronographic images might be usefully compared with 21-cm radio observations where HI bridges have been found.

HAWKINS, M. R. S.:
These particular galaxies are radio quiet in the sense that they do not appear in any of the published catalogues of radio sources. However, one of the most useful aspects of this sort of work should be comparison with radio observations.

URSIES and the Bartol Coudé Observatory

By Arne A. Wyller*
Bartol Research Foundation of the Franklin Institute Swarthmore, Pennsylvania, U.S.A.

With 4 Figures

Abstract

A polar siderostat with a 36-inch flat Cervit mirror is mounted on top of the fifty foot tall building of the Bartol Research Foundation on the campus of Swarthmore College outside Philadelphia. The beam of parallel light is guided through a hole in the roof down along the optical axis of a 24-inch broken Cassegrain reflector, which is mounted stationarily on the third floor of the building. By the use of a plateglass window in the roof hole the circulation of warm air is eliminated and good to excellent solar and stellar image definition results.

The focal plane of the horizontal exit beam of the 24-inch reflector falls on the entrance pinhole plane of the spectrum scanner URSIES (Ultravariable Resolution Single Interferometer Echelle Scanner). The scanner consists of an echelle and/or an interferometer encased in a cylindrical pressure vessel 2 feet in diameter and 4 feet long. The vessel is filled with freongas and the pressure can be varied over a range of 90 inches, permitting a scanrange of 20 Å in the visual and 50 Å in the infrared. In the echelle mode spectral resolution vary from 10^2 to 10^4, while in the interferometer-echelle coupled mode the spectral resolution can be increased up to 10^6 with very high spectral purity. The scanner can also be used as a Halle-type filter with only the interferometer and an interference filter to block out adjacent passbands.

I. Introduction

The echelle-interferometer spectrograph recently developed at Bartol (Fay and Wyller, 1972) enables one, for the first time, to tackle high resolution photoelectric spectrophotometry of stars with modest-sized telescopes in coudé-configurations. This is in large measure due to the very high efficiency of the Bartol spectrograph system in conjunction with the 15-inch siderostat at the Flower and Cook Observatory of the University of Pennsylvania. With this instrument combination it has been possible to scan the HeI λ4922 line in the spectrum of the third magnitude star γ Pegasi with a spectral resolution (0.1–0.2 Å) hitherto reserved for work with much larger telescopes.

The reason for this first successful entry of modest-sized telescopes into the field of high resolution spectrophotometry is the large intrinsic throughput of light

* Permanent address: Stockholm Observatory, Saltsjöbaden, Sweden.

in the Bartol spectrometer and the small angular scale in the focal plane of the small telescope (40″/mm for the 15-inch siderostat). Light loss due to slit limitations simply does not exist for the modest-sized telescope as it does for the large reflectors in their coudé-configurations, which are normally used in high resolution spectroscopic work.

The situation for large reflectors is well characterized by the following remarks of VAUGHAN (1967): "At a resolving power of 80000 with the 12-inch mosaic grating in the coudé spectrograph of the 200-inch telescope, for example, the slit must be only 0.″07 arc in width; with a typical 1-sec atmospheric blurring a few percent of the starlight delivered by the telescope are actually utilized. This is equivalent to a reduction in effective aperture from 200 inches to less than 50 inches!"

In other words, considerably smaller telescopes can effectively compete with the larger ones, if they are equipped with high resolution spectrographs that utilize the light from the entire stellar image, not just a few percent.

II. The Bartol High Resolution Spectrum Scanner

During the past three years the Astronomy and Astrophysics group at Bartol has developed and utilized an exceedingly sophisticated high resolution spectrograph which capitalizes on its ability to convert a smaller telescope effectively into a larger one for a wide range of astronomical studies. There has long been an obvious need for a fresh entry into the field of high resolution spectroscopy with high photometric precision. Observational spectroscopy has not kept pace with the recent computer spurred advances in the theoretical calculations of precise stellar line profiles. Thus, smaller telescopes with high resolution spectrographs of advanced design can now yield significant contributions to modern stellar and solar spectroscopy in a variety of research problems.

The URSIES (Ultravariable Resolution Single Interferometer Echelle Scanner) pressure scanner was developed at Bartol to exploit these advantages. Essential design features were the full utilization of the light in solar surface features (spots) or stellar images (angular diameters 2 to 4 seconds of arc) anywhere in the wavelength range 4000 Å to 12000 Å with spectral resolutions variable from 10^3 to 10^6 and dual channel pulse counting facilities. A vector magnetograph is currently being tested out, which will sample both longitudinal and transverse magnetic fields in sunspots and stars. Recently a two-channel photon counter of novel design has been purchased (Digital Synchronous Computer Model 1110 of SSR Instruments, Inc.) which holds great promise for precision photometry in conjunction with the magnetograph.

The scanner was completed in the fall of 1968 and since then has been in operation in conjunction with the 15-inch siderostat at the Flower and Cook Observatory of the University of Pennsylvania. The observations of sunspots and stars carried out thus far have even exceeded our expectations regarding the usefulness and wide range of applicability of this new instrument.

The use of pinhole slit geometry and dual-channel pulse counters have made possible significant advances in the field of sunspot spectrophotometry. For the

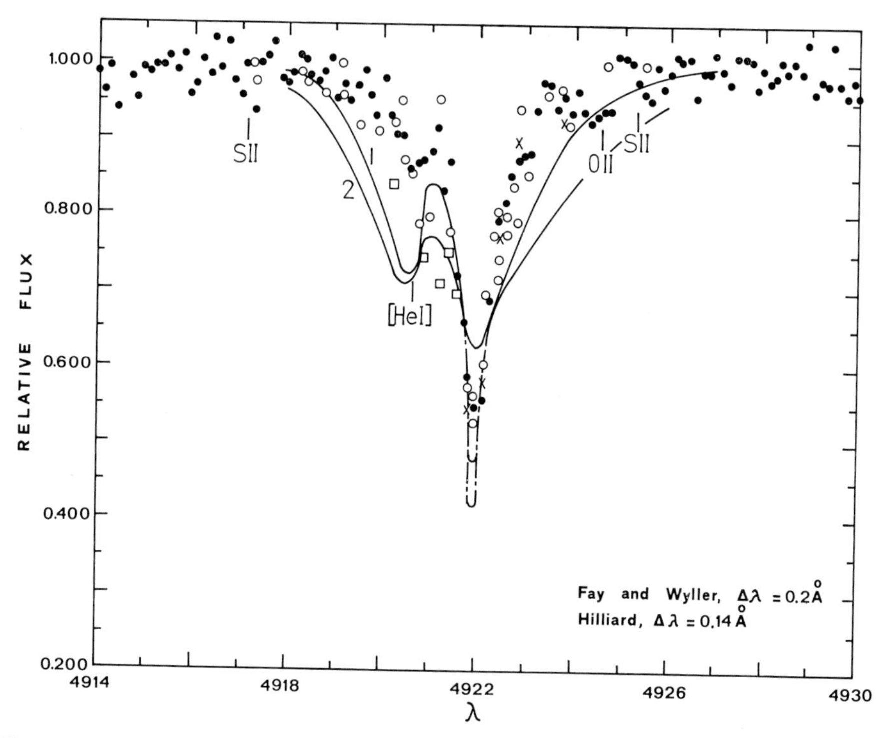

Fig. 1. Profile of the line 4922 Å in γ Peg observed by FAY and WYLLER with URSIES and by HILLIARD with the 36-inch reflector at Kitt Peak with a mechanical multichannel grating scanner

first time photoelectric line profiles of high precision in umbral spectra have been obtained in which the contaminations from scattered disk light have been thoroughly minimized. A vital factor in the success of this venture has been the short and synchronous integration times (0.8 seconds) of the broad-band and narrow-band pulse counters. This enables one, by the broad-band monitor, to single out brief moments of good seeing and record the intensity of the clean umbral line profile at a given wavelength.

The study of the umbral line profile of the Na D_2 line at 5890 (FAY *et al.*, 1972) led Dr. YUN of our group to develop a new umbral model atmosphere for medium-sized spots, which exhibits interesting departures from radiative equilibrium that are attributed to heating effects of the umbral upper atmosphere by the penumbra and photosphere (YUN, 1971). We look forward to new efforts in this field in the coming year with our new vector magnetograph, which has been recently installed.

In the realm of stellar observations significant advances have also been made, which are of special interest in view of the fact that for the first time photoelectric line profiles of high spectral resolution have been obtained with a telescope of such modest size as 15 inches. An important check on the instrumental precision has been made in the case of the star γ Pegasi. The star, which is somewhat brighter than third magnitude visually, was observed in the blue spectral region where the neutral helium line at 4922 Å is of special interest because the line core is a test on deviations from local thermodynamic equilibrium (MIHALAS, private communication).

The line profile was observed in the fall of 1969 with URSIES in the echelle mode (sampling bandpass 0.2 Å) and by Dr. HILLIARD of the University of Arizona, who used the 36-inch reflector at Kitt Peak National Observatory with a mechanical multichannel grating scanner. The concordance of our observed line profiles is very encouraging (see Fig. 1).

On the basis of our experience with stellar scans, we are convinced that with a 24-inch telescope all naked eye stars down to 5th magnitude are accessible to our scanner for high resolution (bandpass 0.1 Å) spectrophotometry of high photoelectric precision; a resolution formerly reserved exclusively for spectral work only with the large reflectors. Projecting the results already obtained with the 15-inch siderostat, it is expected that a 24-inch reflector with our scanner would provide a system that is effectively (and conservatively) the equivalent of a 72- to 100-inch reflector with a conventional high resolution coudé spectrograph. For ultrahigh resolution spectroscopy (sampling bandwidth less than 0.05 Å) the gains would be even greater because of further slit limitations of the large reflectors.

III. Development of Large Siderostat

In order to capitalize on the advantages of the full utilization of the stellar image with a 24-inch reflector, a stationary optical beam (i.e. a Coudé arrangement) is required to feed light to the relatively heavy spectrograph. Generally, a 24-inch telescope is placed on some sort of equatorial mount, and housed in a large domed structure, hence the conventional coudé spectrograph necessitates expensive ancillary facilities (in addition to the cost of the instrument itself). This probably is one reason why the Yale 40-inch, the Columbia 24-inch, and the Princeton 36-inch telescopes, which have conventional coudé foci, are not yet equipped with coudé spectrographs.

The horizontal siderostat arrangement presently in use at the Flower and Cook Observatory is a coudé arrangement in that the reflected ray always passes in a fixed direction independent of the location of the celestial object except for a limited inaccessible region around the celestial pole. This requirement necessitates a fairly complex primary mirror driving mechanism which steers a primary flat mirror in such a manner that a horizontal coudé requirement is satisfied. The image-forming optical unit is usually placed in a stationary horizontal position, and can be an objective lens such as at Flower and Cook Observatory, or it can be a reflecting telescope. The housing then becomes much simpler and less expensive than the conventional dome structure, and the image-forming telescope can be mounted in flexure free position in a thermostated room.

There is a resurgent interest in the construction of siderostats as ancillary feeder telescopes to large coudé spectrographs. In the planning of the European Southern Observatory, the coudé spectrograph of the 150-inch reflector will be fed by a 58–36 inch siderostat, when the large reflector is used with auxiliary instrumentation at the Cassegrain or Newtonian foci. Similar plans are also under way at

Fig. 2. 24-inch Clark reflector mounted on the third floor of Bartol

Kitt Peak National Observatory in conjunction with the Coudé spectrograph of the 84-inch telescope (Hoag, private communication).

For a variety of reasons, the need and desirability for an independent facility at Bartol have become evident, and after detailed consideration of alternative possibilities, the optimum design for the new observatory has been conceived. It embodies the 24-inch Clark Reflector (see Fig. 2) recently retired to make room for a new instrument at the Harvard Observatory, which has most generously been made available to Bartol by Harvard at no cost.

The plan is to mount this telescope in a siderostat arrangement on top of the Bartol building on the campus of Swarthmore College. Such an elevated location is advantageous for several reasons. It affords an inexpensive method for housing the telescope, since no foundation need be poured. It gives complete access to all parts of the sky with no encumbrances by other buildings or trees. Such elevated locations above the ground are currently very much in favor with astronomers, since a large amount of poor seeing is generated by air currents close to the ground. In any event, the seeing requirements for Bartol's program are less stringent than usual because of the special characteristics of the Bartol spectrograph. Preliminary seeing tests on sunspot structure and double stars with a 6-inch telescope and an 800 lb. base structure on the rooftop of our Bartol building indicate absence of severe building vibrations and generally acceptable to good quality seeing. The Bartol building was erected in 1928 and is of an exceptionally sturdy masonry structure to insure stability in the execution, at that time, of delicate electrometer measurements.

The complex steering mechanism for the primary flat in the horizontal siderostat is eliminated in the *polar* siderostat mount. In the latter case the primary flat sends the stationary coudé beam up or down along the polar axis and the mirror turns about this axis at the ordinary diurnal rate. This simplification and also the special advantage of a fixed instrumental polarization plane (important for our

Fig. 3. Bartol 36-inch siderostat

magnetograph) in this arrangement, has led us to adopt the polar siderostat solution for our installation.

As is shown in Fig. 3, a 36-inch flat mirror in a polar siderostat mount will feed the Harvard 24-inch reflector, mounted in a stationary position inside the Observatory with its optical axis along the polar axis. The spectrum scanner is then aligned by laser either with the Cassegrain focus or in the Newtonian position (scale 60″/mm as against 16″/mm in the Cassegrain). The Newtonian mount affords an entirely novel way of coupling a high resolution spectrum scanner with a large scale field. This arrangement will provide a new method for carrying out high resolution studies of extended faint light sources, such as emission nebulae.

On the basis of our accumulated experience in photoelectric line photometry of sunspot spectra the use of an on-line minicomputer as part of our data acquisition system has become highly desirable. A fundamental characteristic of our present observational technique is to observe, in brief moments of good seeing the umbra. This occurs in a rather uncontrollable manner because of seeing excurions, so that about 80% of our printed data have to be rejected. An on-line minicomputer could be programmed to inhibit the print out of spot observations whenever the continuum monitor by its high count rates would indicate that we climb out of the spot. The computer could also be programmed to intermittently reduce the data even to the extent of plotting instantaneously in real time the emerging line profile during the actual scan.

A computer system working along similar lines has been in operation for some years at the Swiss field station of the Solar Observatory of the Oxford University (BLACKWELL *et al.*, 1969). Their system incorporates a PDP-8 minicomputer, which is manufactured by the Digital Equipment Corporation in this country. Since then a more up-to-date version, PDP-11, is being marketed and we feel that this unit

Fig. 4. Roof structure: To the left-entrance aperture to 24-inch Clark reflector. To the right-siderostat pad with 6 clamping bolts

would satisfy our present needs, still leaving room for future extensions by its modular design. This computer could also be used for controlling the stepping motors of the siderostat drive in turn permitting raster scans of the sunspot structure.

While the computer acquisition has not yet materialized, the other developmental plans for the instrumentation of the Bartol Coudé Observatory have been realized. The two-telescope combination is now this spring (April 1972) installed on the roof and third floor of the Bartol building (see Fig. 4).The 36-inch siderostat with Cervit optics has been manufactured by the firm Group 128, Inc. (Waltham, Massachussetts), which previously built for the Smithsonian Astrophysical Observatory a lunar laser-ranging twin telescope.

The Observatory is now considered operational and we are well satisfied with image quality both of sunspots and stars. The seeing-affecting airmass interface between the warmer 24-inch room and the colder exterior (or vice versa), has been resolved by the simple expedient of installing an inexpensive plateglass window, which stops aircurrents through the hole in the roof. The whole Observatory was completed at a cost of $ 65000 in little more than a year's time from the first go-ahead to the final installation of the siderostat. We are extremely grateful to the Harvard Observatory Council for making available to us at no cost their retired 24-inch Clark reflector.

Presently the URSIES spectrum scanner and magnetometer is being refurbished and coupled to the stationary focus of the 24-inch Clark reflector and we expect solar and stellar high resolution spectral scans to start in earnest this fall.

The development of URSIES has been carried out by the generous support of NASA Grant NGR 39-005-066.

Dr. THEODORE FAY participated most valuably in few development while Mssrs. PFEIFFER and SMITH of the Bartol Technical staff carried the major responsibilities for implementary the Coudé Observatory.

References

BLACKWELL, D. E., MALLIA, E. A., PETFORD, A. D.: Monthly Notices Roy. Astron. Soc. **146**, 93 (1969).
FAY, T. D., WYLLER, A. A.: Applied Optics **11**, 1152 (1972).
FAY, T. D., WYLLER, A. A., YUN, H. S.: Solar Phys. **23**, 58 (1972).
YUN, H. S.: Solar Phys. **16**, 379 (1971).
VAUGHAN, A. H. JR.: Ann. Rev. Astron. Astrophys. **5**, 139 (1967).

Discussion

TURLO, Z.:
From my own experience with spectrophotometer similar to one you have just described it is quite evident that there are some specific limitations inherent to the scanning technique by varying gas pressure. First of all it is by no means easy to achieve good reproducibility and linearity of the wavelength scale, unless rather sophisticated system is used for precision controll of the gas pressure. Secondly spectrophotometers of this type are quite slow in that sense that in order to avoid errors caused by turbulence and temperature gradients inside of the Fabry-Perot etalon one can not sample spectrum faster than say two or three samples per minute. In order to make scans through reasonable position of the spectrum one, therefore, would need time interval of the order of hours, which means that overall stability of the instrument as well as stability of the meteorological conditions should be quite remarkable.

UTIGER, H. E.:
Have you considered making slow continuous scans with your gas and measuring the index of refraction with a Michelson interferometer?

WYLLER, A. A.:
We are well aware of this technique pioneered by the PEPSIOS group at Wisconsin. However, for purposes of controlled photometric precision and grating alignment with the interferometer we have preferred to remain at a given pressure level and pulse count at that particular wavelength.

The Problem of Three Bodies

(Invited Lecture)

By V. Szebehely
The University of Texas, Austin, Texas, U.S.A.

Abstract

Analytical and numerical results obtained during the past five years and their astronomical applications are reviewed in the area known as the general problem of three bodies. In this problem the order of magnitude of the masses of the three participating bodies are the same and their distances are arbitrary. Planetary and lunar problems, as well as the several important versions of the restricted problem of three bodies are excluded and the stellar dynamics applications are emphasized.

It is shown that the general problem of three bodies is unstable for most conditions and only special configurations are preserved for astronomically significant periods of time. A dynamical theory supported by numerical experiments is described which offers a possible explanation of binary formation.

I. Introduction

A review of the progress in the stellar three-body problem during the past five years contains some major results of importance in the fields of dynamics, astronomy and mathematics. Some of these results were obtained by analytical and some by numerical tools. Indeed, advanced computer technology assisted the theoretical developments, just as experiments in the physical sciences and observations in astronomy have promoted the discovery of new theories.

The main results of the past five years that represent new knowledge in the solution of the general problem of three bodies are all related to the longtime behavior of this dynamical system. The following results may be listed:

(1) Randomly selected initial conditions and masses lead to an eventual disruption of the system.

(2) Periodic orbits are not dense and families are only formed when mass-variations are combined with changing initial conditions.

(3) Escapes, even with negative total energy, occur with high probability, and are the result of genuine three-body effects and are preceeded by triple close approaches.

(4) Binary and triple collisions occur with measure zero.

(5) The only configurations showing long-time stability are rotational systems without close approaches.

The astronomical consequences of these results are as follows:

(1) Triple stellar systems are inherently unstable except in the case of rotational configurations.

(2) The mechanism of escape is not based on two-body considerations but it requires a triple close approach.

(3) The binary and triple close approaches in general do not introduce tidal effects and, therefore, do not negate the validity of the three-body dynamical model.

(4) Dynamically formed binaries from a triple stellar system consist usually of the two most massive stars of the system; the smallest mass is ejected.

II. Modes of Development of Triple Stellar Systems

In the following sections systems with negative total energy will be discussed only. The analogy between the problem of two bodies and the problem of three bodies is preserved for positive total energy and hyperbolic disruptions of both systems is expected. The gravitational motion of two bodies is bounded for negative total energy but this is not the case for a three-body system with negative total energy. This may be seen considering a simple energy-balance. Let E_t be the total energy of the three-body system and let us assume that at one time during the motion one of the bodies is ejected far enough so that it has only small (perturbing) effect on the remaining two. Let E_e be the energy of the system formed by the ejected star and by the motion of the center of mass of the other stars. Finally let the energy of the binary formed by these two other stars be denoted by E_b. Then we have $E_t = E_b + E_e$. If now the total energy is positive ($E_t \geqq 0$), then E_e must be positive since the energy contained in the binary is always negative. Consequently, if a binary is formed in a system with positive total energy then the third body must escape ($E_e > 0$). This may also be seen from considering the moment of inertia of the system, I, which for $E_t > 0$ goes to infinity as time increases, $I \to \infty$ and since the binary's moment of inertia is bounded, the third body must escape. The case of negative total energy is more complicated. In the energy-balance now we have $E_t < 0$ and $E_b < 0$. Consequently, E_e may be positive (escape) or negative (ejection). For a given total energy, the value of E_b will determine if escape or ejection occurs. When a very close binary is formed E_b is a large negative number since $E_b = -m_1 m_2/(2a)$, where m_1 and m_2 are the masses of the components of the binary and a is its semi-major axis. Consequently, escape requires the formation of a close binary, in which case $E_t - E_b > 0$ or $|E_b| > |E_t|$.

The important experimental fact (obtained by a large number of computer-performed numerical integrations) is that this binary formation does occur with high probability and either an ejection or an escape results sooner or later even for negative total energy.

The occurrence of this hyperbolic-elliptic type motion may be expected sooner or later unless a contraction of the system is prevented by either the existence of high positive total energies (in which case the system disrupts explosively) or by a rotational motion, which is treated next.

By rotational motion or "revolution" we mean a binary surrounded by a halo-type orbit of the third body. Extensive numerical evidence seems to suggest that if the ratio of the semi-major axis of the outer body a, to that of the binary, a_b is larger than approximately 2, then the motion is stable without eccentricities, inclinations and external perturbations. High eccentricity of the orbit of the outer body will introduce perturbations on the binary but once again if the periastron remains larger than twice the semi-major axis of the binary, stability may be expected. The larger the ratio a/a_b is the more stable the triple system is but at the same time it becomes more vulnerable to external perturbations, (for instance the capture of the outer body by another binary may occur).

The question of the existence of periodic orbits comes up naturally at this point since rotational motion is connected with such orbits. Without restrictions on the masses but assuming large values for a/a_b ARENSTORF has shown recently the existence of such periodic orbits. It should be noted that this restriction on the ratio of the distances excludes these orbits from our consideration in this paper, especially since the above-mentioned existence proof does not allow as low ratios as $a/a_b=2$, but in fact it assumes $a/a_b \to \infty$. The search for periodic orbits with arbitrary mutual distances and masses of equal orders of magnitude leads to families of periodic orbits completely different from the families which exist in other dynamical systems of interest in astronomy. The Lagrangean triangular solutions will not be discussed here since they are well described in the existing literature. The fact of considerable importance is that such solutions are unstable for masses of equal order of magnitude. Consequently, the possible existence of other periodic orbits is of interest. Such periodic orbits have been found by STANDISH and SZEBEHELY with the following two common properties: all orbits display collisions or close approaches and no families may be formed by the conventional methods, applicable for instance in the restricted problem. These two results indicate that families of stable periodic orbits in the general problem of three bodies do not exist or in other words stable periodic orbits are not densely distributed in the general problem. The eventual disruption of the system is a consequence of the above statement, since densely distributed periodic orbits would capture (so-to-speak) the motion (see BIRKHOFF).

The various possibilities of the long-time history of triple systems may be summarized as follows. If the angular momentum of the system is small and its total energy is negative the initial motion is called "interplay" or by CHAZY "bounded". Such motion is bounded in the phase space and usually the heaviest mass occupies the central position making alternatingly close binary approaches with the two lighter masses. Eventually one of the lighter masses will be ejected on an elliptic orbit with negative energy while the two remaining bodies form temporarily a loose binary. This type of motion is classified as an "ejection" or by CHAZY as "elliptic-elliptic" type. As the ejected body rejoins the binary it may be once again ejected, it may "escape" or it may participate in further "interplays." Eventually the ejected body will return with high velocity at the proper time and it will be ejected with positive energy on a hyperbolic orbit while the two remaining bodies form a permanent close binary. This type of motion is classified as "escape." The motion is asymptotic and it is termed by CHAZY "elliptic-hyperbolic."

If the total energy of the system is positive the motion is either "hyperbolic-elliptic" as mentioned before or "hyperbolic," that is there is either an "escape" with binary formation or a complete "explosion" when all distances increase linearly with time.

A special "bounded" motion is called "revolution" when an inner binary is enclosed by the orbit of the outside body. This motion can be generated from interplay when sufficient amount of angular momentum is present.

Parabolic motions separate hyperbolic and elliptic types occur but with measure zero.

III. Effect of Masses and Physical Parameters on the Development of Triple Systems

As mentioned before, the body with the smallest mass is usually the escaper. Experimental results indicate that this statement is correct only in a statistical sense since the largest mass may also be ejected or in fact it may escape under special circumstances. Out of 150 experiments with the mass ratios varying from 1:1:1 to 5:5:5 with all possible integer combinations allowed, 87% of the escaping bodies had the smallest mass, 11% the medium mass and 2% the heaviest mass.

By systematically changing the initial velocities of the participating bodies it was found that large angular momentum tends to delay disruption and in fact may result in "revolution." Consequently, increased initial velocity which results in higher total energy will not necessarily increase the probability of escape.

Small changes of the initial conditions offer continuity regarding the outcome of the motion only between the regular beginnings of the motion and the first escape. If an ejection occurs at, say $t=t_1$ and an escape at $t=t_2>t_1$, then a small change in the initial conditions may turn the ejection of t_1 into an escape and the "time to escape" changes discontinuously from t_2 to t_1. Consequently, the above statement regarding the continuity of the outcome with changing initial conditions must be completely and carefully formulated.

As mentioned before, triple close approaches must occur before escape can take place. It is known that if the angular momentum of the system is not zero, total collapse (triple collision) cannot occur. In fact the minimum value of the moment inertia, I, obtainable in a system is governed, among other factors, by the angular momentum. Consequently, large values of the angular momentum will delay or completely prevent the escape. Estimates of the minimum value of I are available and indicate not only the escape characteristics of a system but also the reliability of computational results. Since triple close approaches (or in fact triple collisions) cannot be regularized, computational inaccuracies increase with small values of Min I. We note that binary collisions are regularized today in any sophisiticated computer program.

IV. Remarks on the Modern Literature of the General Problem of Three Bodies

The first numerical experiments verifying escape were performed and published about the same time by AGEKYAN and ANOSOVA (1967) and SZEBEHELY and PETERS

(1967a, b). Randomly selected initial conditions and hundreds of test cases were studied later by AGEKYAN and ANOSOVA (1968), and STANDISH (1971). The stability of the rotational motion has been studied recently by HARRINGTON (1972). Systematic variations of the participating masses and of the initial conditions were studied by SZEBEHELY (1972). The role of the angular momentum may be investigated and estimates for Min I, and consequently for the "closeness" of a triple close approach may be established by using SUNDMAN'S (1912), BIRKHOFF'S (1927) and SIEGEL'S and MOSER'S (1971) work.

Periodic orbits with collision have been found by SZEBEHELY and PETERS (1967a, b) and by STANDISH (1970).

The original classification according to elliptic or hyperbolic type motions was proposed by CHAZY (1929). The terminology of escape, ejection, revolution, etc. was introduced by SZEBEHELY (1971).

Acknowledgement

Partial support of the research covered by this paper was received from the Mathematics Branch of the Office of Naval Research. The presentation of this paper was partially supported by the Astronomy Division of the National Science Foundation and by a research grant from The University of Texas Research Institute, Austin, Texas.

References

AGEKYAN, T. A., ANOSOVA, ZH. P.: Astron. Z. **44**, 1261 (1967).
AGEKYAN, T. A., ANOSOVA, ZH. P.: Akad. Arm. SSR, Astrophys. **4**, 31 (1968).
BIRKHOFF, G. D.: Dynamical System. Providence, R. I.: Am. Math. Soc. Publ. 1927.
CHAZY, J.: J. Math. Pure and Appl. (9), **8**, 353 (1929).
HARRINGTON, R. S.: Celestial Mechanics **6**, 322 (1972).
SIEGEL, C. L., MOSER, J. K.: Lectures in Celestial Mechanics. Berlin-Heidelberg-New York: Springer 1971.
STANDISH, E. M. JR.: In G.E.O. GIACAGLIA (ed.), Periodic Orbits, Stability, and Resonances, p. 375. Dordrecht-Holland: D. Reidel 1970.
STANDISH, E. M. JR.: Celestial Mechanics **4**, 44 (1971).
SUNDMAN, K. F.: Acta Math. **36**, 105 (1912).
SZEBEHELY, V.: Celestial Mechanics **4**, 116 (1971).
SZEBEHELY, V.: Celestial Mechanics **6**, 84 (1972).
SZEBEHELY, V., PETERS, C.F.: Astron. J. **72**, 876 (1967a).
SZEBEHELY, V., PETERS, C.F.: Astron. J. **72**, 1187 (1967b).

Discussion

MARKELHS, V. V.:
What was your method of integration?

SZEBEHELY, V.:
Runge-Kutta-Fehlberg in most cases but in some cases we used a recurrent power series method.

MARKELHS, V. V.:
In those cases what was the order of expansion?

SZEBEHELY, V.:
Usually about 15.

HADJIDEMETRIOU, J.:
Have you made the computations using the same initial conditions but going backwards, to follow the past history of the triple system? Is it always the star which comes from infinity, the same star which escapes?

SZEBEHELY, V.:
If all initial velocities are zero, then the motions are identical, going forwards or back in time. If the initial velocities are not zero then the backwards integration is identical to a new set of initial conditions preserving the total energy and the angular momentum magnitude. Therefore, it would be expected *statistically* that the two integrations would result in the escape of the smallest mass.

BAKOS, G. A.:
1. Among visual binaries there are triple systems with a close system in the center and one star much further away. However, there are also types where the central star is single and the distant one a double massive system. How can you explain this?

2. Would you by extrapolation say that quadruple systems are also unstable?

SZEBEHELY, V.:
1. It is possible that such systems are not stable for long time periods. These systems are presently being investigated by numerical integration.

2. Yes, but this answer is an extrapolation based on three-body results only.

Periodic Orbits of the Restricted Problem for Various Values of the Mass-Ratio

By G. Contopoulos and M. Zikides
University of Thessaloniki and University of Athens, Greece

With 5 Figures

Abstract

The evolution of the families k, j, t, z, v is given by varying the mass-ratio.

A few years ago a project started at the University of Thessaloniki, aiming at finding the evolution of the main families of periodic orbits in the restricted three-body problem with mass-ratio μ. In the case of orbits starting perpendicularly to the x-axis a family is represented by its characteristic in the $x-C$ plane for every value of μ; thus, in the three-dimensional space (x, μ, C) (Fig. 1) a family is represented by a surface.

A first step in this direction is Bozis' study (1970) of collision periodic orbits for the entire range (0,1) in μ. His $\mu-C$ diagram gives the intersections of the surfaces, representing the various families, by the plane $x=\mu$ (Fig. 1). A similar study could have been made on the plane $x=\mu-1$, which represents collisions with the second body. However, in the restricted problem we can replace μ, x, y by $1-\mu$, $-x$,

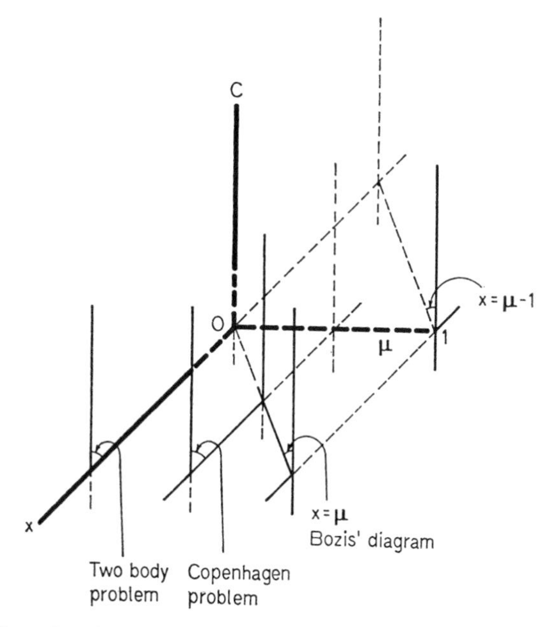

Fig. 1. The three-dimensional space (x, μ, C), in which the characteristics of the various families of periodic orbits are represented as surfaces

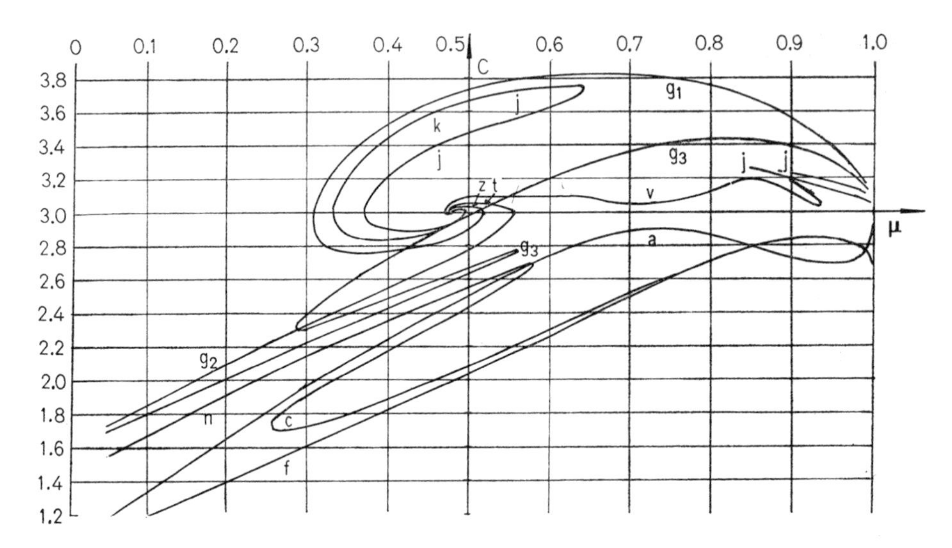

Fig. 2. BOZIS' diagram, giving the values (μ, C) of collision periodic orbits. In the space (x, μ, C) BOZIS' plane is represented as $x=\mu$

$-y$ and recover again periodic orbits. The correspondence of the families is as follows (HÉNON, 1965):

μ	a	c	f	g	k	j	l	m	n	o	r	t	v	w	z
$1-\mu$	b	c	h	i	k	j	l	m	n	o	r	s	u	x	y

Thus, if we study the intersections of all families by the plane $x=\mu$ we have, by symmetry, its intersections by the plane $x=\mu-1$.

BOZIS' diagram (Fig. 2) gives some new connections between a number of families of periodic orbits. Thus:

(1) Family c is, on one hand, a continuation of family n, and, on the other hand, it branches off from a collision periodic orbit of family f ($\mu\simeq 0.72$). Thus the characteristics of families c and f for this value of μ coincide. Therefore, the families n, c, f (and h because of the symmetry) form one surface reaching $\mu=0$ (two-body problem) and $\mu=1$.

(2) Family z is a continuation of family g_1, and family t a continuation of family g_2. Thus, the families g, t, z form a complex surface reaching $\mu=0$ and $\mu=1$. Because of symmetry the families i, s, y form a similar surface.

(3) Families k and j form one complex surface. We will see below that this does not reach $\mu=0$, or $\mu=1$.

Thus, the number of independent families is reduced by at least 8.

Recently a systematic study has been made of the families k, j, t, z, v. All these families have one common characteristic, namely their $\mu-C$ curves spiral around the point S^* ($C=3$, $\mu=0.4755$). This point corresponds to a collision-asymptotic-periodic orbit, which is a limiting case of both asymptotic periodic orbits I and II.

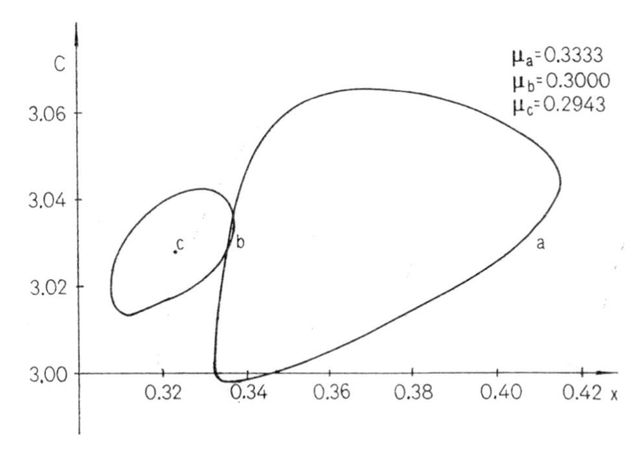

Fig. 3. Characteristic curves for various values of μ of the family k. For $\mu=\mu_c$ the whole characteristic is reduced to a point

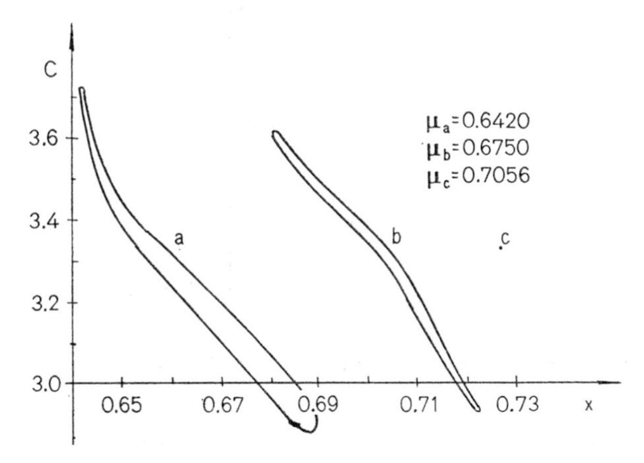

Fig. 4. Characteristic curves for various values of μ of the family j. For $\mu=\mu_c$ the whole characteristic is reduced to a point

The $\mu-C$ curves of the k and j families have minima at $\mu=0.3333$ and $\mu=0.3753$ respectively and a maximum at $\mu=0.642$. The question is what happens to these families beyond these limits.

We found that for $\mu<\mu_{\min}$ the characteristic curves are closed (Fig. 3), becoming smaller in size, as μ becomes smaller. For $\mu=0.2943$ the characteristic curve of family k shrinks to one point. Similarly for $\mu>\mu_{\max}$ we have closed characteristic curves (Fig. 4) which become smaller as μ increases and reach a point for $\mu=0.7056$.

Thus, the surface (k, j) extends from $\mu=0.2943$ to $\mu=0.7056$. This proves that these families cannot be continued to the two-body problem.

Similarly the $\mu-C$ curves of the families z, t, v have minima at $\mu=0.47614$, $\mu=0.4675$, and $\mu=0.4595$ respectively. The characteristic surfaces terminate at values of μ a little smaller than the above.

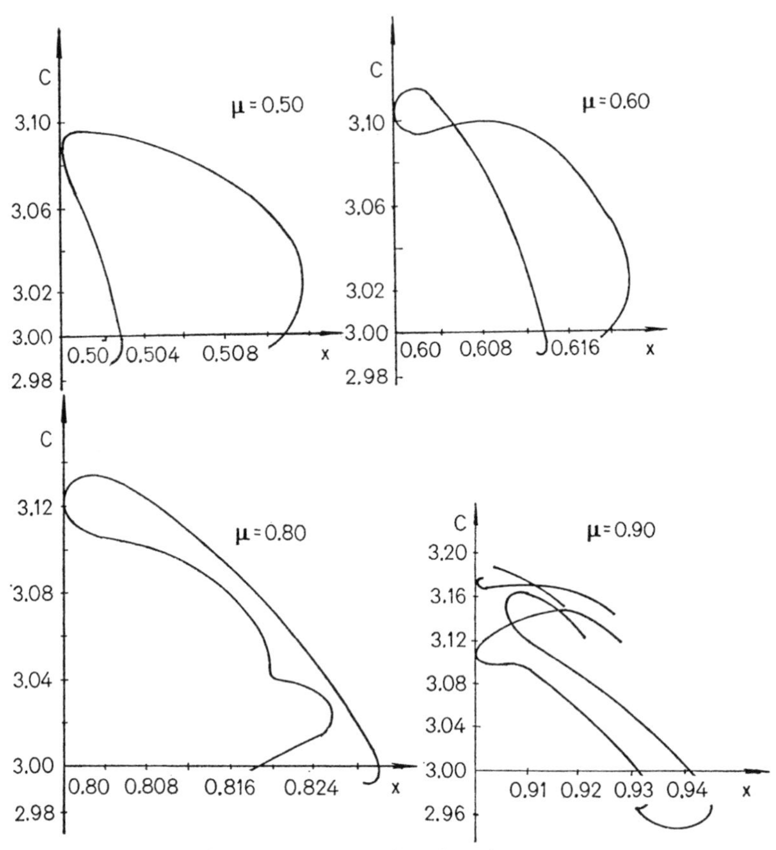

Fig. 5. Characteristic curves for various values of μ of the family v

The $\mu - C$ curves of the families z and t have maxima at $C=3$ and $\mu=0.518$ and $\mu=0.548$, where they join the $\mu - C$ curves of families g_1 and g_2, respectively.

Family v has a more complicated bahaviour. Its $\mu - C$ curve extends up to $\mu=0.935$ and then starts making some loops. We found that the characteristic curves of the family v for μ up to 0.80 end in loops around I' and V (Fig. 5). However, for μ approaching 0.90 the characteristic curves become quite complicated (Fig. 5).

The termination of the v-type $\mu - C$ curve is still uncertain.

BOZIS marks also some j-type branches of $\mu - C$ curves in the region between $\mu=0.85$ and $\mu=0.999$. However, our present study established that these families are not connected with the $k-j$ families studied above, therefore, their indication as j families is misleading. Further work is being made to find the termination of these families.

References

BOZIS, G.: In G. E. O. GIACAGLIA (ed.), Periodic Orbits, Stability and Resonances, p. 176. Dordrecht-Holland: D. Reidel 1970.

HÉNON, M.: Ann. Astrophys. **28**, 499, 992 (1965).

On the Possibility of a Resonance Capture of the Asteroid Toro by the Earth

By Lars Danielsson and Ramesh Mehra
Royal Institute of Technology, Stockholm, Sweden

With 8 Figures

Abstract

Among the very few tabulated asteroids with perihelion inside the Earth's orbit (Apollo asteroids) some have periods nearly commensurable with the Earth's. The perturbed orbit of one of these, 1685 Toro, has been integrated over 1200 years. The results show that Toro is at present captured in resonance with the Earth and that also Venus has a drastic influence on the orbital elements.

I. Introduction

The present orbital elements of the asteroid 1685 Toro shows that its period is almost exactly 1.6 years. Further the relative positions of Toro and the Earth exhibit a certain symmetry including close approaches. This quasi-resonance – if we may call it so – has been investigated more in detail by integrating the perturbed Toro-orbit from 1580 AD to 2800 AD, i.e. totally 1200 years. The perturbations by Venus, Earth, Mars, Jupiter and Saturn are included in the orbit integration according to Cowell's method. The accuracy of the computation is extremely high; after 100 years which requires 10^4 integration steps the relative error is about 10^{-8}; i.e. the absolute error in the position of the Earth is 10^{-5} AU. A minor part of these results, leading to some false conclusions, were published earlier (Danielsson and Ip, 1972).

The standard method of studying variations in orbit elements, etc. is based on secular variations due to the slowly changing forces from the major planets. However, since we know that Toro comes at least as close as 0.06 AU to the Earth this method is considered to be rather unreliable. The time-span – 1200 years – covered by the present accurate integration is admittedly too short to determine the relative importance of the secular variations and the resonance features caused by the inner planets.

II. Present Orbits

The orbits of Toro and the terrestrial planets are at present oriented as shown in Fig. 1 (projection to the ecliptic plane). The possible positions of the Earth and Venus when Toro is at perihelion are marked. These positions are quasi-stable. The orbit of Toro relative to the Earth, i.e. in a coordinate system rotating with

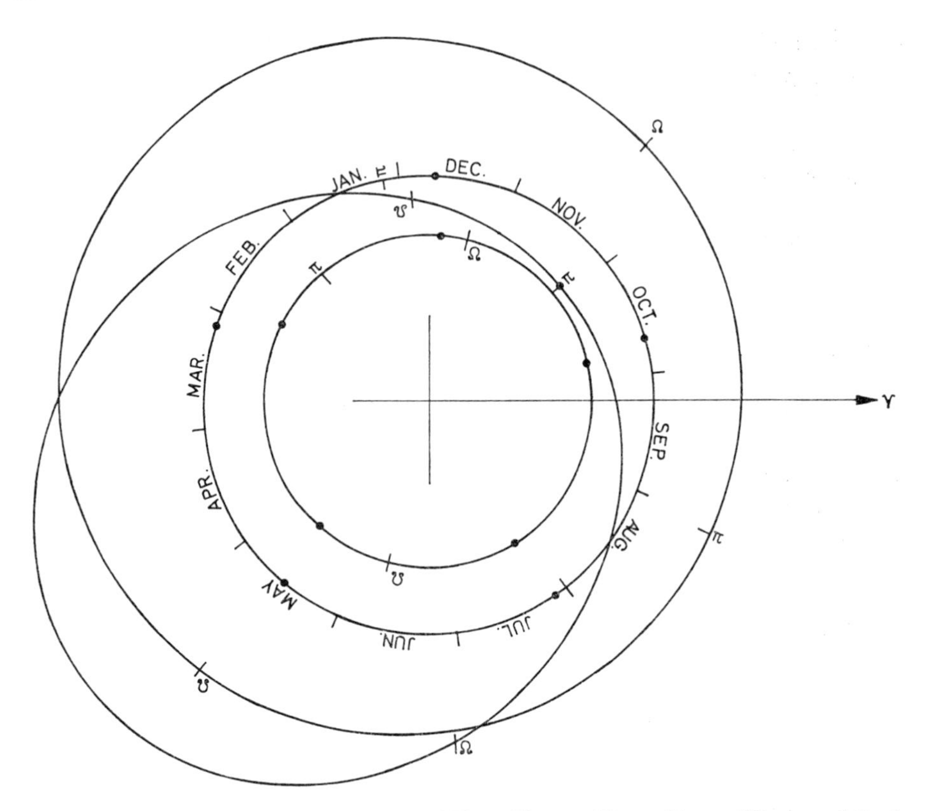

Fig. 1. The orbits of the terrestrial planets and Toro. The possible positions of Venus and Earth when Toro passes its perihelion are marked

the Earth, is shown in Fig. 2. It takes Toro eight years to complete this loop. The loop is not exactly closed instead it can be said to oscillate around its center (the Sun). In the figure this is best illustrated by keeping the 8-year Toro loop fixed and letting the Earth oscillate along the arc (projected to the orbital plane of Toro) marked in the figure. At the turning points the distance of Toro from the Earth varies between 0.07 and 0.15 AU. At these close encounters the Earth is in the longitudes corresponding to the months of August and January. At the August (every 8th year) close encounter, in the vicinity of its ascending node, Toro is typically 2° ahead of the Earth and 6° above the ecliptic plane in the present mode of libration. Thus Toro is retarded, i.e. the angular momentum and the period are decreased, which tends to move it away from the Earth and the encounter will be more and more distant. Instead Toro will approach closer to the Earth in the vicinity of its descending node (January every 8th year). At the January encounter Toro is typically 3° behind the Earth and 2° below the ecliptic plane. Toro is accelerated, its period increased and again it tends to move away from the Earth at this point to approach it in the vicinity of the ascending node (August). This process can be seen in the variation of the period, Fig. 3. (Due to the high latitude

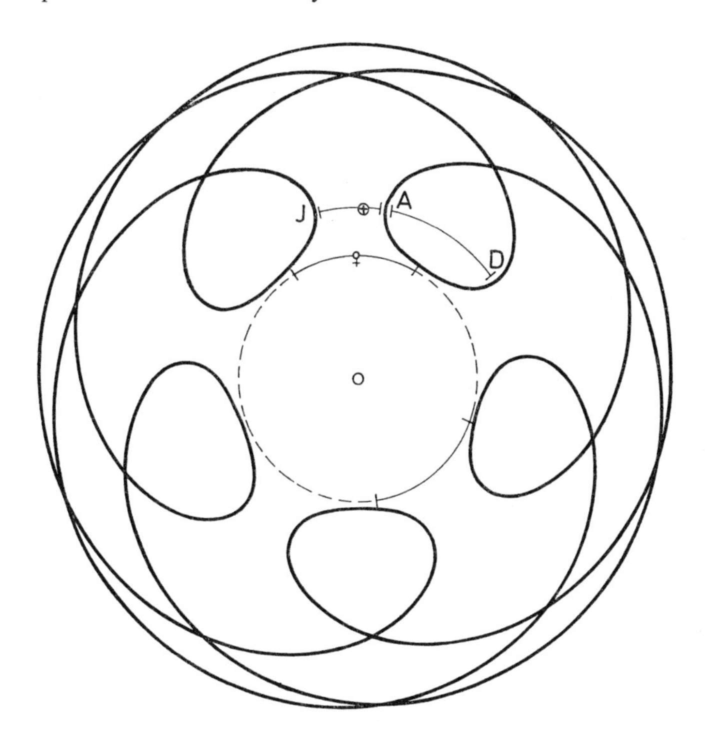

Fig. 2. The orbit of Toro in a coordinate system rotating with the Earth. The fact that this loop oscillates is accounted for by letting the Earth oscillate along the arc J-A-D. At the same time Venus moves along the inner circle, having a temporarily stable phase relative to Toro on the solid parts of the line

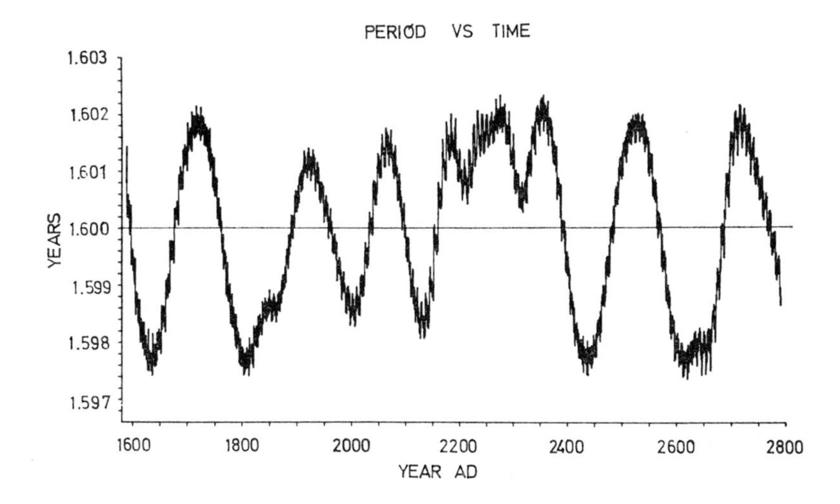

Fig. 3. Variation of the period with time

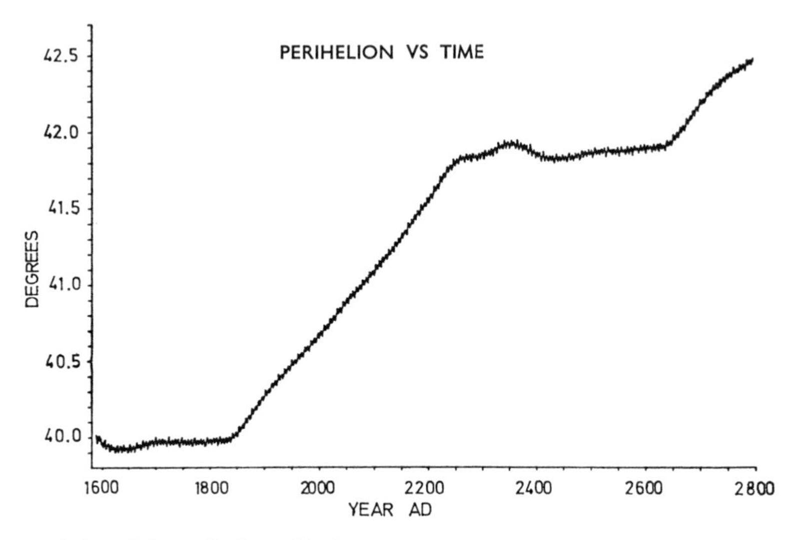

Fig. 4. Variation of the perihelion with time

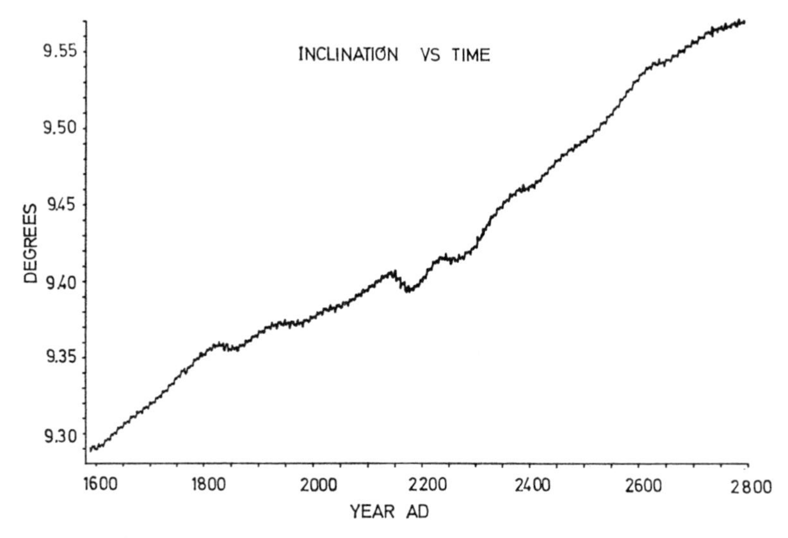

Fig. 5. Variation of the inclination with time

at the August encounter Toro can slip to the wrong side of the Earth there, see below.)

In the encounters with the Earth Toro approaches closest in the acceleration phase (January, Fig. 1). This fact causes the libration curves to be slightly unsymmetric (see e.g. Fig. 3), the parts with increasing period are often of shorter duration than those with decreasing period. Further, the period of the libration seems to vary with its amplitude and be between 140 and 200 years. The effect of the close encounters can be seen in all the orbital elements except the node (see Fig. 4–7).

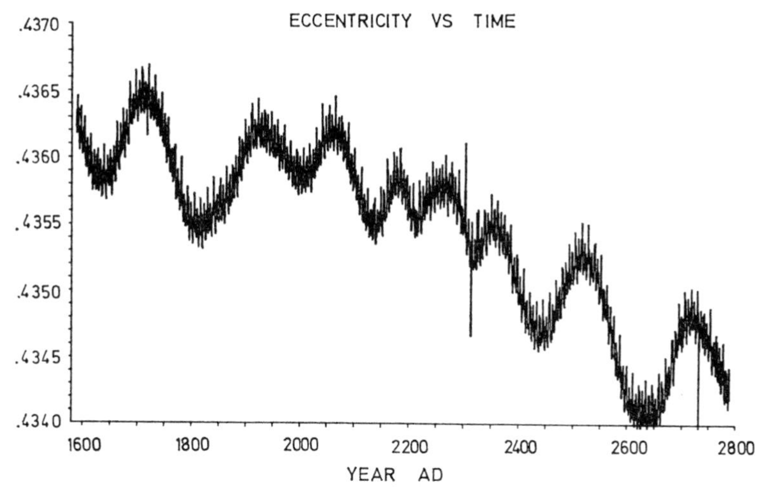

Fig. 6. Variation of the eccentricity with time

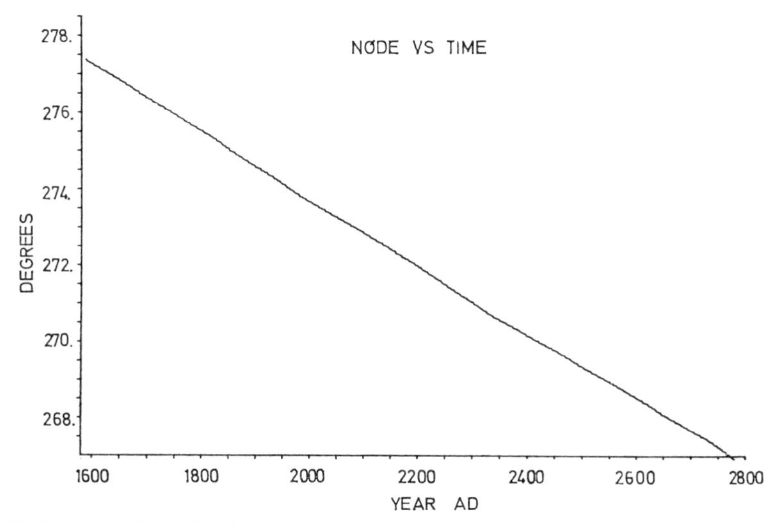

Fig. 7. Variation of the node with time

III. Libration

The libration of the Toro orbit is in fact rather complicated. As indicated in Fig. 2 there are two semi-stable regions where an oscillation can take place, i.e. *J–A* and *A–D*. The phase variation in Fig. 8 is based on the longitudinal distance from Toro to the Earth and Venus when Toro is at perihelion. The data are reduced so that the positions of the planets always refer to the points which are at $\lambda \simeq 13°$ (October) in Fig. 1, i.e. $n \times 72°$ has been added $n=0, 1, 2, 3$, or 4. For zero inclination the equilibrium phase angle would be expected to be 36° and the resonance is expected to disappear if the amplitude (relative to the equilibrium) of the libration exceeds 36° (i.e. if the phase angle passes 0° or

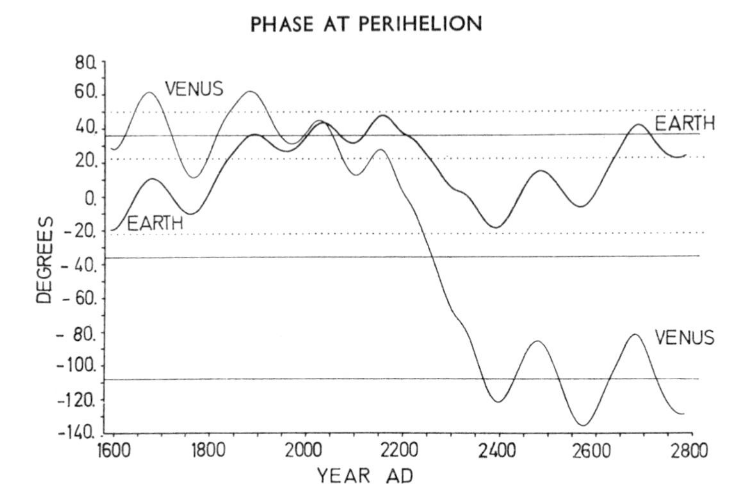

Fig. 8. The difference in longitude between Toro and the Earth and Venus at the October perihelion every 8th year

72°) and to be drastically changed already when it exceeds 13°. In reality, due to the finite inclination, Toro can rather easily penetrate to the other side of the Earth at the August encounter. In Fig. 8 the corresponding phase variation for Venus is also shown. It exhibits a certain resonance-like oscillation superimposed on an average phase drift of 18° per century. The average phase drift is explained by the fact that Venus period is 0.11 days too short for an exact 5:13 commensurability with Toro. However, it is remarkable that Toro does seem to "lock" temporarily on the period of Venus, during which time the Earth's phase has an average drift in the opposite direction, so that the difference between the Earth's and Venus' phase in Fig. 8 is a straight line, inclined 18°/century (that is how close the Earth and Venus are to a 8:13 commensurability).

It can be seen in Fig. 8 that the phase of the Earth is constant on the average and that there is a time-span of 800 years between two points with identical phase-relation. Toro's average period of revolution computed over this 800 years period of libration differs from 1.6 years by less than 10^{-5} year. (An error of two days; the average integration step is three days.)

Toro's longitude of perihelion exhibits some remarkable features too. It stays constant for part of the investigated time and has otherwise a constant drift relative to the vernal equinox, see Fig. 4. The changes in the slope of this curve occur simultaneously with clearly visible irregularities in period and inclination, cf. Fig. 3 and 5. The changes from the horizontal parts to the inclined are again 800 years apart and thus occur at the same phase relative to the Earth and Venus, cf. Fig. 8.

IV. Discussion

In summary we can state that the 1200 year accurate orbit integration has given good evidence that the asteroid 1685 Toro is in a 5:8 resonance with the Earth.

The average period of Toro is 1.6 years with a maximum error of 10^{-5} years. The period oscillates around this average with a period of 140 to 200 years. Also the eccentricity and the phase relative to the Earth and Venus show the same oscillation. Due to periodically repeated interactions with Venus there is also a superimposed oscillation with a period of 800 years, which is best seen in phase variation, see Fig. 8. Toro is also close to a 5:13 resonance with Venus and for a part of the 800-year period it is "locked" on Venus instead of the Earth. However, in average the phase relative to the Earth is constant, while Venus' phase relative to both the Earth and Toro drifts with a speed of 18°/century. The "peak-to-peak" amplitude of phase libration relative to the Earth is 65°.

The implication of the present results is that a completely new type of orbital resonance in the planetary system has been discovered. While in previously known resonances close encounters are avoided – the best known example is Neptune-Pluto – the resonance between the asteroid 1685 Toro and the Earth and Venus entirely relies on close encounters. This is of course a natural consequence of the very small sphere of influence of any terrestrial planet so that no resonances have been anticipated in this region.

There is no reason to believe that Toro is a unique case; also a few other Apollo asteroids have periods very nearly commensurable with the Earth's, others show a very regular variation of the period, indicating an orbital resonance of some kind.

The stability of the Toro resonance over longer periods of time cannot be judged with certainty from the present material. Although the secular variation of the orbital elements can change the situation – the stability being particularly sensitive to the inclination – it is quite possible that the Jupiter interaction will be overruled by the strong periodic interactions with the Earth and Venus.

One of the objectives for studying orbits like Toro's in detail is that this is one of the possibilities for a pre-capture orbit for the Moon (Alfvén and Arrhenius, 1969). Another is that the existence of orbits like Toro's may be important for determining the lifetimes of Apollo asteroids.

Acknowledgement

The authors are indebted to Mr. N. Carlborg of Stockholm Observatory at Saltsjöbaden for kindly making his Cowell program available for the orbit integration and to Professor Hannes Alfvén for many stimulating discussions.

References

Alfvén, H., Arrhenius, G.: Science **165**, 11 (1969).
Danielsson, L., Ip, W.-H.: Science **176**, 906 (1972).

Commensurable Mean Motions as a Tool in Solar System Dynamical Studies

By A. E. Roy
Department of Astronomy, University of Glasgow, United Kingdom

Abstract

The orbits of the planets about the Sun and of the satellites about their primaries are in general well-behaved so that extended numerical integrations are required before significant changes can be expected to be observed. For example the possibility of secular changes in the semimajor axes of the planetary orbits has been suggested but such changes are so slow that integrations of order 10^7–10^8 years may be required. The mechanism producing such changes is such that the unaveraged equations of motion must be integrated numerically with vigour if the mechanism's effect is to be detected. Such integrations are unpractical at present. Apart from the large amount of machine time required round-off error accumulation would remove all accuracy if use were made of the usual methods such as Cowell's, Encke's or any of the available perturbational methods.

Roy and Ovenden (1954, 1955) pointed out that the Solar system is richly endowed with commensurabilities in pairs of mean motions. It is suggested by the present author that use may be made of these commensurabilities to provide a rigorous Encke-type method of special perturbations to enable with moderate expenditure of machine time and a minimum of accumulation of round-off error greatly extended numerical integrations to be carried out. The method makes use of the fact that in closely commensurable systems, the dynamical geometry (relative positions and velocities) is repeated to a high degree of approximation every commensurability period. The behaviour of the bodies in the system is, therefore, highly similar in the new period to their behaviour in the previous period. The positions throughout the previous period are in the computer and used as reference data in an Encke-type precedure for calculating the positions in the new period. There is no reason in principle in fact for believing that unlike the standard "first difference" Encke method the new method cannot be taken to higher orders so that the "step" ultimately becomes the commensurability period for example in the main lunar problem, of length 18 years 11 days (or one Saros) or in the Jupiter-Saturn-Sun problem a step of 900 years (the length of the Grand Inequality).

References

Roy, A. E., Ovenden, M. W.: Monthly Notices Roy. Astron. Soc. **114**, 232 (1954).
Roy, A. E., Ovenden, M. W.: Monthly Notices Roy. Astron. Soc. **115**, 296 (1955).

Galactic Models and Stellar Orbits

(Invited Lecture)

By Jaan Einasto
W. Struve Astrophysical Observatory, Tôravere, Estonia, USSR

With 19 Figures

Abstract

Methods of construction composite models of galaxies are outlined. The available observational information renders it possible to distinguish the following populations in nearby galaxies: nucleus, core, bulge, halo, disk, young population. Parameters of models are presented on six galaxies (our Galaxy, M 31, M 32, M 87, and the dwarf galaxies in Fornax and Sculptor). New data indicate that the degree of mass concentration to the centre of the galaxies is much higher than adopted in previous models, and that giant galaxies may be surrounded by massive coronae of very large dimensions.

Recent developments in the derivation of stellar orbits and kinematical characteristics of various stellar populations in galaxies are also summarized.

In the last section models of the physical and dynamical evolution of galaxies are discussed.

I. Introduction

The construction of models is the most effective tool for a synthesis of various observational data and for a quantitative study of the physical and dynamical structure and evolution of stellar systems.

In earlier model studies the goal has been more specific, since the models have been considered mainly as mass distribution models only (Perek, 1962).

Classical models of spiral galaxies are based on rotational velocities, which are identified with circular velocities. The models of elliptical galaxies are based on luminosity distribution. Mass distribution is derived on the assumption that the mass-to-luminosity ratio is constant over the whole system. The latter quantity is determined from the virial theorem by means of the velocity distribution in the centre of the system.

Some years ago it became apparent that a new approach to the construction of models of stellar systems is necessary. First of all, recent observational and theoretical data have raised serious doubts about the validity of the assumptions mentioned above. Secondly, present observational techniques permit us to study not only the general distribution of mass in galaxies, but also the dynamical and physical structure of different stellar populations in them. And finally, the study of the physical and dynamical evolution of galaxies has become an important new branch of investigation. All these aspects must be taken into consideration in developing new methods for the modelling of galaxies.

In the following we shall give a review of the problems connected with the construction of spatial and kinematical models of galaxies, and shortly describe the corresponding procedures. New results on the structure and evolution of galaxies will be presented.

The spatial structure of galaxies and their populations is closely associated with properties of stellar motions. This aspect will also be discussed in connection with the problem of stellar orbits.

II. New Observational Data on Galaxies

As mentioned above, new observational data enable us to study the distribution of stars of different populations in galaxies. By a population we mean the family of stars or other objects having similar physical properties (age, chemical composition, etc.) and similar parameters of spatial distribution and kinematics. The population problem in galaxies has been recently reviewed by King (1971).

The most important new observational data on which the study of populations is based are the spectrophotometric and colorimetric data concerning the stellar content of galaxies.

Pioneering work in this direction has been made by Morgan and Mayall (1957) and Spinrad (1962). The first two authors discovered that the spectra of the centres of M 31 and of the giant elliptical galaxies have cyanogen bands of normal strength. This indicates that the majority of stars in the central region of these galaxies are old population I stars rather than metal-poor population II stars of the globular cluster type. Population II is also present there, but it is not the dominating population. Spinrad has found that in the centres of giant elliptical and spiral galaxies the D lines of sodium are very strong, which indicates the presence of red dwarfs in large numbers. Later McClure (1969) and Spinrad and Taylor (1971 a) found that the lines of metals in the spectra of some old open clusters and nuclei of giant galaxies are also very strong. They interpret the results in terms of super-metal-rich nuclei of giant galaxies and old open clusters (Spinrad *et al.*, 1970). Spinrad and his collaborators have developed methods for the derivation of the mass-to-luminosity ratios of galaxies on the basis of spectrophotometric data (Spinrad *et al.*, 1971; Spinrad, 1971; Spinrad and Taylor, 1971 b).

Spectrophotometric methods can be applied for central bright parts of galaxies only. Valuable information on the star content of galaxies, including faint peripheral regions, can be obtained by photometric methods, using the metallicity parameter Q (van den Bergh, 1967) or other combinations of U, B, V, R, I, J, K, L, colours.

By means of large optical and radio telescopes the distribution of individual bright objects in nearby galaxies can also be studied. Among these objects we mention globular clusters studied by Vetešnik (1962), Kinman (1963), Racine (1968), Sharov (1968 a), van den Bergh (1969), and others, novae (Sharov, 1968 b), supernovae (Tammann, 1970), emission nebulae (Baade and Arp, 1964;

DEHARVENG and PELLET, 1969; COURTES *et al.*, 1968), stellar associations (VAN DEN BERGH, 1964), and neutral hydrogen (s. VAN WOERDEN, 1967).

Parameters characterising the physical and dynamical structure of populations cannot be determined directly from observations in all cases. For example, the mass-to-luminosity ratios of the bulge, the disk and the halo of a galaxy cannot be derived from spectrophotometric observations. These quantities can be estimated from the models of the evolution of galaxies or from other data (see below).

III. Methods for the Construction of Models of Stellar Systems

By a model of a stellar system we mean a set of functions and parameters describing the main features of the system, including the presence of populations in it. The main descriptive functions are the following: the gravitational potential Φ and its radial and vertical derivatives $-K_R$, $-K_z$, the spatial density of mass ρ, the projected (along the line of sight) density of mass P, the projected luminosity density L, velocity dispersions in a cylindrical system of coordinates σ_R, σ_θ, σ_z, the centroid velocity V_θ, and the inclination angle α of the major axis of the velocity ellipsoid to the plane of the galaxy. All these functions are connected with each other by means of various equations; the form of the equations depends on certain assumptions on the shape of equidensity surfaces and on some other conditions.

To construct a model of a galaxy, three problems are to be solved:

1. The choice of equations connecting descriptive functions.

2. The choice of an analytic form for the initial descriptive function.

3. The determination of parameters of the descriptive functions.

To establish the form of connection equations, we make two assumptions. First, we suppose that galaxies are in a steady state and are symmetrical in respect to the rotational axis and equatorial plane. Secondly, we suppose that galaxies consist of ellipsoidal components of different flattening that represent different stellar populations in them. These assumptions are a good approximation to real galaxies. The assumption of an ellipsoidal distribution of mass has the advantage that the form of the connection equations is the simplest one (EINASTO, 1969a, b; 1970a, b). For the projected density of a component we have

$$P(A)=fL(A)=\frac{1}{2\pi E}\int_A^\infty \frac{\mu(a)\,da}{a\sqrt{a^2-A^2}}, \tag{1}$$

where $a=\sqrt{R^2+\varepsilon^{-2}z^2}$ is the semimajor axis of the equidensity ellipsoid with the axial ratio ε; R, z are cylindrical co-ordinates, and A is the semimajor axis of the projected density ellipsis with the apparent axial ratio E,

$$E^2=\sin^2 i+\varepsilon^2\cos^2 i, \tag{2}$$

i being the angle between the plane of the system and the line of sight. Further, f is the mass-to-luminosity ratio of the component. According to our assumption,

f is constant over the whole component. Finally the expression

$$\mu(a)=4\pi\varepsilon a^2\rho(a) \tag{3}$$

is the mass distribution function (the mass of an equidensity layer of unit thickness at the equator).

The gravitational potential and its gradients can be found from the following expressions

$$K_R(R,z)=RG\int_0^1\frac{\mu(a)u^2\,du}{a^2[1-(eu)^2]^{1/2}}, \tag{4}$$

$$K_z(R,z)=zG\int_0^1\frac{\mu(a)u^2\,du}{a^2[1-(eu)^2]^{3/2}}, \tag{5}$$

$$\Phi(R,z)=-\int_R^\infty K_R(R,z)\,dR=-\int_z^\infty K_z(R,z)\,dz, \tag{6}$$

where a is connected with the integration variable u by the formula

$$a^2=u^2(R^2+z^2[1-(eu)^2]^{-1}). \tag{7}$$

If the galaxy consists of more than one component, the corresponding sums are to be written at the right-hand side of (4) and (5).

Kinematical functions can be calculated by means of hydrodynamical equations of stellar dynamics

$$\frac{1}{R}(\sigma_R^2-\sigma_\theta^2)+\frac{1}{\rho}\frac{\partial}{\partial R}(\rho\sigma_R^2)+\frac{1}{\rho}\frac{\partial}{\partial z}[\rho\gamma(\sigma_R^2-\sigma_z^2)]-\frac{V_\theta^2}{R}=-K_R, \tag{8}$$

$$\frac{1}{R}\gamma(\sigma_R^2-\sigma_z^2)+\frac{1}{\rho}\frac{\partial}{\partial R}[\rho\gamma(\sigma_R^2-\sigma_z^2)]+\frac{1}{\rho}\frac{\partial}{\partial z}(\rho\sigma_z^2)=-K_z, \tag{9}$$

where

$$\gamma=1/2\,\mathrm{tg}\,2\alpha. \tag{10}$$

If the galaxy has a composite structure, the total number of functions to be determined is large and the numerical integration of equations is impossible. The procedure used by most investigators is the following: an analytic expression is adopted for one of the descriptive functions—this function can be called the *initial descriptive function*. Other functions will be calculated by means of the formulae given above. The problem of determining the model of a galaxy is reduced to the determination of the parameters of the initial descriptive function.

The spatial density ρ ,the projected density P or the gravitational potential Φ can be used as the initial descriptive function. In practice, if we wish to determine all the descriptive functions mentioned, the most convenient initial function is the spatial density ρ.

A large variety of analytic expressions has been proposed for the initial descriptive function. To select the most suitable expressions, the following conditions can be used (KUTUZOV and EINASTO, 1968; EINASTO, 1969b, 1971):

1. the mass and luminosity densities must always be non-negative and finite;

2. the mass density and its derivative must have no breaks, not even at the centre of the systems;

3. the density must decrease outwards;

4. the moments of the mass distribution function must be finite

$$M_j\{\mu\} = \int_0^\infty \mu(a)\, a^j\, d a < \infty, \quad j \geqq -2; \tag{11}$$

5. the descriptive functions are to form a mutually consistent system of functions;

6. the model of the galaxy must allow of stable circular motions.

It should be noted that in special types of galaxies some of these conditions may not be fulfilled.

A discussion of the analytical expressions used earlier in galactic models has shown (EINASTO, 1968a, b, 1969b) that almost all expressions do not satisfy some of the conditions mentioned. Therefore, we have proposed a new density law (EINASTO, 1970b; EINASTO and EINASTO, 1972a)

$$\rho(a) = \rho(0) \exp\{x - [x^{2N} + a^2 (k a_0)^{-2}]^{1/(2N)}\}, \tag{12}$$

where

$$\rho(0) = h \mathfrak{M} (4\pi \varepsilon a_0^3)^{-1} \tag{13}$$

is the central density of the component, $\mathfrak{M}$ is the mass of the component, a_0 its effective (mean) radius, for the definition of a_0 see EINASTO and EINASTO (1972a)[1], x and N are structural parameters, h and k dimensionless normalizing parameters, depending on x and N. Our experience has shown that by an appropriate choice of structural parameters the law (12) can be used in all cases.

The practical procedure of determining the model of a galaxy consists of three steps.

First, a photometric model of the galaxy is to be derived. To determine a representative photometric model, the profiles of the galaxy are needed in several photometric systems and both along the major and the minor axis. The preliminary values of model parameters can be found from graphs by a comparison of the observed curves B, B–V, and U–B versus log A with theoretical ones. The final values of model parameters can be determined by a trial-and-error procedure; for this purpose a special computational program is composed. In photometric models the following parameters can be found for the components of the galaxy: structural parameters, effective radii, axial ratios (flattening), luminosities and colours. The colours give us the relations between the mass-to-luminosity ratios, for instance

$$B - V = (B - V)_\odot + 2.5 \log(f_B / f_V). \tag{14}$$

The second step is the determination of the mass distribution model. For this purpose absolute values of the mass-to-luminosity ratios for all components are

[1] The physical meaning of the effective radius a_0 is clear from the expression $2a_0 = \mathfrak{M}/S_0$, where S_0 is the mass of a meridional plane-parallel layer of unit thickness, going through the centre of the system.

to be determined. For the central bright components they can be found from spectrophotometric data by the Spinrad method, and independently by the virial theorem from velocity dispersion. For fainter components the absolute value of f can be estimated theoretically, on the basis of the results of the calculations of the physical evolution of galaxies and their populations.

In the case of galaxies with a composite structure we obtain from the virial theorem (EINASTO 1971)

$$\langle\sigma_r^2\rangle_k = \beta_r\, G\, a_{0k}^{-1} \sum_{l=1}^{n} \mathfrak{M}_l H_{kl}. \tag{15}$$

In this expression $\langle\sigma_r^2\rangle_k$ is the mean dispersion of the galactocentric radial velocities of the component k of the galaxy (this quantity can be determined from observations),

$$\beta_r = \beta_R \cos^2 i + \beta_z \sin^2 i, \tag{16}$$

where β_R, and β_z are the coefficients defined in EINASTO and EINASTO (1972a). Further, a_{0k} is the mean radius of the component k, $\mathfrak{M}_l$ the mass of the component l and

$$H_{kl} = \int_0^\infty \mu_k^0(\alpha)\, m_l(\alpha)\, \alpha^{-1}\, d\alpha, \tag{17}$$

where μ_k^0 is the dimensionless mass distribution function of the component k, $\alpha = a/a_{0k}$, and

$$m_l(\alpha) = \mathfrak{M}_l^{-1} \int_0^\alpha \mu_l(\alpha)\, d\alpha \tag{18}$$

is the dimensionless inner mass of the component l.

The final step in the construction of a model of the galaxy is the determination of its dynamical and kinematical descriptive functions. Knowing the mass distribution function, the gravitational potential of the whole galaxy and its gradients can be calculated. Thereafter the kinematical functions can be established for different components of the galaxy. The difficulty lies in the fact that the system of hydrodynamical equations is not closed: for the five functions σ_R, σ_θ, σ_z, V_θ, γ we have only two equations: (8), and (9).

An approximate solution can be obtained, supposing $\gamma = 0$ and $\sigma_R^2, \sigma_\theta^2, \sigma_z^2 \ll V_\theta^2$. In this case we obtain from the first hydrodynamical equation for $z=0$ the expression

$$V_\theta^2 = R\, K_R \equiv V^2, \tag{19}$$

where V is the circular velocity; and from the second equation (Jeans approximation, subscript zero)

$$\rho(z)[\sigma_z^2(z)]_0 = \int_z^\infty K_z(R, z)\, \rho(z)\, dz. \tag{20}$$

Formulae (19) and (4) are the basic ones in the classical method for the determination of the models of oblate galaxies.

To obtain a more accurate solution, three additional equations are to be used. Recently we proposed (EINASTO, 1970a) to close the system of equations by means of the formulae (KUZMIN, 1953),

$$\gamma = R z (R^2 + z_0^2 - z^2)^{-1}, \tag{21}$$

the Lindblad formula

$$k_\theta \equiv \sigma_\theta^2 / \sigma_R^2 = 1/2(1 + \partial \ln V_\theta / \partial \ln R), \tag{22}$$

and the KUZMIN (1961) formula

$$k_z \equiv \sigma_z^2 / \sigma_R^2 = k_\theta (1 + k_\theta)^{-1}. \tag{23}$$

Formula (21) follows from the theory of the third integral of motion, z_0 is a constant (in a more general case z_0 varies with R and z, EINASTO and RÜMMEL, 1970). It should be noted that formulae (22) and (23) are valid near the plane of symmetry of a galaxy and for the flat components of a galaxy only. In the general case the problem of closing hydrodynamical equations is more complicated and has presently no final solution.

IV. The Structure of Galaxies

a) Populations in Galaxies. The first classification (BAADE, 1944) recognized only two types of populations. Population I consists of hot and intrinsically bright stars and forms spiral arms in the galaxies, population II consists of stars of the globular cluster type and is the main constituent of elliptical galaxies and the central regions of spiral galaxies. Later KUKARKIN (1949) in his investigation of the stars in our Galaxy proposed to divide the various stellar subsystems into three components of the Galaxy: the spherical, the intermediate, and the flat component according to the shape of their equidensity surfaces. At the Vatican Conference on Stellar Populations (OORT, 1958) a more detailed population classification (hereafter VPC) was proposed, based also on the study of stars in our Galaxy. As at that time the knowledge of the spatial structure and composition of stellar populations in external galaxies was in an early stage, the VPC is not complete for external galaxies. Now we have much more information on the structure of the external galaxies and it seems reasonable to refine the VPC to some extent.

Let us start with the nuclear regions of galaxies, which are not represented in the VPC.

According to the modern data on the majority of galaxies, the nuclear region has a composite structure. In the centre of most galaxies a small very dense star cluster is located. Usually this cluster is called the *nucleus* of the galaxy. Its colour is red, and its metal-content and mass-to-luminosity ratio are high.

At the very centre of some galaxies there exists a body, which must be responsible for various forms of nuclear activity in galaxies (explosions as in the case of M 82, jets of the type of M 87, the existence of bright and broad emission lines as in the Seyfert galaxies, etc.). Sometimes this body is also called the nucleus of the galaxy, but this may lead to confusion with the central dense cluster of stars. Therefore, the introduction of a new term for the active body would be necessary. At the present time the physical nature of this body is not fully understood, and it seems too early to select a term characterizing its physical nature, such as

Table 1. Galactic populations

Population	ε	$-\log Z$	$\log \frac{a_0}{a_{0\,\text{halo}}}$	$\log t$ years
Nucleus	1	1	-3	10
Core	1	1	-1.5	10
Bulge	1	1.5	-0.5	10
Halo	0.1–1	2–5	0	10
Disk	0.04–0.1	1.5	0.5	9–10
Young	<0.04	1.5	0.5	<9

magnetoid (OZERNOY and USOV, 1971; OZERNOY, 1972) or dead quasar (LYNDEN-BELL, 1969). We propose to use for this body the term *kernel* of the galaxy[2].

The main body of an elliptical galaxy and the central body of a spiral galaxy is usually called the *bulge* of the galaxy. Available data indicate that the bulge consists of stars of normal chemical composition and that the mass-to-luminosity ratio is of the order 10–20. Recent spectrophotometric (SPINRAD *et al.*, 1971) and colorimetric (SANDAGE *et al.*, 1969) data indicate, however, that the central part of the bulge has a different physical structure: its metal-content and mass-to-luminosity ratio are much higher than those for the bulge in general. According to our definition, a population is a family of objects with similar physical properties. Therefore, we can speak of a new population of stars, which may be called the *core* of the galaxy. This term was used earlier by KING (1962) in a geometrical sense to designate the central part of a stellar system with a given density distribution. We attribute to this term physical meaning by relating it to a new recently recognized population of stars in giant galaxies.

The largest "spherical" stellar population in galaxies is the *halo*, which is characterized by a large deficit of metals and a low value of the mass-to-luminosity ratio.

All populations considered above consist of old stars, which in galaxies form subsystems of nearly spherical form. These populations differ in their chemical composition, mean radii and densities (Table 1).

In SO and spiral galaxies there exist some additional populations. The *disk* consists of stars of normal metal content, its flattening parameter ε is smaller and the mean radius a_0 larger in comparison with "spherical" populations. The age dispersion of stars seems to be larger than in the bulge. The *young population*, also called extreme I (OORT, 1958) and flat (KUKARKIN, 1949) population, consists of young stars and the star-generating medium (interstellar gas and dust). Objects belonging to the young population are distributed irregularly or they form the spiral arms of galaxies.

Observations of stars of different subsystems of stars in our Galaxy indicate that between the extreme population I and the extreme halo population there exist objects which form a continuous transition of subsystems. The transitional

[2] Originally this term was proposed by VORONTSOV-VELYAMINOV (1972) for the central star cluster (the nucleus). We adopted it in the preliminary version of this paper (EINASTO, 1972c). A discussion with some astronomers studying the nuclei of galaxies has convinced us that the terminology given above is more convenient.

Table 2. Some general data on galaxies

Galaxy	$(m-M)_0$	Ref.	A_B	Ref.	i	σ_r km·s^{-1}	Ref.	f_V	Ref.
M31	24.2	1	0.60	1	12.5	225	7	43	9
M32	24.2	1	0.60	1	0	98	7	6	9
M87	30.9	2	0.21	6	0	550	8	—	
Fornax	21.8	3, 4	0.20	3	0	—		—	
Sculptor	19.4	5	0.20	3	0	—		—	

References: 1. VAN DEN BERGH (1968). 2. SANDAGE (1968). 3. DE VAUCOULEURS and ABLES (1968). 4. HODGE (1961 a). 5. HODGE (1961 b). 6. SHAROV (1963). 7. MINKOVSKY (1962). 8. BRANDT and ROOSEN (1969). 9. SPINRAD (1971).

subsystems in the VPC are divided into three populations: intermediate population I, disk, and intermediate population II (OORT, 1958). In external galaxies we cannot observe individual stars belonging to a transitional population. But there exists indirect evidence that in external galaxies such transitional populations do occur, at least between the extreme halo and the disk populations. The mean axial ratio of equidensity ellipsoids found from photometric data for the Andromeda galaxy is $\varepsilon=0.30$ for the halo and $\varepsilon=0.08$ for the disk (EINASTO, 1971). Since the extreme halo population objects (globular clusters) in M31 have $\varepsilon=0.54$ (EINASTO, 1972a), there must exist intermediate subsystems with $0.08\leqq\varepsilon\leqq0.3$ in order to obtain for the halo the mean value $\varepsilon=0.3$. Kinematical data on stars in our Galaxy indicate that subsystems within this range of the axial ratio ε are associated with RR Lyrae stars of different mean periods. RR Lyrae stars belong to the metal-deficient stars, therefore one may conclude that subsystems in this range of ε are *intermediate halo subsystems* (EINASTO, 1971).

b) Photometric Models. Detailed models are now available on six galaxies: our Galaxy (EINASTO, 1970b; EINASTO and EINASTO, 1972b), the Andromeda galaxy, M31 (EINASTO, 1971; EINASTO and RÜMMEL, 1972) and its dwarf elliptical companion, M32, the giant elliptical galaxy M87 and the very faint dwarf elliptical galaxies in Fornax and Sculptor (present paper). The data on these galaxies are given in Tables 2 and 3. Table 2 presents the adopted true distance modulus $(m-M)_0$[3], the blue absorption $A_B=B-B_0$, the inclination angle i, the velocity dispersion in the centre σ_r, and the spectrophotometrically determined mass-to-luminosity ratio f_V. Table 3 gives the mass $\mathfrak{M}$, the blue luminosity L_B, the effective radius a_0, the axial ratio of equidensity surfaces ε, the mass-to-luminosity ratio f_B, two colours, B–V, U–B, the mean density $\langle\rho\rangle$ defined by the expression

$$\langle\rho\rangle=\mathfrak{M}(4\pi\varepsilon a_0^3\chi)^{-1}, \tag{24}$$

[3] During the First European Astronomical Meeting in Athens Dr. G. A. TAMMANN informed us that according to the new distance scale (SANDAGE, 1972) the distance of M87 is 20 Mpc instead of 15 Mpc adopted in the present paper. In this case the masses of the components of this galaxy are to be increased by a factor of 4/3 and the luminosities by a factor of $(4/3)^2$ but the mass-to-luminosity ratios are to be reduced by a factor of 3/4 and the mean densities by a factor of $(3/4)^2$.

Table 3. Parameters of galactic models

Galaxy	Populations	$\mathfrak{M}$ $10^9_\odot$	L_B $10^9_\odot$	a_0 kpc	ε	f_B	$B-V$	$U-B$	$\log\langle\rho\rangle$ $\odot$ pc^{-3}	$\log P(0)$ $\odot$ pc^{-2}	$B(0)$	N	x
Galaxy	Spheroid	35		0.9	0.60							4	10.5
	Disk	95		6.4	0.10							1	1.5
	Young	13		8	0.02							1	1.5
	Total	143											
M 31	Kernel	0.1											
	Nucleus	0.3	0.006	0.005	0.90	48	0.89	0.67	5.19	6.64	14.6	2	3
	Core	27	0.6	0.15	0.90	48	0.89	0.65	2.73	5.78	16.8	4	7
	Bulge	46	3.3	0.8	0.90	14	0.85	0.47	0.78	4.57	18.5	4	7
	Halo	20	6.2	3.0	0.30	3	0.65	0.16	−0.83	3.05	20.7	4	7
	Disk	85	7.5	9.2	0.08	11	0.82	0.43	−1.27	2.46	23.5	1	1
	Young	6	2.5	8.0	0.02	3	0.33	−0.17	−1.55	1.40	24.6	0.5	0
	Total A	185	20			9							
	B	370	20			18							
M 87	Nucleus	15	0.1	0.03	1.00	110	1.03	0.71	4.49	6.77	15.2	2	3
	Core	1090	10	1.4	0.98	110	1.03	0.71	1.41	5.47	18.4	4	7
	Bulge	1510	50	10	0.90	30	0.88	0.58	−0.99	3.89	21.0	4	7
	Halo	226	70	76	0.35	3	0.60	0.05	−4.05	1.30	25.0	4	7
	Total A	2840	130			22							
	B	23000	130			170							
M 32	Nucleus	0.05	0.01	0.003	0.8	6	0.87	0.38	4.05	6.25	13.3	2	3
	Bulge	0.91	0.19	0.04	0.8	5	0.84	0.34	3.03	5.77	15.1	4	7
	Halo	0.65	0.20	0.20	0.8	3	0.65	0.05	0.79	3.92	18.5	4	7
	Total	1.61	0.40			4							
Fornax	Halo	0.072	0.024	0.90	0.65	3	0.65	0.05	−2.18	1.34	24.9	0.5	0
Sculptor	Halo	0.004	0.0012	0.36	0.65	3	0.65	0.05	−2.21	1.02	25.0	1.5	1

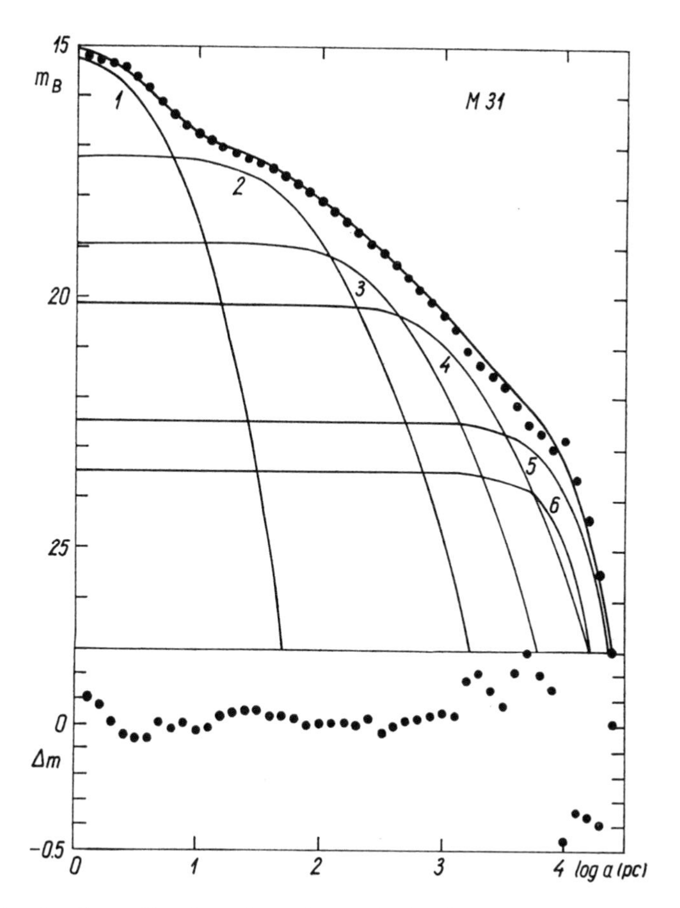

Fig. 1. The luminosity profile along the major semi-axis of M31. Dots represent observations compiled by EINASTO (1969a), curves represent various components (1 nucleus, 2 core, 3 bulge, 4 halo, 5 disk, 6 young population). In the lower part of the figure the deviations of the model from observations are shown

where χ is a dimensionless constant of the order 1 (for the definition of χ see EINASTO, 1972b). Finally, $B(0)$ is the central blue brightness of the component seen face-on and corrected for the effect of interstellar reddening, and $P(0)$ is the central value of the surface density (perpendicular to the plane of symmetry of the galaxy), N and x are structural parameters.

In Fig. 1–5 photometric profiles are given along the major axis for all the 5 external galaxies studied. The curves represent components of galaxies and total luminosity profiles, observations of various authors are plotted with different symbols. The agreement of models with observations is very good – the deviations are much smaller than the observational errors. The region of the spiral arms in the Andromeda galaxy presents the only exception: the model does not represent individual spiral arms. In Fig. 6 and 7 the run of the apparent axial ratio of equi-density ellipses is given for M87, and M31, in Fig. 8 the colour variation for

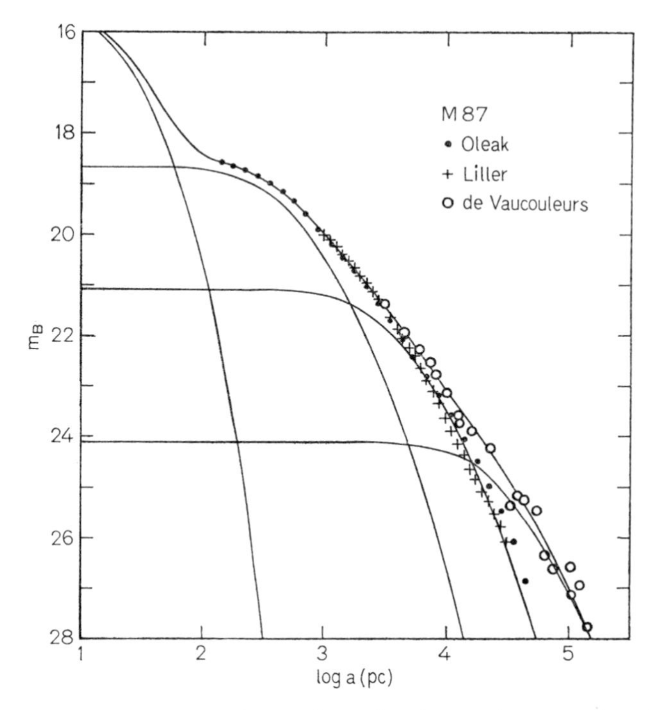

Fig. 2. The same as Fig. 1 for M 87. Observations of various authors are plotted by different signs: OLEAK (personal communication), LILLER (1960), DE VAUCOULEURS (1969). At large distances from the centre photographic observations (OLEAK, LILLER) deviate from photoelectric ones (DE VAUCOULEURS), which indicate the presence of scale errors in photographic observations. Curves indicate the following components (left to right): nucleus (not visible due to large distance of M 87), core, bulge, and halo

M 87. As can be seen, these photometric functions are also satisfactorily reproduced by the model.

c) Mass-to-Luminosity Ratio of Components. The critical point in model constructing is the determination of mass-to-luminosity ratios for individual galactic components.

For the nucleus and the core this ratio can be determined from observations by two methods, namely from spectrophotometric data and from the virial theorem. The necessary observational data are given in Table 2. Both kinds of data are available for M 31 and M 32 and they give mutually consistent results. The adopted values for f_B are given in Table 3.

For another extreme case, the halo, we have no direct determinations of f_B. On the analogy of globular clusters we may assume that f_B is approximately equal to 1 (SCHWARZSCHILD and BERNSTEIN, 1955). This value, however, represents a lower limit for the halo. First, the low-mass stars with high values of f_B may evaporate from clusters, which is not the case with halos. Secondly, GRIFFIN (1972) has shown on the basis of his new radial velocity determinations that the masses of globular clusters are somewhat larger than previously adopted. Finally,

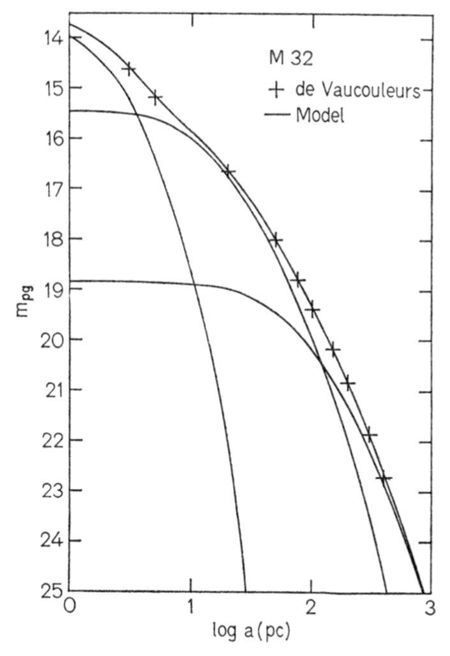

Fig. 3. The same for M32. The observed profile has been taken from DE VAUCOULEURS (1953), the components (left to right) are: nucleus, bulge, and halo

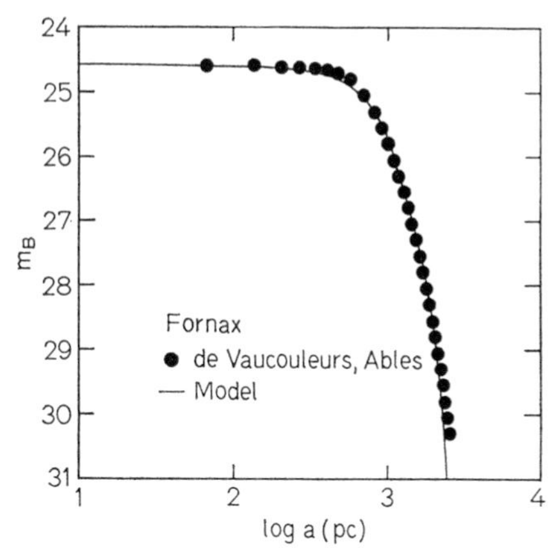

Fig. 4. The same for the Fornax galaxy; observations have been taken from DE VAUCOULEURS and ABLES (1968)

a halo population in general has a smaller ε than the subsystem of globular clusters, which indicates a larger value of the mass-to-luminosity ratio. Therefore, we have adopted a preliminary value $f_B=3$ for the halos of all the galaxies considered. There may exist differences in f_B for the halos of galaxies of different mass and morphological types, but more information is needed to draw quantitative conclusions.

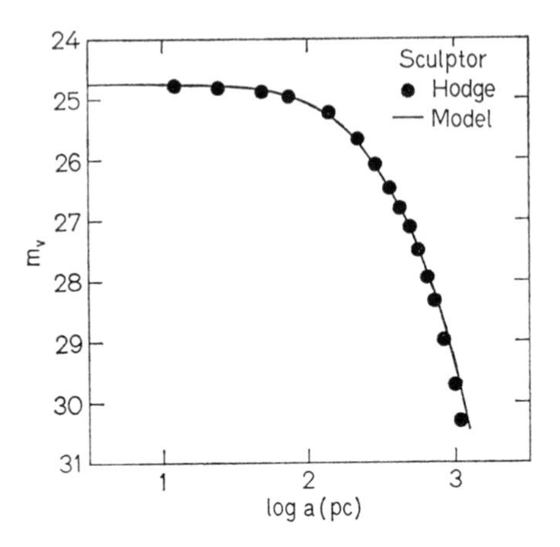

Fig. 5. The same for the Sculptor galaxy. The luminosity profile has been derived from star counts (HODGE 1961a, b)

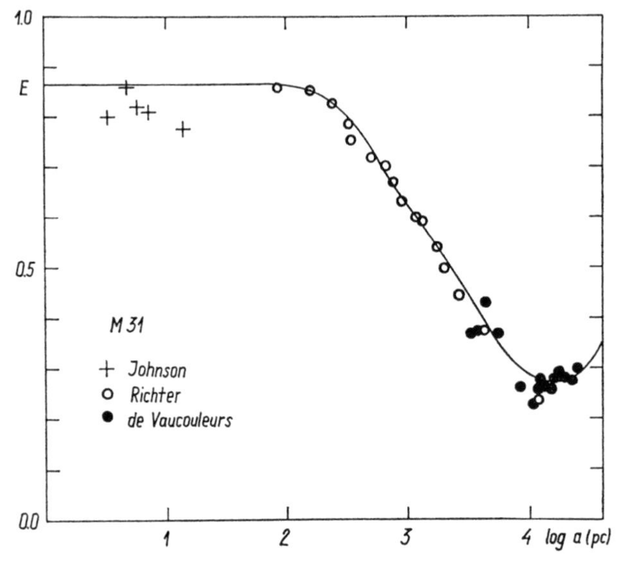

Fig. 6. The apparent axial ratio of equidensity ellipses as a function of the major semi-axis for M31. The curve represents the model, observed values have been taken from JOHNSON (1961), RICHTER and HÖGNER (1963) and DE VAUCOULEURS (1958)

The mass-to-luminosity ratio of the young population is estimated from theoretical calculations of the physical evolution of galaxies (EINASTO, 1971). The mass of interstellar gas is also included in the mass of this component.

For the bulge and the disk the mass-to-luminosity ratio can be found from the total mass of the galaxy, which is known from rotational velocity data (M31), and from physical evolution calculations. We assume that in M31 the bulge and the disk have the same chemical composition and that mass-to-luminosity ratio

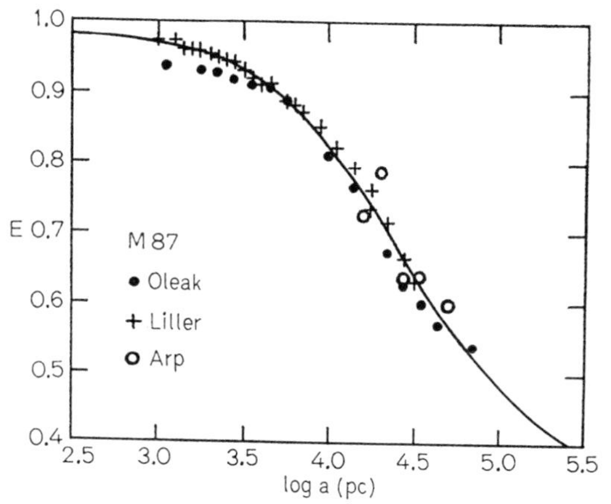

Fig. 7. The same as in Fig. 6 for M 87. Observed axial ratios have been taken from OLEAK (private communication), LILLER (1960) and ARP and BERTOLA (1969)

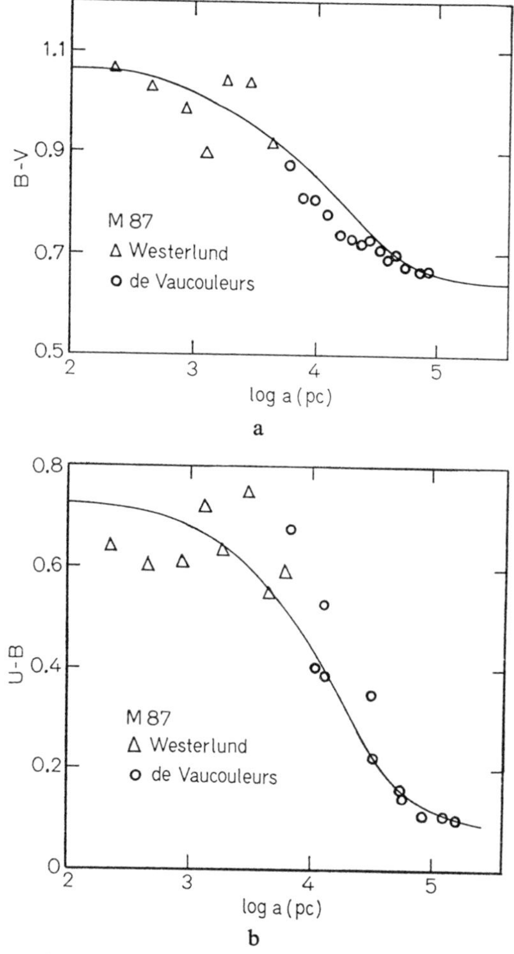

Fig. 8a and b. Colour variations in M 87. Curves represent the model; observed values have been given according to WESTERLUND and WALL (1969) and DE VAUCOULEURS (1969)

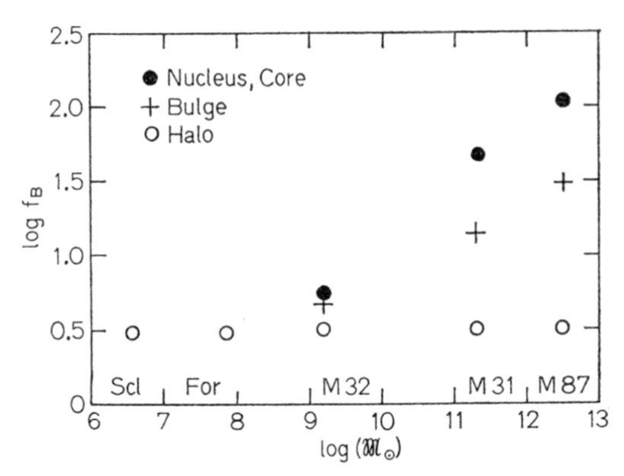

Fig. 9. Dependence of mass-to-luminosity ratios f_B of old galactic components on the total mass of the galaxy $\mathfrak{M}$

differences are caused by differences in the age distribution of stars. The latter can be estimated by physical evolution calculations. The absolute values of f_B are estimated from the mass of M31 (the masses of all other components are calculated from luminosities and mass-to-luminosity ratios).

For M87 and M32 f_B of the bulge can be estimated from physical evolution calculations. These estimates are, however, rather rough, since the chemical composition is known only approximately. A better estimate can be found from the relation between $B-V$, $U-B$ colours and the mass-to-luminosity ratios for the old populations.

The dependence of $\log f_B$ of individual galactic components on the total mass of the galaxy $\mathfrak{M}$ and on the $B-V$ and $U-B$ colours is shown in Figs. 9 and 10. We see that there exists a close correlation between these quantities. Since all the populations plotted in Figs. 9 and 10 have approximately the same age, differences in f_B can be attributed to differences in composition.

d) Mass, Mass-to-Luminosity Ratio, and Circular Velocity Distribution. For all the external galaxies studied the general mass, mass-to-luminosity ratio and circular velocity distributions have been calculated. The general mass distribution has been found from the formula

$$M(a)=\sum_{j=1}^{n}\int_{0}^{a}\mu_j(a)\,da, \tag{25}$$

where n is the number of components in a given galaxy. The general mass-to-luminosity ratio has been calculated from the expression

$$f_B(a)=P(a)/L_B(a), \tag{26}$$

where $P(a)$ and $L_B(a)$ are the face-on projected mass and luminosity densities (summed over all components), the projected luminosity being corrected for the interstellar absorption.

In Fig. 11 the general mass distribution has been plotted for 5 galaxies, the lines of the constant mean inner densities

$$\langle\rho(a)\rangle=3M(a)(4\pi a^3)^{-1} \tag{27}$$

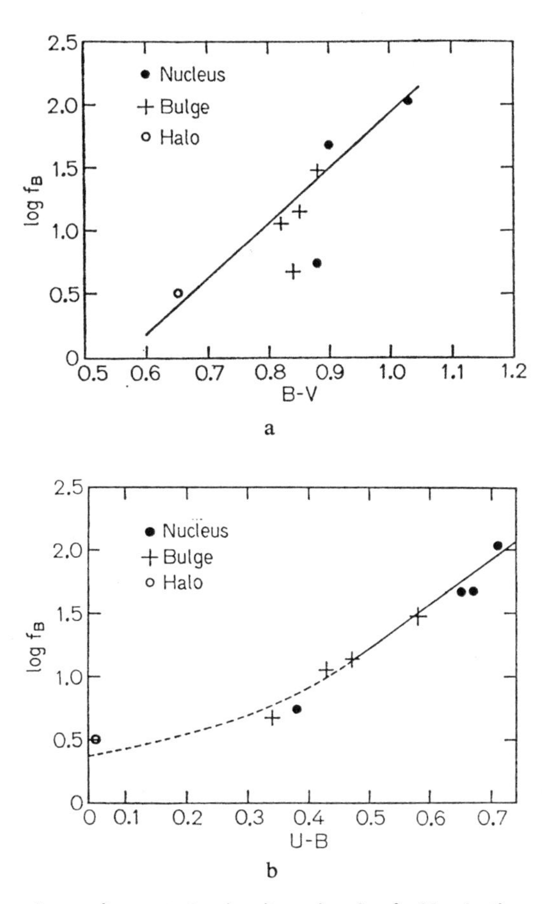

Fig. 10a and b. Dependence of mass-to-luminosity ratios f_B of old galactic components on their *B-V* and *U-B* colours

are also given. These densities correspond to the spherical distribution of mass. In Fig. 12 the run of the mass-to-luminosity ratio is given for the same 5 galaxies. In Fig. 13 the circular velocity curve is given for the Andromeda galaxy, M31. For the sake of comparison the corresponding functions are given for the Andromeda galaxy according to the Roberts (1966) model, a typical representative of models constructed by the classical method.

There are two main differences between classical and new models. First, the new possess a much higher mass concentration to the centres of galaxies. Secondly, in comparison with the models using the generalized Bottlinger law of circular velocities (as the Roberts model), new models have considerably lower densities in the outer region. The second aspect is connected with the problem of the "missing mass" in galaxies and will be considered in the next section.

The reason for the difference in the central mass concentration lies in the main assumptions of model construction. The new models are based on photometric data and various independent data on the mass-to-luminosity ratio on galactic

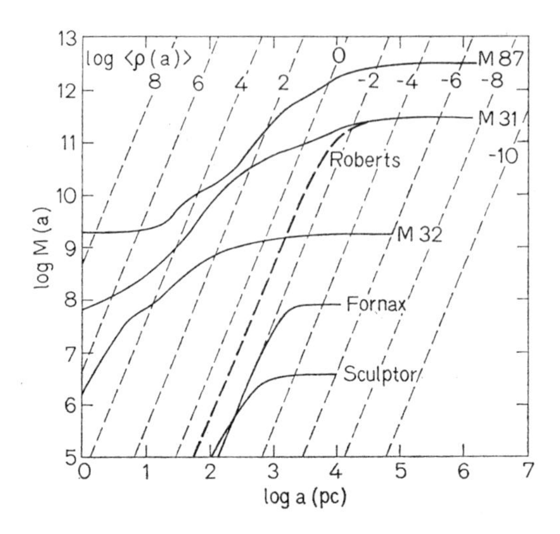

Fig. 11. General mass distribution of galaxies M 87, M 31, M 32, Fornax and Sculptor. For the Andromeda galaxy, M 31; for comparison, mass distribution is given according to the ROBERTS (1966) model. Lines of constant mean inner density are given for the spherical distribution of mass. Mass is given in solar units, distance in parsec

populations. The basic assumption of the classical method of model construction, is this – the rotational velocities can be identified with the circular ones over the *whole range of distances R*. In other words, it is supposed that the pressure term in the first hydrodynamical Eq. (8) is negligible in comparison with the rotational velocity term. In fact, in the central regions of galaxies the opposite situation takes place. E.g., in the nucleus of M 31 the observed rotational velocity amounts only to 7% of the circular velocity calculated from the virial mass of the nucleus. Consequently, the pressure term exceeds the rotational one by a factor of 200!

Sometimes the identification of rotational velocities with circular ones has been justified by the fact that interstellar gas, having a small velocity dispersion, also rotates slowly in the central parts of galaxies (RUBIN and FORD, 1970). In this case, however, the model has negative densities in some parts. Apparently the slow rotational velocity of the gas is of a different origin, which may be connected with the nuclear activity of the galaxy, as supposed by OORT (cited by RUBIN and FORD, 1970).

A practical conclusion from this situation is the following: in the inner parts of galaxies the rotational velocities cannot be used for the determination of the mass distribution in galaxies [formulae (4) and (19)].

As a result of differences in circular velocities the density distribution in classical and new models is very different. Combining the virial theorem and the spectrophotometric data we obtain for the central density of M 31 $\rho(0) = 10^6\ \mathfrak{M}_\odot\ \mathrm{pc}^{-3}$; for the Roberts model the central density is $1.4\ \mathfrak{M}_\odot\ \mathrm{pc}^{-3}$. A high central density has recently been found for our Galaxy and M 31 also by OORT (1971). However, the presence of dense nuclear regions in galaxies has not been incorporated in earlier overall galactic models.

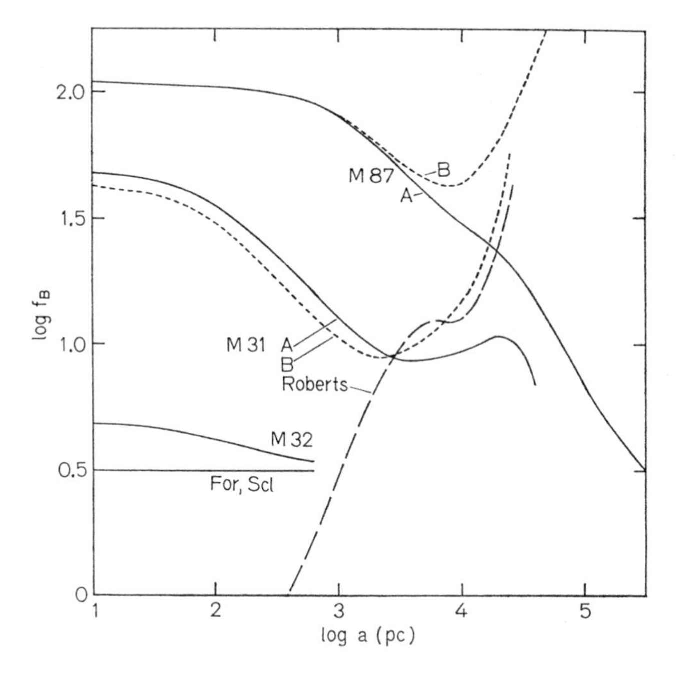

Fig. 12. Distribution of the projected mass-to-luminosity ratio for galaxies M 87, M 31, M 32, Fornax, and Sculptor. For galaxies M 87 and M 31 the distribution has been calculated for two models: A (without corona), and *B* (with corona). For M 31 the distribution is given also according to the ROBERTS (1966) model

The distribution of the mass-to-luminosity ratio in classical and new models is also very different. Due to the rapid increase of the luminosity density and the slow increase of mass density, the mass-to-luminosity ratio decreases towards the centre in old models (Fig. 12). In the centre, according to previous models, $f(0) \approx 10^{-2}$. This is in contradiction with direct spectrophotometric estimates (SPINRAD, 1971). On the other hand, using physically reasonable values of the parameters of the star formation function it is imposible to obtain theoretically a low mass-to-luminosity ratio for old metal-rich stellar populations.

An additional argument in favour of new models is the similarity of the physical characteristics of the central regions of spiral galaxies with the corresponding characteristics of elliptical galaxies. The latter have high mass-to-luminosity ratios in central regions.

As mentioned in the Introduction, classical models of elliptical galaxies are constructed, assuming the mass-to-luminosity ratio to be constant over the whole galaxy. Our results show that this is correct for dwarf elliptical galaxies only. The higher the mass of the galaxy, the larger is the difference between the mass-to-luminosity ratio in central and peripheral regions. Since a higher mass-to-luminosity ratio indicates a higher metal-content of stars, we conclude that in the central regions of massive galaxies metal-enrichment has been more effective than in the low mass galaxies.

The general conclusion is that both elliptical and spiral galaxies have a much higher degree of mass concentration towards the centres as assumed earlier. For instance, 25% of the total mass of the Andromeda galaxy is located inside a spheroid with a major semiaxis 1.8 kpc, the corresponding mean inner density being $\langle \rho(a) \rangle = 2\, \mathfrak{M}_\odot\, \mathrm{pc}^{-3}$.

e) The Problem of "Missing Mass" in Galaxies. This problem has two aspects, i.e. the local and the overall one.

The local problem of the "missing mass" arises about 10 years ago. OORT (1960) and HILL (1960) have investigated vertical motions of stars in the solar vicinity and determined the vertical gradient of the gravitational acceleration characterized by the Kuzmin parameter C,

$$C^2 = -(\partial K_z / \partial z)_{z=0}. \tag{28}$$

The total dynamical density of matter ρ in the solar vicinity has been calculated by using the Poisson equation (KUZMIN, 1952)

$$4\pi G \rho = C - 2(A^2 - B^2), \tag{29}$$

where A and B are Oort constants. The result is $C = 90\ \mathrm{km \cdot s^{-1} \cdot kpc^{-1}}$ and $\rho = 0.15\, \mathfrak{M}_\odot\, \mathrm{pc}^{-3}$, which is about twice the observed star and interstellar gas density $\rho = 0.09\, \mathfrak{M}_\odot\, \mathrm{pc}^{-3}$ (EINASTO and KUTUZOV, 1964, OORT, 1965). The difference has been attributed to stars of unknown origin.

KUZMIN (1955), EELSALU (1958) and EINASTO (1966) have found a considerably smaller value $C = 70\ \mathrm{km \cdot s^{-1} \cdot kpc^{-1}}$, which is in good agreement with direct estimates of mass density. Recently JÔEVEER (1972) developed a new method for the determination of C. The result is again $C = 70\ \mathrm{km \cdot s^{-1} \cdot kpc^{-1}}$. Therefore, the hypothesis of the presence of unknown matter in the solar vicinity in a great amount seems to have no strong observational support[4].

The overall problem of "missing mass" has arisen newly in connection with a different phenomenon. In the outer regions of spiral galaxies calculated from new models circular velocities possess a considerably steeper radial gradient than the observed gradient of rotational velocities. That can be seen in Fig. 13 from a comparison of our model with the one by Roberts—the latter reproduces direct observations at $R \geqq 10$ kpc fairly well. These differences may by explained in two ways.

1. In the outer regions of spiral galaxies there exist large systematical deviations of rotational velocities from circular ones. In this case our models are correct and no "missing mass" exists.

2. Spiral galaxies have coronae of great masses and dimensions. In this case our models must be respectively corrected to account for the "missing mass".

The first hypothesis is supported to some extent by the discovery of ROBERTS (1967) and COURTES *et al.* (1968) that many spiral galaxies have large differences

[4] MURRAY and SANDULEAK (1972) bring evidence that the missed matter in the solar vicinity is in the form of very cold low-mass dwarfs. However, a preliminary analysis made in Tartu by JÔEVEER, has shown that the number of red dwarfs has been overestimated by these authors.

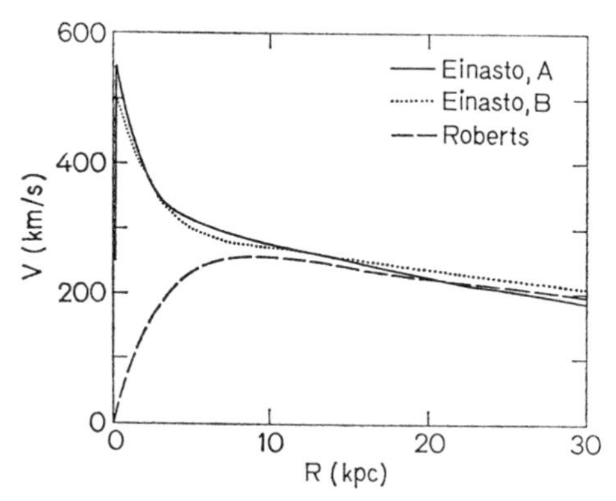

Fig. 13. Distribution of circular velocity of the Andromeda galaxy M 31 according to the Einasto models (A without and B with the corona) and the ROBERTS (1966) model

in the rotation curves of the opposite sides which can hardly attributed to differences in circular velocities. However, to explain the systematic nature of these deviations, additional assumptions are required.

The second hypothesis would lessen the discrepancy between the total cosmological density of matter and the mean density of observed matter (SHAPIRO, 1971) and the discrepancy between the virial and the count masses of the clusters of galaxies. In which form the "missing mass" may exist? The matter in question cannot be in the form of neutral gas, since this gas would be observable. The matter cannot be in the form of stars too. Luminosity decreases in outer galactic regions rapidly, therefore, if the matter is in the form of stars, the latter must be of very low luminosity to be invisible. The presence of low-luminosity stars in outer galactic regions without bright ones would require a powerful process of a large-scale segregation of stars according to mass (low-luminosity stars have smallest masses), but this is highly improbable. There remains the possibility that the unknown matter exists in the form of rarefied ionized gas. In connection with this we refer to a recent investigation by DE YOUNG (1972), who has given evidence for the presence of an intergalactic medium in clusters of galaxies. To prove the consequence of this hypothesis, we have added the corona in our model B, its mass is estimated to be equal to the stellar component of the galaxy ($185 \times 10^9 \mathfrak{M}_\odot$), the effective radius $a_0 = 20$ kpc, and the mean density about $10^{-3} \mathfrak{M}_\odot \text{ pc}^{-3}$. Respective descriptive functions are shown in Figs. 12 and 13.

There exists some evidence that very massive gaseous coronae may surround giant elliptical galaxies too. If we suppose that the outer limiting radii of galaxies, containing practically all their masses, do not change with time and that the corresponding mean densities characterize the initial densities of protogalaxies at the time of galaxy formation, we obtain for the initial density of M 87 the value of $3 \times 10^{-5} \mathfrak{M}_\odot \text{ pc}^{-3}$, which is about 1.5 orders of magnitudes smaller than the initial density of M 31, $10^{-3} \mathfrak{M}_\odot \text{ pc}^{-3}$. This result is rather surprising, since it is usually accepted that elliptical galaxies are denser than the spiral galaxies (HOLMBERG, 1964). The initial densities of both galaxies would be equal if the galaxy M 87 had a corona with the mass $2 \times 10^{13} \mathfrak{M}_\odot$ and with the effective radius $\alpha_0 = 100$ kpc (model B). We see that the total mass of the corona may be an order of magnitude greater than the mass of the stellar component of the galaxy!

However, evidence for the presence of massive coronae around giant galaxies is not very strong and further studies are necessary.

V. Stellar Orbits and Kinematical Properties of Galactic Populations

The knowledge of stellar orbits in galaxies is important in several respects. To establish the possible existence and form of the third integral of motion, it is necessary to compute three-dimensional orbits of stars. To develop a theory of evolution of stars the places of star formation must be known, which can be found by backward orbit calculations. Statistics of orbital elements are needed both for the theory of galaxy evolution and for the construction of representative models of galaxies.

a) Stellar Orbits and the Third Integral of Motion. According to the Jeans theorem (JEANS, 1915) the phase density of stars depends on the velocity components v_R, v_θ, v_z through isolating integrals of motion I_i. In the case of steady-state stellar systems, symmetrical in respect to a plane and an axis of rotation, there exist in general two independent isolating integrals of motion: the energy integral I_1 and the angular momentum integral I_2. If there exist no more isolating integrals, the star density in the velocity space would be a function of I_1 and I_2 only, and the velocity distribution would be axially symmetrical, the dispersions σ_R and σ_z being equal. This contradicts observations: in the solar vicinity the dispersion of radial motions for all subsystems of the stars studied exceeds the dispersion of vertical motions.

This discrepancy has sometimes been interpreted as an indication of the non-steady state of the Galaxy. However, KUZMIN (1953) has pointed out that the inequality of the velocity dispersions can be explained within the framework of the theory of a stationary galaxy, provided we suppose the existence of a third isolating integral of motion, I_3. As a first approximation, a quadratic dependence of I_3 on velocity components has been supposed. The existence of a third quadratic integral of motion imposes a definite restriction for the gravitational potential of the galaxy. The corresponding form of the potential has already been proposed by EDDINGTON (1915). KUZMIN (1953, 1956) has shown that this form of potential enables us to construct reasonable models of galaxies.

In the case of the quadratic third integral, the galactic orbits of stars are located inside the toroids bordered by confocal hyperboloids and ellipsoids. Two projections of such an orbit, calculated in 1953 by EELSALU, are given in Fig. 14. Later these orbits have been called box-orbits (OLLONGREN, 1962).

The family of confocal hyperboloids and ellipsoids has another important property. Together with the meridional planes they are tangent to the axes of velocity ellipsoids, and are, therefore, called main velocity surfaces (EDDINGTON, 1915).

The gravitational potential of real galaxies does not exactly coincide with the potential following from the theory of the quadratic third integral. Therefore, if

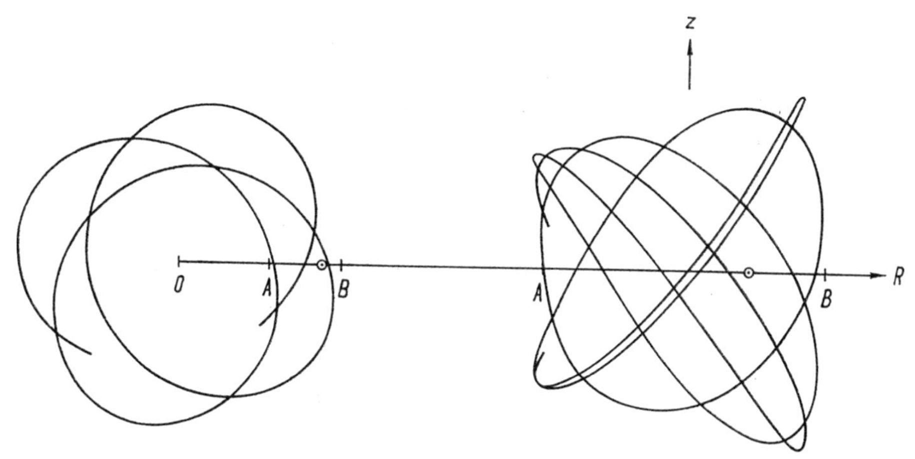

Fig. 14. Vertical (left) and meridional projections of the three-dimensional orbit of a star, calculated by EELSALU in 1953. Points A and B correspond to the minimal and the maximal distance of the orbit from the galactic centre at the equatorial plane of the Galaxy (Reproduced from KUZMIN, 1969)

there exists a third isolating integral, it does not have a quadratic form and the main velocity surfaces are not confocal hyperboloids and ellipsoids. To verify the possible existence of a third integral, OLLONGREN (1962) has computed a series of three-dimensional orbits of stars for the SCHMIDT (1956) model of the Galaxy. As in the previous case the orbits are located in toroids, but the borderlines of meridional sections have a more complicated form. This seems to show that the third isolating integral exists, but does not have a quadratic form.

Recent studies, reported in the IAU Symposium No. 25 (CONTOPOULOS, 1966), indicate that the problem is much more complicated. For some regions of the galaxy meridional projections of stellar orbits are enveloped by boxes, indicating the isolating character of the third integral. In other regions stellar orbits fill all the space allowed by the energy integral, in this region the third integral is not isolating (ergodic orbits).

From the standpoint of galaxy modelling an important problem to be solved is the determination of the tilt $\alpha(R, z)$ of the velocity ellipsoid outside the equatorial plane. This problem has been solved so far only for the case when the third integral has a quadratic form and the main velocity surfaces are confocal hyperboloids and ellipsoids.

b) Stellar Orbits and Places of Star Formation. The possibility of backward orbit calculations and the knowledge of stellar ages enables us to find the places of star formation. For this purpose CONTOPOULOS and STRÖMGREN (1965) have published extensive tables of plane galactic orbits. Recently GROSSMAN and YUAN (1970) have calculated new tables of plane galactic orbits for the potential with a spiral component.

Orbit calculations have been used for the derivation of places of star formation by DIXON (1967a, b, 1968). He has found that stars are born near the apocentres of their orbits and then fall slightly towards the centre of the galaxy.

According to a suggestion made by OORT (1964), the reason for this is that interstellar gas is supported partly by non-gravitational forces. This problem has also been studied by HUBE (1970), who has found no clustering of star formation points near apocentres.

Vertical oscillations of stars have been studied also by JÔEVEER (1968, 1972). He has shown that groups of young stars oscillate synchronously with the frequency $\omega_z = C = 70\ \mathrm{km \cdot s^{-1} \cdot kpc^{-1}}$. This is the case if stars were born above or below the galactic plane with small random velocities, as suggested by OORT. HUBE has failed to prove the existence of such vertical oscillations since he has adopted too large a value $C = 90\ \mathrm{km \cdot s^{-1} \cdot kpc^{-1}}$ for the Kuzmin constant.

c) Statistics of Stellar Orbits and Kinematical Properties of Galactic Populations. Statistics of the elements of stellar orbits as well as velocity dispersions and centroid velocities yield valuable information on the kinematical properties of galactic populations. In our Galaxy kinematical characteristics can be determined for solar vicinity for all populations of stars, but the parameters of the spatial distribution can be estimated only for the subsystems of bright stars. This makes kinematical characteristics very useful for the determination of the mean age of a given star group.

For cosmogonic studies both the statistics of stellar orbital elements and the statistics of velocity components can be used.

The first procedure has been used by EGGEN *et al.* (1962) to study the initial collapse phase of the Galaxy. Individual orbits of a large number of stars have been calculated and then orbital elements have been determined. The same results could be obtained more directly from the three-dimensional Bottlinger diagram, which gives the orbital elements directly from the components of the velocity vectors of stars (KUZMIN, 1956).

Centroid velocities and elements of the velocity dispersion tensor have been used for the study of galactic dynamical evolution by VON HOERNER (1960) and by EINASTO (1970b, 1971). In both studies absolute ages have been estimated for some subsystems, which allows one to accomplish age calibration for all kinematic groups. All three studies showed that at the first stage of its evolution the Galaxy was rapidly contracting both in radial and vertical directions.

If the mass density and the gravitational potential of a galaxy are known, it is possible to calculate kinematical descriptive functions for various test-populations by means of the formulae given in Section III. Such calculations have been carried out for seven test-populations of our Galaxy and six test-populations of the Andromeda galaxy (EINASTO, 1971). The results on one test-population of our Galaxy are given in Fig. 15. In addition to kinematical functions, the density of the population and its gradients have also been plotted.

Our calculations show that the dispersion of vertical velocities σ_z for disk populations depends largely on R and only slightly on z. Earlier a strong dependence of σ_z on z has been obtained by INNANEN and FOX (1967), which, however, is in contradiction with observations (WAYMAN, 1961). The reason for this difference lies in the use by INNANEN and FOX of the Schmidt density law with a sharp outer boundary.

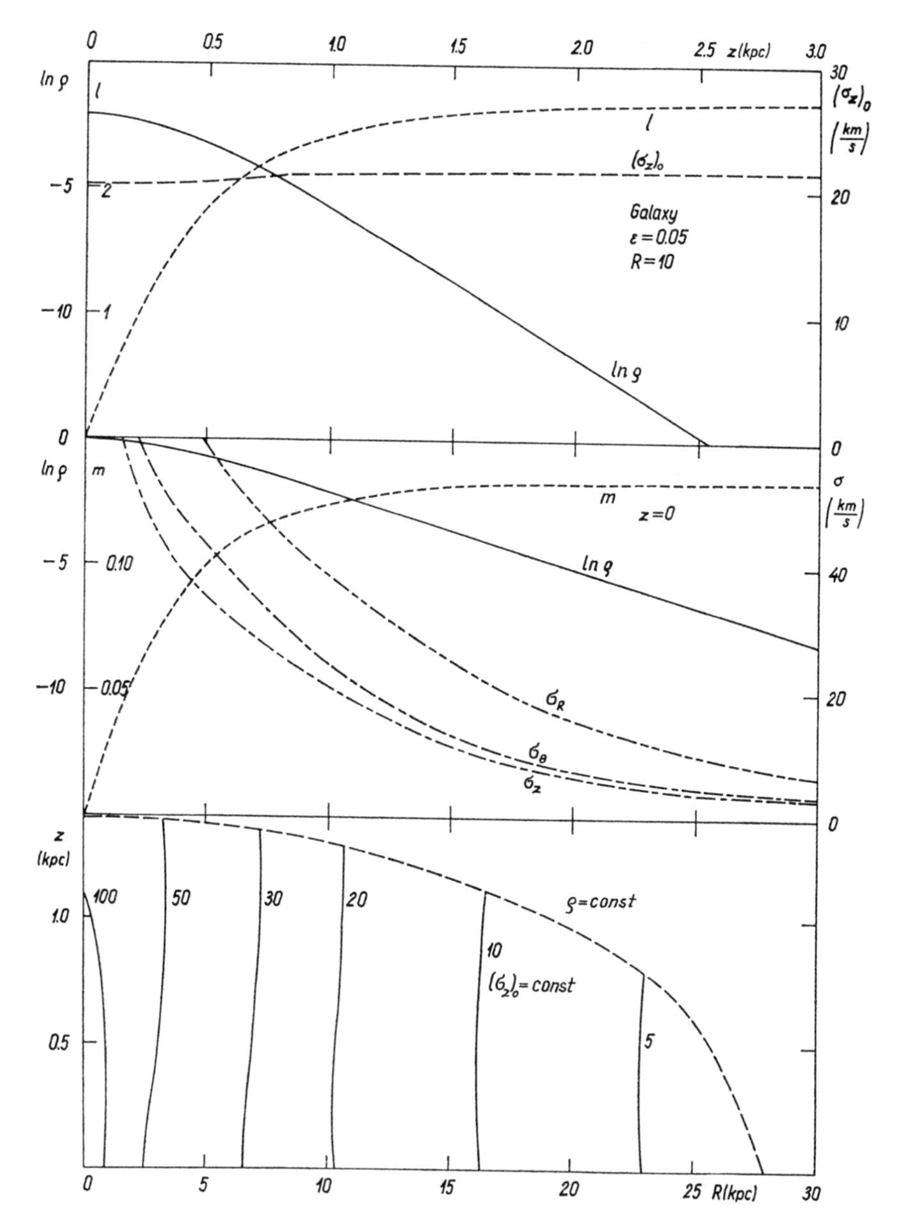

Fig. 15. Various descriptive functions of a test-population of the Galaxy with the effective radius $a_0 = 7.4$ kpc, axial ratio $\varepsilon = 0.05$ and structural parameters $N = 1$ and $x = 1.5$. In the upper part of the figure the z-dependence of the density ρ [in units of the central density $\rho(0)$], the vertical density gradient $l = -\partial \log \rho / \partial z$ and the velocity dispersion $(\sigma_z)_0$ are given for $R = 10$ kpc. In the central part of the figure the R-dependence of the density ρ, of the radial density gradient $m = -\partial \log \rho / \partial R$, and of the three velocity dispersions σ_R, σ_θ, σ_z are shown for the galactic plane $z = 0$. In the lower part of the figure the lines of the constant density ρ and the velocity dispersion $(\sigma_z)_0$ are given for the meridional plane of the Galaxy. Velocities are expressed in $\mathrm{km \cdot s^{-1}}$

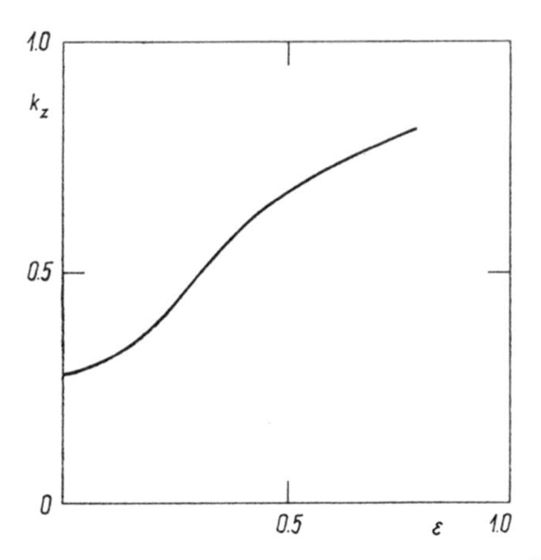

Fig. 16. Dependence of the axial ratio of the velocity ellipsoid $k_z = \sigma_z/\sigma_R^2$ on the flattening parameter ε of galactic subsystems for solar vicinity

Another interesting result is that the dispersion σ_R of galactocentric radial velocities increases considerably towards the centre of the Galaxy. The classical theory of a steady-state galaxy predicts σ_R to be independent of R (EDDINGTON, 1915). This theory is based on the assumption that the velocity distribution exactly follows the Schwarzschild law. Observations, however, show that this assumption is valid only approximately.

As mentioned in Section III, one important kinematical parameter is the ratio of velocity dispersions k_z. The Kuzmin formula (23) gives for the solar vicinity $k_z = 0.285$, which is in good agreement with the observed value for flat subsystems, $k_z = 0.278$. For intermediate and halo subsystems the direct determination of k_z is difficult due to large observational errors. This quantity, however, can be determined indirectly by the integration of the hydrodynamical Eqs. (8) and (9), adopting the observed relation between the centroid velocity and the mean velocity dispersion. The resulting dependence of k_z on the flattening of test-populations has been reproduced in Fig. 16 (EINASTO, 1971).

VI. The Evolution of Galaxies

Now we shall consider the application of the method of modelling to the study of the evolution of galaxies.

The evolution of galaxies is caused by four different processes: the formation of stars out of interstellar matter, the enrichment of interstellar matter by the products of stellar evolution, the change of physical parameters of populations as a result of stellar evolution, and the change of the spatial distribution of stars and interstellar matter. All processes are proceeding simultaneously and are mutually affecting one another. For instance, the rate of star formation depends on the density of interstellar matter, the contraction of interstellar matter is controlled by the number of young stars, etc. Therefore, a perfect theory of galaxy evolution must take into account all processes simultaneously. However, the problem is very complicated and such a theory does not exist at present. A separate study of different processes seems to be a realistic approach.

To construct models of the physical evolution of galaxies, the following data are needed: the rate of star formation as a function of time, the distribution of newly-born stars according to their mass (the initial mass function), the evolution tracks of stars of different mass and composition, the rate of the production of various isotopes by stars of different mass, the bolometric corrections and intrinsic colours of stars as functions of their chemical composition, absolute temperature and luminosity. The last-named data are necessary to calculate observable photometric quantities for model galaxies.

Dynamical parameters of stellar populations are rather conservative, therefore, the present dynamical state of the populations preserves information on the conditions in the galaxy at the time of their formation. By comparing the dynamical parameters of populations of different ages, it is possible to reconstruct the past dynamical history of the galaxy.

a) The Rate of Star Formation. OORT (1958) supposed that the rate of star formation is proportional to the interstellar gas density in a rather high power. This was confirmed by SCHMIDT (1959), who deduced from the study of star and gas densities in the solar vicinity that

$$R \equiv d\rho_s/dt = \gamma \rho_g^n. \tag{30}$$

For the power of this empirical law SCHMIDT obtained the value $n=2$. Practically the same result $n=1.84$ has been found by SANDULEAK (1969) for the Small Magellanic Cloud. Recently HARTWICK (1971) has found for the Andromeda galaxy, M31, a much higher value, $n=3.5$, and has supposed that this parameter may depend on local conditions of star formation. However, HARTWICK has used hydrogen densities, uncorrected for the antenna smearing effect. A rediscussion of the data on M31, taking the antenna smearing effect into consideration, gives $n=2$ (EINASTO, 1972b). In the last-mentioned study the coefficient of proportionality, γ, has also been found. The value $n=4$ has been obtained (time being expressed in 10^9 years and densities in solar masses per cubic parsecs). The result corresponds to the normal chemical composition of interstellar gas. The available data indicate that star formation does not take place at very low gas densities ($\rho_g < 0.01\ \mathfrak{M}_\odot\ \mathrm{pc}^{-3}$).

Of course, the rate of star formation depends not only on gas density, but on the gas temperature and composition, may be also on some other factors. An attempt to take into account the temperature and the composition of gas has been made by MATSUDA (1970). The adopted dependence, however, is rather arbitrary.

If the total density of matter $\rho = \rho_s + \rho_g$ is supposed to be independent of time, the differential equation (30) can be integrated analytically. The result for the mass of gas $\mathfrak{M}_g$ in the galaxy is in case $n=2$

$$\mathfrak{M}_g = \mathfrak{M}(1+\tau)^{-1}, \tag{31}$$

where $\mathfrak{M}$ is the total mass of the galaxy,

$$\tau = t/t_0 \tag{32}$$

denotes dimensionless time, and

$$t_0=(\gamma\langle\rho\rangle)^{-1} \tag{33}$$

designates the characteristic time of star formation, $\langle\rho\rangle$ being the mean total density of matter in the galaxy. It should be mentioned that the widely used exponential law (TINSLEY, 1968)

$$\mathfrak{M}_g=\mathfrak{M}\,e^{-\tau} \tag{34}$$

corresponds to the case $n=1$ and contradicts the Schmidt law.

b) The Initial Luminosity Function. This function was derived first by SALPETER (1955), who found that the distribution of the number of newly-born stars according to their mass m can be satisfactorily expressed by the law

$$F(m)=a\,m^{-p}, \tag{35}$$

where a is a normalizing constant and $p=2.35$. This law has been confirmed by SANDAGE (1957), VAN DEN BERGH (1957) and others. REDDISH (1965) called attention to the case $p=2.33$, which corresponds to an equal distribution of the potential energy among stars of different mass. SAAR (1972) has shown that this particular case is valid within a very wide range of masses of celestial bodies from stars to clusters of galaxies. The law (35) cannot be applied to very small masses since the integral $\int F(m)\,dm$ does not converge. Therefore, there must exist an effective lower limiting mass m_0.

c) Models of the Physical Evolution of Galaxies. Similar models have been constructed by TINSLEY (1968) and EINASTO (1971), using evolutionary tracks of stars, bolometric corrections, and intrinsic colours. The main results of these studies are the following.

TINSLEY has made an attempt to reproduce the observed integrated photometric properties of galaxies by a corresponding choice of parameters of the stellar birth function. The chemical composition of stars has not been varied. In this case the necessary variations in other parameters become too large. In particular, the observed high values of the mass-to-luminosity ratios of giant elliptical galaxies have been obtained by adding arbitrarily red dwarfs in an unprobably large amount.

Our study has shown that the most important factor in galactic evolution is the initial chemical composition of stars. In Fig. 17 we reproduce the dependence of f_B on time for three different compositions. Following a suggestion by TRURAN and CAMERON (1970), we suppose that the lower limiting mass of the stellar birth function depends on the composition of gas. For the normal composition $Z=0.02$ we adopt $m_0=0.03\,\mathfrak{M}_\odot$ (REDDISH 1965), for $Z=0.10$ $m_0=0.001\,\mathfrak{M}_\odot$ and for $Z=10^{-5}$ $m_0=0.1\,\mathfrak{M}_\odot$ to obtain for old populations the observed values of mass-to-luminosity ratios (Fig. 10).

Another important parameter determining the evolution of galaxies is the mean density. Differences in the mean density may be responsible for the formation of galaxies of different types: ellipticals, spirals and irregulars (HOLMBERG, 1964). However, serious difficulties arise, if we wish to explain the formation of different populations by differences in their mean density alone.

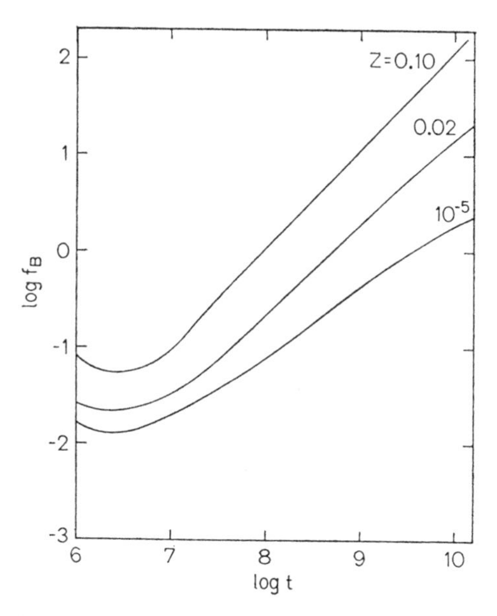

Fig. 17. Dependence of the mass-to-luminosity ratio of stellar populations with a different heavy-elements content Z on the age t of the population (in years)

The characteristic times of star formation t_0, calculated for various components of galaxies from data given in Table 3, vary within very broad limits. In particular, the characteristic time for the formation of the halo is too large. To overcome this difficulty, one may suppose that the parameter γ of the Schmidt law (30) is not a constant, but depends on the chemical composition of interstellar gas. However, this assumption cannot explain the presence of halos of different mean densities, since the halo stars in different galaxies have approximately the same composition. There remains another explanation – to suppose that halo stars were formed at the final stage of the contraction of the protogalaxy just before gas clouds lost their high kinetic energy by collisions. In other words we suppose that halo stars formed near the pericentres of their first revolutions. At that time the mean density of the protogalaxy was high and a rapid formation of halo stars was possible.

d) Dynamical Evolution of Galaxies. Differences in kinematical properties and the corresponding parameters of the spatial distribution of stars of different spectral type have been known for a long time (Rootsmäe, 1943; Kukarkin, 1949; Parenago, 1951). Rootsmäe has interpreted these data as an age effect. He supposed that the star-generating medium (gas) was in gradual contraction. A different explanation has been proposed by Spitzer and Schwarzschild (1953), Gurevič (1954) and Kuzmin (1961), namely acceleration of stars by interaction with massive clouds of interstellar matter and stars (the effect of irregular gravitational forces).

Von Hoerner (1960) showed that both processes may be active in galaxies. In the initial stage of the galactic history the first process (rapid contraction of interstellar matter towards the centre and the plane of the galaxy) dominates, thereafter

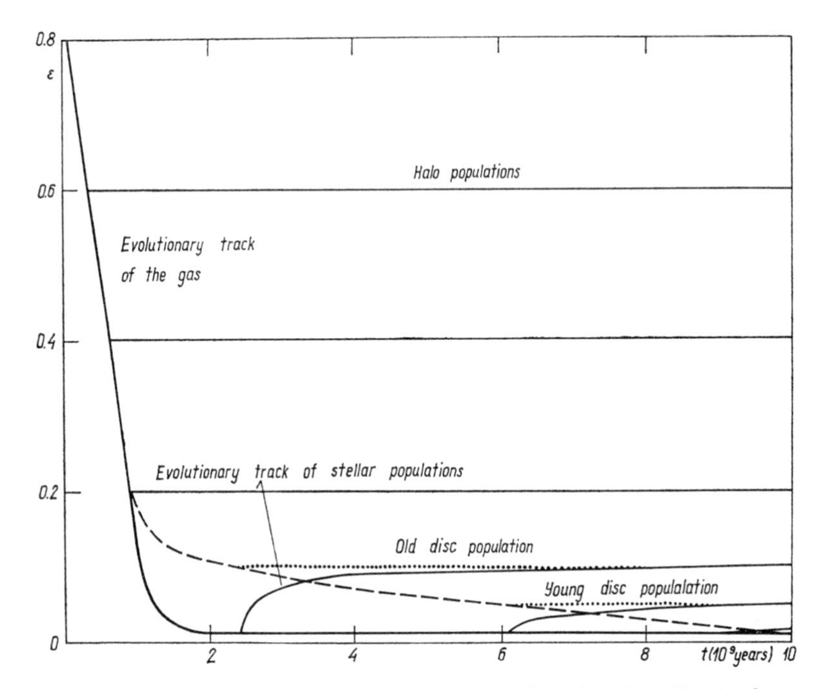

Fig. 18. Possible evolutionary changes of the axial ratio ε of equidensity ellipsoids for gas and stellar populations of different ages. Halo subsystems ($\varepsilon > 0.10$) have probably preserved their initial ε values, but the axial ratios of disk subsystems have increased due to the action of irregular forces. An alternative picture (the independence of ε for all stellar populations) is shown by dashed (gas) and dotted (stars) lines. (Reproduced from EINASTO, 1971)

the second process (acceleration) takes over. The contraction or collapse phase has been studied in detail by EGGEN *et al.* (1962) and recently by EINASTO (1971). The results of VON HOERNER have been confirmed. In the central parts of the Galaxy contraction proceeds very quickly (in 10^7 years), the bulge and the inner halo are formed in a few 10^8 years, the formation of the disk is completed in 10^9 years. The degree of contraction of the gas is very large in the centre of the galaxy (about 100); in the solar vicinity the degree of radial contraction of the gas is about 5; and the degree of vertical contraction is about 50 (Fig. 18).

The consequences of a rapid initial contraction of the galaxy have been investigated by LYNDEN-BELL (1967b). He showed that violent relaxation takes place as the result of a fast change of the gravitational field. This relaxation explains the similarity of the density distribution of all elliptical galaxies. Recent numerical studies have confirmed these results (see CONTOPOULOS, 1970).

The effect of the angular momentum of the protogalaxy on its subsequent evolution is also to be mentioned. This problem has been investigated by many authors (LYNDEN-BELL, 1967a; BROSCHE, 1970; SANDAGE *et al.*, 1970). Protogalaxies with a small relative angular momentum lead to the formation of elliptical galaxies, protogalaxies with a large momentum lead to the formation of spiral and irregular galaxies.

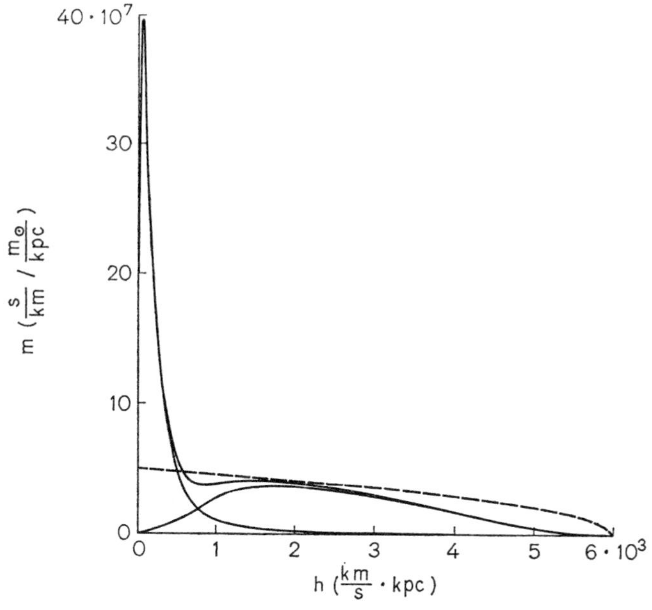

Fig. 19. Distribution of mass versus angular momentum h for the Andromeda galaxy M31. Thin lines correspond to spherical (core, bulge, halo) and to disk populations, the thick line to the galaxy in general. The respective distribution for a homogeneous spheroid rotating with constant angular velocity is indicated by a dashed line. The masses and the maximal values of angular momenta of both models are identical

It is usually accepted that the angular momentum distribution in galaxies is identical with the corresponding distribution of an uniformly rotating homogeneous spheroid (CRAMPIN and HOYLE, 1964; INNANEN, 1966). This result, however, is based on old models of spiral galaxies, where the central mass concentration is underestimated. The only spiral galaxy for which a new model is available is the Andromeda galaxy. The angular momentum distribution of M 31 is reproduced in Fig. 19 (EINASTO, 1971). Thus, we see that there exists a large excess of mass of low momentum as compared with the homogeneous spheroidal model. This indicates that the turbulent mixing of the gas may have an important role in the early history of the Galaxy.

VII. Conclusions

The main new results reported first in this review article are the following:

1. Galaxies have a much higher degree of central mass concentration than generally adopted.

2. The rotational velocities in the central parts of spiral galaxies cannot be used for the determination of mass distribution, since the circular velocities are considerably higher than the rotational ones.

3. The mass-to-luminosity ratio increases considerably towards the centres of galaxies.

4. Massive galaxies may be surrounded by coronae of large mass and dimensions.

5. There exists a close correlation between the colour and the mass-to-luminosity ratio of old populations. Both the colour and the mass-to-luminosity ratio depend on the initial chemical composition of stellar populations.

6. The dispersion σ_z of the vertical velocities of the disk population stars is almost independent of the height above the galactic plane. The dispersion σ_R of the galactocentric radial velocities increases towards the centre of the Galaxy.

7. The ratio of velocity dispersions $k_z = \sigma_z^2/\sigma_R^2$ increases with the increasing thickness of the population.

8. The angular momentum distribution of spiral galaxies differs considerably from the corresponding distribution of uniformly rotating homogeneous spheroids.

Acknowledgements

The work reported in this review has been partly carried out in accordance with an agreement between the Academies of Sciences of the GDR and the Estonian SSR. The author is deeply grateful to Prof. H.-J. Treder, Director of the Zentralinstitut für Astrophysik, and Dr. H. Oleak for sending photometric data on M87 prior to publication, and to Mrs. Liia Einasto, Mrs. Urve Rümmel, Messrs. A. Kaasik, P. Kalamees, P. Traat and J. Vennik for taking part in data processing and preparation of the manuscript. Fruitful conversations with Prof. G. Kuzmin, Prof. D. Lynden-Bell, Dr. L. Ozernoy and Dr. E. Saar are also to be mentioned.

Note added in proof: Recently Ostriker and Peebles (1973) showed that flattened galaxies are stable against the formation of bars only if the rotational energy of the galaxy is smaller than 14 per cent of its total gravitational energy. This condition requires that normal spiral galaxies must have spherically distributed halos or coronae, exceeding in mass the known halo population stars considerably. This is an independent strong argument in favour of the presence of mass of unknown origin in spiral galaxies.

References

Arp, H., Bertola, F.: Astrophys. Letters **4**, 23 (1969).
Baade, W.: Astrophys. J. **100**, 137 (1944).
Baade, W., Arp, H.: Astrophys. J. **139**, 1027 (1964).
Bergh, S. van den: Astrophys. J. **125**, 445 (1957).
Bergh, S. van den: Astrophys. J. Suppl. **9**, 65 (1964).
Bergh, S. van den: Astron. J. **72**, 70 (1967).
Bergh, S. van den: Comm. David Dunlap Obs. No. 195 (1968).
Bergh, S. van den: Astrophys. J. Suppl. **19**, 145 (1969).
Brandt, J. C., Roosen, R. G.: Astrophys. J. **156**, L 59 (1969).
Brosche, P.: Astron. Astrophys. **6**, 240 (1970).
Contopoulos, G. (ed.): The Theory of Orbits in the Solar System and in Stellar Systems. IAU Symp. No. 25. London-New York: Academic Press 1966.
Contopoulos, G.: Contr. Astron. Department, University of Thessaloniki No. 53 (1970).
Contopoulos, G., Strömgren, B.: Tables of Plane Galactic Orbits. New York: Inst. Space Studies 1965.

Courtes, G., Carranza, G., Georgelin, Y., Monnet, G., Pourcelot, A.: Ann. Astrophys. **31**, 63 (1968).
Crampin, D. J., Hoyle, F.: Astrophys. J. **140**, 99 (1964).
Deharveng, J. M., Pellet, A.: Astron. Astrophys. **1**, 208 (1969).
Dixon, M. E.: Monthly Notices Roy. Astron. Soc. **137**, 337 (1967a).
Dixon, M. E.: Astron. J. **72**, 429 (1967b).
Dixon, M. E.: Monthly Notices Roy. Astron. Soc. **140**, 287 (1968).
Eddington, A. S.: Monthly Notices Roy. Astron. Soc. **76**, 37 (1915).
Eelsalu, H.: Publ. Tartu Astron. Obs. **33**, 153 (1958).
Eggen, O. J., Lynden-Bell, D., Sandage, A. R.: Astrophys. J. **136**, 748 (1962).
Einasto, J.: In J. C. Pecker (ed.), Transactions IAU, vol. XIIB, p. 436. London-New York: Academic Press 1966.
Einasto, J.: Publ. Tartu Astron. Obs. **36**, 396 (1968a).
Einasto, J.: Publ. Tartu Astron. Obs. **36**, 414 (1968b).
Einasto, J.: Astrofizika **5**, 137 (1969a).
Einasto, J.: Astron. Nachr. **291**, 97 (1969b).
Einasto, J.: Astrofizika **6**, 149 (1970a).
Einasto, J.: Teated Tartu Astron. Obs. No. 26, 1 (1970b).
Einasto, J.: Structure and Evolution of Regular Galaxies. Tartu (1971) (unpublished).
Einasto, J.: In D. S. Evans (ed.), External Galaxies and Quasi-Stellar Objects. IAU Symp. No. 44, p. 37. Dordrecht-Holland: D. Reidel 1972a.
Einasto, J.: Astrophys. Letters **11**, 195 (1972b).
Einasto, J.: Teated Tartu Astron. Obs. No. 40 (1972c).
Einasto, J., Einasto, L.: Teated Tartu Astron. Obs. No. 36, 3 (1972a).
Einasto, J., Einasto, L.: Teated Tartu Astron. Obs. No. 36, 46 (1972b).
Einasto, J., Kutuzov, S. A.: Teated Tartu Astron. Obs. No. 10, 1 (1964).
Einasto, J., Rümmel, U.: Astrofizika **6**, 241 (1970).
Einasto, J., Rümmel, U.: Teated Tartu Astron. Obs. No. 36, 55 (1972).
Griffin, R. F.: Observatory **92**, 28 (1972).
Grossman, K., Yuan, C.: Tables of Plane Galactic Orbits with Spiral Field. New York: Inst. Space Studies 1970.
Gurevič, L. E.: Voprosy Kosmogonii **2**, 151 (1954).
Hartwick, F. D. A.: Astrophys. J. **163**, 431 (1971).
Hill, E. R.: Bull. Astron. Inst. Neth. **15**, 1 (1960).
Hodge, P. W.: Astron. J. **66**, 249 (1961a).
Hodge, P. W.: Astron. J. **66**, 384 (1961b).
Hoerner, S. von: Mitt. Astron. Rechen-Inst. Heidelberg, Ser. A, No. 13 (1960).
Holmberg, E.: Uppsala Astr. Obs. Medd. No. 148 (1964).
Hube, D. P.: Astron. Astrophys. **9**, 142 (1970).
Innanen, K. A.: Astrophys. J. **143**, 150 (1966).
Innanen, K. A., Fox, D. R.: Z. Astrophys. **66**, 308 (1967).
Jeans, J. H.: Monthly Notices Roy. Astron. Soc. **76**, 70 (1915).
Jôeveer, M.: Publ. Tartu Astron. Obs. **36**, 84 (1968).
Jôeveer, M.: Teated Tartu Astron. Obs. No. 37, 3 (1972).
Johnson, H. M.: Astrophys. J. **133**, 309 (1961).
King, I. R.: Astron. J. **67**, 471 (1962).
King, I. R.: Publ. Astron. Soc. Pacific **83**, 377 (1971).
Kinman, T. D.: Astrophys. J. **137**, 213 (1963).
Kukarkin, B. V.: The Investigation of the Structure and Evolution of Stellar Systems on the Basis of the Study of Variable Stars. Moscow: Gostechizdat 1949.
Kuzmin, G. G.: Publ. Tartu Astron. Obs. **32**, 5 (1952).
Kuzmin, G. G.: Publ. Tartu Astron. Obs. **32**, 332 (1953).
Kuzmin, G. G.: Publ. Tartu Astron. Obs. **33**, 3 (1955).
Kuzmin, G. G.: Astron. Z. **33**, 27 (1956).
Kuzmin, G. G.: Publ. Tartu Astron. Obs. **33**, 351 (1961).
Kuzmin, G.: Astronomy and Geodesy in Estonian SSR. Tartu, p. 52 (1969).
Kutuzov, S. A., Einasto, J.: Publ. Tartu Astron. Obs. **36**, 341 (1968).

Liller, M. H.: Astrophys. J. **132**, 306 (1960).
Lynden-Bell, D.: In H. van Woerden (ed.), Radio Astronomy and the Galactic System. IAU Symp. No. 31, p. 257. London-New York: Academic Press 1967a.
Lynden-Bell, D.: Monthly Notices Roy. Astron. Soc. **136**, 101 (1967b).
Lynden-Bell, D.: Nature **223**, 690 (1969).
Matsuda, T.: Prog. Theor. Phys. **43**, 1491 (1970).
McClure, R. D.: Astron. J. **74**, 50 (1969).
Minkowsky, R.: In G. G. McVittie (ed.), Problems of Extra-Galactic Research, p. 112. New York: MacMillan 1962.
Morgan, W. W., Mayall, N. U.: Publ. Astron. Soc. Pacific **69**, 291 (1957).
Murray, C. A., Sanduleak, N.: Monthly Notices Roy. Astron. Soc. **157**, 273 (1972).
Ollongren, A.: Bull. Astron. Inst. Neth. **16**, 241 (1962).
Oort, J. H.: Ric. Astron. Specola Astr. Vatic. **5**, 415 (1958).
Oort, J. H.: Bull. Astron. Inst. Neth. **15**, 45 (1960).
Oort, J. H.: In F. J. Kerr and A. W. Rodgers (eds.), The Galaxy and the Magellanic Clouds. IAU-URSI Symp. No. 20, p. 1. Canberra: Australian Academy of Science 1964.
Oort, J. H.: In A. Blaauw and M. Schmidt (eds.), Galactic Structure, p. 455. Chicago: Chicago University Press 1965.
Oort, J. H.: Pont. Acad. Scient. Scripta Var. **35**, 321 (1971).
Ostriker, J. P., Peebles, P. J. E.: Astrophys. J. (1973) (in press).
Ozernoy, L. M.: In D. S. Evans (ed.), External Galaxies and Quasi-Stellar Objects. IAU Symp. No. 44, p. 290. Dordrecht-Holland: D. Reidel 1972.
Ozernoy, L. M., Usov, V. V.: Astrophys. Space Sci. **13**, 3 (1971).
Parenago, P. P.: Trudy gos. Astr. Inst. Sternberga **20**, 26 (1951).
Perek, L.: Adv. Astron. Astrophys. **1**, 165 (1962).
Racine, R.: J. Roy. Astron. Soc. Can. **62**, 367 (1968).
Reddish, V. C.: In A. Beer (ed.), Vistas in Astronomy, vol. 7, p. 173. London: Pergamon Press 1966.
Richter, N., Högner, W.: Astron. Nachr. **287**, 261 (1963).
Roberts, M. S.: Astrophys. J. **144**, 639 (1966).
Roberts, M. S.: Oral presentation to the IAU Commission 33 Meeting, Prague, 1967.
Rootsmäe, T.: Tartu Astron. Obs. Kalender p. 74 (1943).
Rubin, V. C., Ford, W. K., Jr.: Astrophys. J. **159**, 379 (1970).
Saar, E.: In preparation (1972).
Salpeter, E. E.: Astrophys. J. **121**, 161 (1955).
Sandage, A.: Astrophys. J. **125**, 422 (1957).
Sandage, A.: Astrophys. J. **152**, L 149 (1968).
Sandage, A.: Quart. J. Roy. Astron. Soc. **13**, 282 (1972).
Sandage, A. R., Becklin, E. E., Neugebauer, G.: Astrophys. J. **157**, 55 (1969).
Sandage, A., Freeman, K. C., Stokes, N. R.: Astrophys. J. **160**, 831 (1970).
Sanduleak, N.: Astron. J. **74**, 47 (1969).
Schmidt, M.: Bull. Astron. Inst. Neth. **13**, 15 (1956).
Schmidt, M.: Astrophys. J. **129**, 243 (1959).
Schwarzschild, M., Bernstein, S.: Astrophys. J. **122**, 200 (1955).
Shapiro, S. L.: Astron. J. **76**, 291 (1971).
Sharov, A. S.: Astron. Z. **40**, 900 (1963).
Sharov, A. S.: Astron. Z. **45**, 146 (1968a).
Sharov, A. S.: Astron. Z. **45**, 335 (1968b).
Spinrad, H.: Astrophys. J. **135**, 715 (1962).
Spinrad, H.: Pont. Acad. Scient. Scripta Var. **35**, 45 (1971).
Spinrad, H., Greenstein, J. L., Taylor, B. J., King, I. R.: Astrophys. J. **162**, 891 (1970).
Spinrad, H., Gunn, J. E., Taylor, B. J., McClure, R. D., Young, J. W.: Astrophys. J. **164**, 11 (1971).
Spinrad, H., Taylor, B. J.: Astrophys. J. **163**, 303 (1971a).
Spinrad, H., Taylor, B. J.: Comments Astrophys. Space Phys. **3**, 40 (1971b).
Spitzer, L. Jr., Schwarzschild, M.: Astrophys. J. **118**, 106 (1953).
Tammann, G. A.: Astron. Astrophys. **8**, 458 (1970).
Tinsley, B. M.: Astrophys. J. **151**, 547 (1968).

TRURAN, J. W., CAMERON, A. G. W.: Nature **225**, 710 (1970).
VAUCOULEURS, G. DE: Monthly Notices Roy. Astron. Soc. **113**, 2 (1953).
VAUCOULEURS, G. DE: Astrophys. J. **128**, 465 (1958).
VAUCOULEURS, G. DE: Astrophys. Letters **4**, 17 (1969).
VAUCOULEURS, G. DE, ABLES, H. D.: Astrophys. J. **151**, 105 (1968).
VETEŠNIK, M.: Bull. Astron. Inst. Czech. **13**, 180 (1962).
VORONTSOV-VELYAMINOV, B. A.: Extragalactic Astronomy. Moscow: Nauka 1972.
WAYMAN, P. A.: Roy. Observ. Bull. No. 36 (1961).
WESTERLUND, B. E., WALL, J. V.: Astron. J. **74**, 335 (1969).
WOERDEN, H. VAN (ed.): Radio Astronomy and the Galactic System. IAU Symp. No. 31. London-New York: Academic Press 1967.
YOUNG, D. S. DE: Astrophys. J. **173**, L 7 (1972).

Discussion

MAVRIDIS, L. N.:
Which is the value of the mass-to-luminosity ratio of the gaseous corona surrounding, according to your model, the galaxies?

EINASTO, J.:
Very high.

BOULESTEIX, J.:
About the problem of "missing mass" in galaxies you said that in outer regions of spiral galaxies there exist large systematical deviations of rotational velocities from circular ones. In the case of M 33, which was studied by COURTES *et al.* (1968), the recent observations and results show that the large difference in rotation curves of opposite sides seem to disappear in outer parts (5 kpc from the center), while they are very important between 0.5 kpc and 2 kpc. Is that coherent with your model?

EINASTO, J.:
To solve the problem of the existence of "missing mass" it is necessary to find reasons for the systematical differences between circular and rotational velocities. The differences of rotational velocity curves of opposite sides of galaxies only show that differences between circular and rotational velocities may exist. Our models are made in two variants in the one case we suppose that there exist a massive corona, responsible for the differences in velocity curves. In the another case, we have no corona and suppose that the velocity curve differences are caused by systematic deviations of rotational velocities from circular ones. Presently the problem is open.

The Gravitational N-Body Problem for Star Clusters

(Invited Lecture)

By Roland Wielen
Astronomisches Rechen-Institut, Heidelberg, GFR.

With 11 Figures

Abstract

We summarize recent numerical studies on the gravitational N-body problem for small stellar systems containing up to 500 stars. The principal aims and fundamental problems of this experimental approach to the dynamical evolution of stellar systems are discussed. The properties of the investigated star cluster models are described and the general results of numerical N-body experiments are reviewed. A quantitative comparison between the experimental results and the predictions of statistical theories is carried out for the evolution of the density distribution and for escaping stars of isolated, spherical star clusters. The simulation of the dynamical evolution of open clusters, with special emphasis on the resulting total lifetimes, and of clusters of galaxies are discussed.

I. Introduction

The dynamical evolution of small stellar systems containing a few hundred stars can be studied by numerically integrating the equations of motion of all the stars as an N-body problem. The equations of motion of N stars under their mutual gravitational attraction are given by

$$\ddot{\boldsymbol{r}}_i = \sum_{\substack{j=1 \\ j \neq i}}^{N} \frac{G m_j}{|\boldsymbol{r}_j - \boldsymbol{r}_i|^3} (\boldsymbol{r}_j - \boldsymbol{r}_i); \qquad i = 1, \ldots, N. \tag{1}$$

Here, r_i is the position vector of the star no. i, m_i the mass of this star, and G the gravitational constant. These equations of motion may be extended by adding external gravitational fields or by considering a change of the masses m_i with time.

From a methodic point of view, the numerical N-body experiments belong to two branches of dynamical astronomy, namely to celestial mechanics and to stellar dynamics. As in celestial mechanics, the basic procedure is the computation of the individual orbits of all the stars in the exact gravitational field of the other stars. However, contrary to celestial mechanics, these individual orbits are not the information at which we aim. Generally in stellar dynamics, we are interested in the evolution of the system as a whole. Hence, we finally deduce from all the individual orbits the desired statistical information about the evolution of the whole system. A typical example for such a "macroscopic" property of a system is the average mass density as a function of position in the system. Of course, the individual orbits also give valuable insight into the "microscopic" behaviour of a stellar system; e.g., we can analyse in detail the dynamical mechanism by which escaping stars gain their positive energy.

What can we hope to learn from numerical N-body experiments? There are two different applications of the results: First, we can test the statistical theories of stellar dynamics. For that purpose, the physical situation should be chosen as simple as possible in order to test the basic assumptions and predictions of the theories. For example, we should study isolated spherical clusters of stars of equal mass. Only if the theories can successfully describe the results obtained from such simple experiments, should we introduce additional complications like different masses of the stars, external fields, etc. Second, we can simulate the dynamical evolution of real stellar systems. At present, we are able to handle systems of up to 500 stars over an astronomically significant period in the numerical experiments. Such a number N is typical for many open star clusters and clusters of galaxies.Therefore, we can compare the N-body experiments directly with these types of astronomical objects without any further theory, provided that our numerical models contain all the physically relevant effects. For example, for open clusters we have to include the actual spectrum of stellar masses, perhaps a mass loss due to the internal evolution of the stars, the tidal field of the Galaxy, and gravitational shocks by passing HI-clouds.

II. Fundamental Problems

The numerical N-body experiments have the advantage that they are, as far as possible, free from mathematical assumptions which are not physically inherent in the problem. In contrast, all the presently available statistical theories of stellar dynamics have to introduce such additional assumptions. However, there also exists an hitherto unsolved fundamental problem in the interpretation of N-body experiments. The problem is caused by the fact that the solutions of the general N-body problem for star clusters are in most cases highly unstable. This instability of the individual orbits of the stars in a cluster is basically a physical phenomenon; the instability is not caused by the numerical technique of integration. The instability of the solutions of the N-body problem for star clusters has first been pointed out by MILLER (1964), who discusses this problem in detail in a recent paper (MILLER, 1971).

Consider a cluster with some given initial values for the coordinates and velocities of all the stars, and take a second cluster which differs from the first one by very small differences in the initial values. If we compare then the orbits of the corresponding stars in the two clusters, we find that these orbits diverge rapidly from each other. The differences grow on the average exponentially with time. The physical reason for this instability is the amplification of any small perturbation by encounters with other stars, or, more generally, by fluctuations of the gravitational field on a small scale in space and time. For instance, a slight variation of the impact parameter for a close encounter produces a large change in the angle of deflection. This behaviour of stellar systems differs drastically from the expected stability of the solar system. The planetary system is stabilized by the large relative mass of the sun compared to all the planets and by the highly ordered motions of the planets.

The unavoidable errors during the numerical integration start the instability of the individual orbits. Hence we cannot trace the individual orbits over long periods of time by any numerical technique. Our numerical star cluster model deviates more and more from the "true" system during the course of its evolution. The numerical experiments are reversible in time only over short periods. Furthermore, the N-body experiments are not reproducable in detail, e.g., if one changes the method of integration or the computer, because of the instability.

The fundamental problem is now whether or not the instability on the microscopic level (individual orbits) destroys also the validity of the statistical results for macroscopic properties derived from N-body experiments. Unfortunately, this problem is still unsolved, in spite of its fundamental importance for the applicability of the results of the numerical experiments. This shows that the experiments are not as free from additional assumptions as one might expect at the beginning. It is generally argued, and supported by the comparison of different experimental results, that the derived statistical properties of the dynamical evolution of a stellar system as a whole are not biased by the microscopic instability for the following reasons: (a) There is no indication that either the basic instability itself or the numerical integration introduce any significant systematic error into the statistical behaviour of the computed individual orbits. (b) The additional "numerical" relaxation caused by the instability is much slower than the physical relaxation in the star cluster. For instance, the change of the total energy of a star during a close encounter and the angle of deflection are always very accurately computed. Hence each "elementary event" in the evolution of the cluster is correctly reproduced by the experiments. It is only the detailed sequence of such events which is affected by the instability. On the other hand, the computed system always looses information on the long-period correlations between the motions of the stars, and the consequence of this is unknown at present.

The microscopic instability seems to be a general property of ergodic or quasi-ergodic systems. For example, a gas or a plasma show a similar microscopic instability which does not affect their statistical behaviour. Furthermore, the ergodic orbits in the restricted three-body problem show the same type of instability as the N-body experiments. One should investigate the statistical properties of those orbits by numerical and analytical methods. A comparison of both results would give valuable information whether such an orbital instability affects the statistical properties. After this warning about an unsolved problem in the basic concept of N-body experiments, we shall now discuss the general properties of star cluster models investigated by numerical N-body experiments.

III. Properties of Star Cluster Models

a) General References. During the last decade, since the pioneering work of VON HOERNER (1960), the computer simulation of stellar systems has been developed quite extensively. Many symposia and colloquia were partly or fully devoted to this subject. The proceedings of those conferences, listed at the end of this paper, may be used as convenient general references. We shall not try to summarize here all those investigations, but we shall concentrate on star cluster models which contains about 100 stars or more. While the studies of very small systems, containing between 10 and 50 stars, give also interesting results (e.g. VAN ALBADA, 1968; HAYLI 1970), it seems to be safer to use the larger systems in order to derive results which are more representative for actual star clusters. Such larger systems, containing up to 500 stars, have been extensively studied by AARSETH (1963, 1966, 1968, 1969, 1971a, b, c) and by WIELEN (1967a, b, 1968, 1969, 1971a). AARSETH has summarized his investigations in a concise form in a forthcoming review paper (AARSETH, 1973). This allows us to concentrate here on the star cluster models which I have studied, and to perform a comparison between AARSETH's and my models. It is very encouraging that the conclusions derived from both sets of models are generally in good agreement.

b) Numerical Integration. We shall not attempt here a presentation of the numerical techniques of integrating the gravitational N-body problem. AARSETH (1971c) has reviewed the direct integration methods including regularization of close two-body encounters and binaries. A general derivation of the basic polynomial method used in the integration is given by WIELEN (1967a). An essential feature of these methods is the use of a step size which varies with time and differs from star to star (variable and individual time steps). A meeting on the numerical solution of ordinary differential equations with special emphasis on the gravitational N-body problem will be held at Austin, Texas in October 1972. It may give a fresh impetus to the improvement of the numerical techniques. The main problem is always the most economic use of computer time. For a standard computer (IBM 7090), the typical integration time per mean crossing time is of the order of one hour for $N=100$ and increases with N^2, e.g., to one day for $N=500$. In order to cover a significant period of the dynamical evolution, a star cluster model has to be integrated for several crossing times. Hence the time requirements obviously limit severely the progress in the numerical studies of star cluster models.

c) Basic Time Scales. The total gravitational field in a stellar system may be formally divided into (a) the regular field, which is caused by the smoothed-out mass distribution of the whole system, and (b) the irregular field, which is mainly due to the near neighbours of a star. A star describes a smooth orbit in the regular field. This orbit is, however, perturbed by encounters with other stars, or, more generally stated, by the irregular field which fluctuates rapidly in space and time.

The basic time unit, which is now generally used in the N-body experiments, is the crossing time T_{cr}. This is the time which an average star needs to cross the cluster once; or quantitatively:

$$T_{cr}=2\bar{R}/v_{rms}, \tag{2}$$

where $\bar{R}$ is the harmonic mean distance between all pairs of stars in the cluster, and v_{rms} is the root mean squared velocity of the stars. Using the virial theorem, the crossing time can be expressed in terms of the total energy $\mathscr{E}$ and the total mass $\mathscr{M}$ of the cluster (e.g. VAN ALBADA, 1968):

$$T_{cr}=G\mathscr{M}^{5/2}/(-2\mathscr{E})^{3/2}. \tag{3}$$

Table 1. Typical values of the basic time scales

Quantity	Open cluster	Globular cluster	Cluster of galaxies
N	10^3	$2 \cdot 10^5$	10^2
$\mathcal{M}$ [$\mathcal{M}_\odot$]	500	10^5	10^{13}
R [pc]	1	5	$5 \cdot 10^5$
T_{cr} [years]	$4 \cdot 10^6$	$3 \cdot 10^6$	$10 \cdot 10^9$
T_{rh} [years]	$2 \cdot 10^7$	$2 \cdot 10^9$	$9 \cdot 10^9$

The crossing time T_{cr} measures essentially the mean orbital period of a star in the regular field, or, in macroscopic terms, the hydrodynamical time scale of the cluster ($T_{cr} \propto (G\langle\rho\rangle)^{-1/2}$).

For comparison with statistical theories of stellar dynamics, we shall adopt the relaxation time T_{rh}, as defined and used by SPITZER in his recent papers (e.g. SPITZER and HART, 1971a). We find the following relation between T_{rh} and T_{cr}:

$$T_{rh} = (0.0143\, N/\log_{10}(0.4\, N))\, T_{cr}. \tag{4}$$

The relaxation time T_{rh} measures the rate with which encounters between the stars can effectively change the regular orbits of the stars and the overall state of the cluster. According to Eq. (4), the ratio of the two time scales, T_{rh}/T_{cr}, depends on the number N of stars in the cluster only. For $N=100$, T_{rh} and T_{cr} are about equal; for $N=1000$, the ratio T_{rh}/T_{cr} is 5.5. Hence it is always difficult to clearly separate the effects of a general mass motion (affecting the regular field) and of relaxation (irregular field) in small clusters. The two time scales become well separated only for globular clusters ($T_{rh}/T_{cr} \sim 1000$) and richer systems (e.g. galaxies). In Table 1, we list typical values for the basic time scales in actual clusters.

The general scheme of the dynamical evolution of a star cluster, expected from theoretical considerations (see e.g., HÉNON, 1964; MICHIE, 1964) and essentially confirmed by the numerical N-body experiments, is the following: Any bound cluster which may not be initially in a dynamical equilibrium, settles into a quasistationary state after a few crossing times. Then, this dynamical equilibrium is changed by the relaxing effect of encounters on a time scale characterized by the relaxation time. No statistical equilibrium can strictly be reached. The cluster evolves towards a final state in which almost all stars have escaped, i.e. the cluster disperses in space, while a few remaining stars (at least one binary) absorb the binding energy of the whole cluster.

d) Initial Conditions. The initial condition (positions, velocities, masses of the stars) may be chosen under various aspects. The state of real stellar systems at the epoch of formation is unknown at present. Even the present state of an actual cluster is not completely known, since we cannot measure the detailed velocity distribution in clusters. Hence, we may either choose very simple initial conditions, e.g., a homogeneous sphere for the positions and velocities, as is done by AARSETH. On the other hand, we may adopt certain initial conditions which are appropriate from a theoretical point of view. For example, we know stationary and stable solutions of the encounterless Liouville equation for the distribution function of the stars in the 6-dimensional phase space. The dynamical state of such a stellar system would be constant in time, if there were no relaxation by encounters. In other words, any dynamical evolution which we observe for such systems in numerical N-body experiments, is due to relaxation alone. In this way, we can easily separate the effects of relaxation and overall mass motions, in spite of $T_{rh} \sim T_{cr}$. We use as a special stationary and stable initial state the Plummer model (WIELEN, 1967a, b). This is a spherical system with a density distribution $\rho(r)$ as in a polytropic gas sphere of Emden index 5:

$$\rho(r) = \frac{\mathcal{M}}{\frac{4\pi}{3} R^3} \frac{1}{\left(1+\left(\frac{r}{R}\right)^2\right)^{5/2}}. \quad (5)$$

The Plummer model has no finite limiting radius, but finite mass. For describing the length scale of the model, we use the median radius in projection, R. The velocity distribution is isotropic, depends on r and differs from a Maxwellian one significantly.

Table 2 summarizes the initial properties of the star cluster models which I have studied. The models contain between 50 and 500 stars, and are labeled by characters for identification. They differ either in the spectrum of stellar masses (equal masses or a realistic spectrum), or by the amount of overall rotation, or by the presence of the tidal field of the Galaxy, or by an assumed mass loss of stars due to internal evolution, or simply by different random numbers used for producing the actual initial coordinates. For the "realistic" spectrum of stellar masses, given quantitatively in Table 5, we have basically adopted the standard Salpeter function, $dN \propto m^{-2.35}\, dm$, but have formed discrete mass groups.

Initially, the stars in our models are distributed independently of each other, by using a one-particle distribution function. In real stellar systems, correlations between the stars do exist, either from the epoch of formation (e.g. proto-binaries), or these particle correlations are build up during the dynamical evolution. Since we do not now how long it takes to build up representative correlations, we prefer to integrate relatively few models over a longer time rather than to study many models over very short periods. The typical time of integration covers 10 crossing times.

e) General Results. We shall now summarize the general results of the numerical N-body experiments, which have been performed with star cluster models containing up to 500 stars. Most of the following qualitative statements are confirmed by independent investigations of different authors. A selection of the main results will be discussed quantitatively in the following chapters.

(1) The central density in a cluster increases systematically with time. (2) A halo of stars is steadily formed in the outer regions of a cluster. (3) The rate of dynamical evolution is strongly increased by unequal masses of the stars. (4) The most massive stars segregate towards the center of the cluster. (5) In the core of a cluster, one or a few binaries are formed which absorb most of the binding energy of the cluster. (6) During close encounters, some stars gain enough energy for escape from the cluster. (7) The mean velocity of the stars in a cluster decreases

Table 2. Properties of star cluster models

Initial state	Spectrum of stellar masses	External field		Model and time of integration							
				$N=50$		$N=100$		$N=250$		$N=500$	
Plummer's model	Realistic	None		HP	$16\,T_{cr}$	P P2	$10\,T_{cr}$ $6\,T_{cr}$	DP DP2	$7\,T_{cr}$ $11\,T_{cr}$	FP FP2	$11\,T_{cr}$ $10\,T_{cr}$
		Galactic field	$R=0.08\,\xi_L$			G2	$10\,T_{cr}$	DG2	$8\,T_{cr}$		
			$R=0.15\,\xi_L$			G3	$21\,T_{cr}$	DG3	$10\,T_{cr}$		
			$R=0.3\,\xi_L$			G	$10\,T_{cr}$				
	Equal masses	None		HE	$13\,T_{cr}$	E	$18\,T_{cr}$	DE	$30\,T_{cr}$		
Plummer's model with differential rotation	Realistic	None		HR	$38\,T_{cr}$	R	$13\,T_{cr}$	DR	$7\,T_{cr}$		
Plummer's model with initial mass segregation	Realistic	None				M	$16\,T_{cr}$				
Evolved state of models P, G or G2 ($10\,T_{cr}$)	Realistic but with mass loss of evolving stars	None				L L2	$19\,T_{cr}$ $19\,T_{cr}$				
		Galactic field	$R=0.08\,\xi_L$			G2L	$13\,T_{cr}$				
			$R=0.3\,\xi_L$			GL	$13\,T_{cr}$				

with the distance from the center. Clusters are strongly "non-isothermal". (8) In the core of a cluster, the velocity distribution is nearly isotropic. (9) In the outer parts of a cluster, the motions of the stars are primarily in the radial direction and hence strongly nonisotropic. (10) No equipartition of the kinetic energy takes place among the stars of different masses. (11) After a possible early phase of adjustment, the virial theorem is, on the average, nicely fulfilled for bound clusters. (12) A slow overall rotation of a cluster does not significantly affect the dynamical evolution as outlined above.

IV. Comparison with Statistical Theories

a) Evolution of the Density Distribution. The evolution of the density distribution in spherical stellar systems can be conveniently described by plotting the radii $r_M(t)$ of those spheres which contain a fixed fraction M_r of the total mass of the cluster, e.g. 10%, 50% and 90% of $\mathcal{M}$, as a function of time. We shall compare the results for isolated, spherical star cluster models with the predictions of statistical theories given recently by Hénon (1971a, b, 1972) and by Spitzer and his collaborators (Spitzer and Hart, 1971a, b; Spitzer and Shapiro, 1972; Spitzer and Thuan, 1972). In these statistical treatments, it is assumed that the relaxation of the cluster is only due to independent two-body encounters. The effect of these encounters is numerically calculated by a Monte-Carlo technique for quite realistic cluster models (e.g. with non-isotropic velocity distributions). The procedure corresponds formally to a solution of the Fokker-Planck equation which describes the diffusion of stars in the velocity space. Another technique for studying the dynamical evolution of clusters on the basis of Fokker-Planck encounter terms has been applied by Larson (1970a, b). He uses a hydrodynamical approach based on the moments of the velocity distribution.

In Fig. 1 we compare the results for isolated spherical clusters of stars of equal mass. The two N-body models E ($N=100$) and DE ($N=250$), shown as solid lines, agree very well with each other, if the radii of the spheres are plotted as a function of the elapsed relaxation times. The Monte-Carlo model of Hénon (1972) has the same initial conditions as the N-body models. The agreement between Hénon's model and the N-body results is rather good. However, Hénon has used for that model rather large time steps in the numerical integration. A revised program with smaller time steps leads, for other models, to a faster evolution (Hénon, 1972). If this trend applies also to Hénon's model shown in Fig. 1, then the present agreement may be destroyed (see Hénon, 1973). Spitzer and Thuan (1972) have used different initial conditions for their models. We have plotted in Fig. 1 their model E1/E2 which is most similar to the N-body models E and DE. Spitzer's and Thuan's model evolves faster than the N-body models. Especially dramatic is the complete collapse of the core of their cluster model, as shown by the sphere containing 10% of the total mass, after 21 T_{rh}.

The question whether a finite fraction of a stellar system can really collapse to "infinite" density after a finite period of its dynamical evolution, is of interest for a possible explanation of the active nuclei of galaxies or of quasars (see e.g., Spitzer, 1971). On the average over many models with various initial conditions,

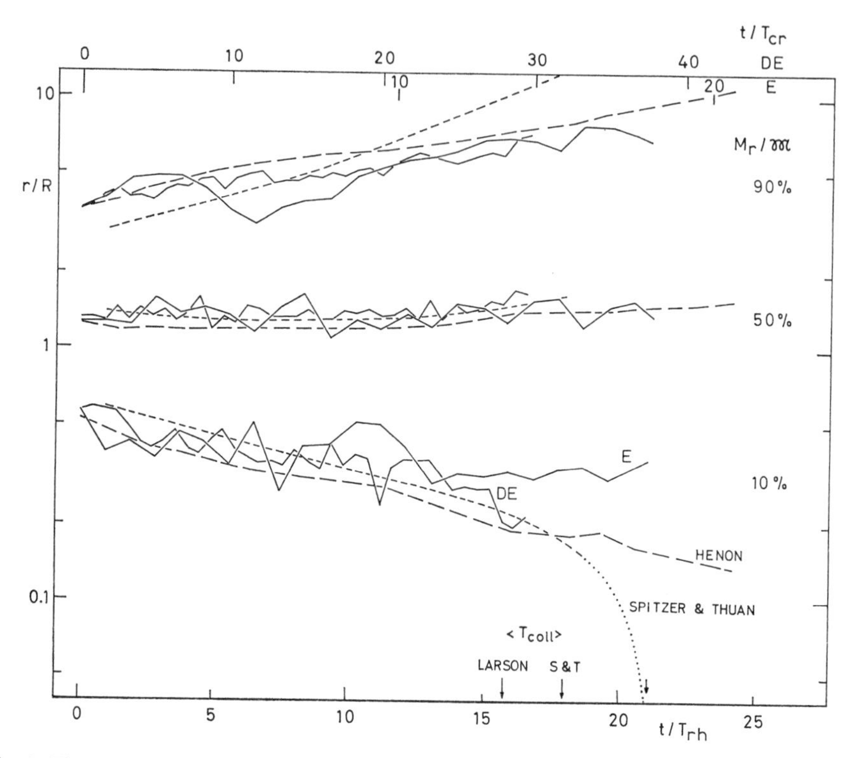

Fig. 1. The evolution of the density distribution in isolated spherical clusters of stars of equal masses, as shown by the change with time of radii of spheres which contain a fixed fraction of total mass. Comparison between N-body experiments (solid lines) and statistical theories (dashed lines)

SPITZER and THUAN find a mean collapse time of about 18 T_{rh}. LARSON predicts even an earlier collapse, after only 16 T_{rh}. HÉNON finds also a infinite central density after a finite time. However, this happens in Henon's model much later ($>50\ T_{rh}$) and does not involve a finite fraction of the mass. In Henon's model, the mass of the collapsing core becomes zero when the central density becomes infinite. Expecially this latter discrepancy between the results of HÉNON and of SPITZER and THUAN is unexplained at present. Unfortunately, the presently available N-body models cannot really clarify the problem of possible infinite central density: The contraction of the core of a cluster is clearly shown by the N-body experiments; an early collapse of a shell containing about 10% of $\mathcal{M}$ before 20 T_{rh} is not compatible with the experimental results, but a collapse after a longer time ($T_{coll} > 25\ T_{rh}$) cannot be ruled out. Furthermore, it is unclear whether the dominant central binary usually formed in N-body experiments can be essentially identified with the extremely high central density predicted by the statistical theories.

We shall now study the increase in the rate of dynamical evolution, if we replace the equal masses of the stars by a realistic spectrum of stellar masses. First, we compare in Fig. 2 the evolution of the density distribution for the N-body models E (equal masses) and P and R (unequal masses). The models E and P

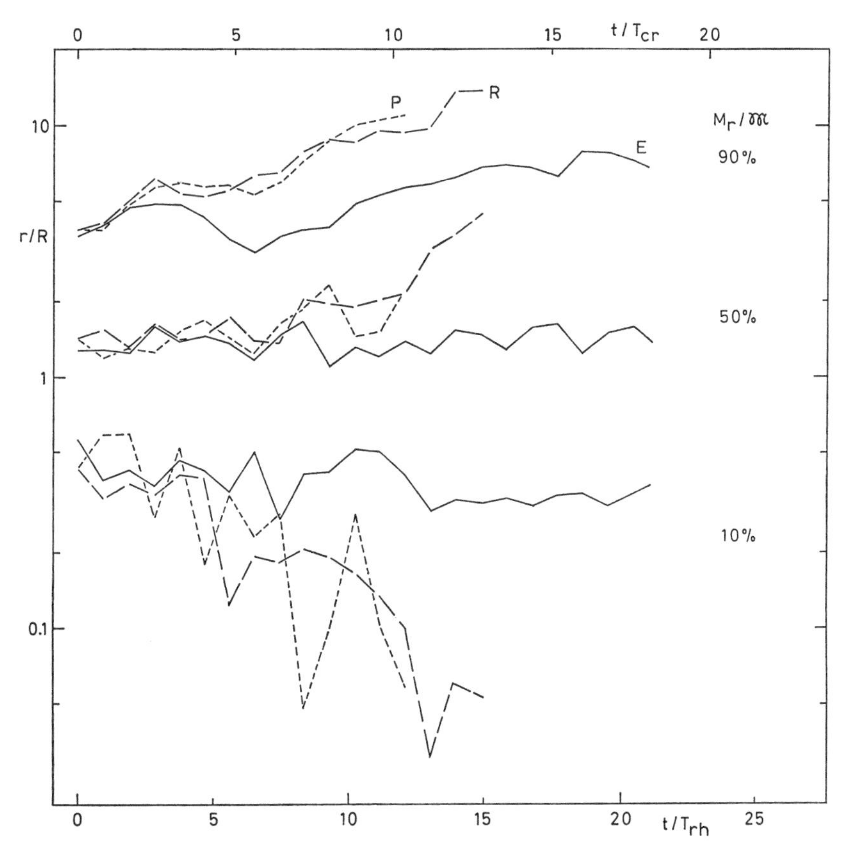

Fig. 2. Evolution of the density distribution in isolated spherical systems. Comparison between clusters containing stars with equal masses (solid line) and those with a realistic mass spectrum (dashed lines)

differ in the mass spectrum, but are initially otherwise strictly identical. A comparison of the models P or R with model E indicates that a star cluster with a realistic mass spectrum evolves by a factor of at least five, perhaps ten, faster than a cluster containing identical stars. After 15 T_{rh}, the mean density inside the sphere containing 10% of $\mathcal{M}$ differs by a factor of almost 10^3 for equal and unequal masses. The nice agreement between the curves for the models P and R proves that the overall rotation of the model R has no significant effect on the evolution of density distribution, especially not on the increase of the central density.

In Fig. 3, we compare the evolution of the N-body models P and R with a corresponding Monte-Carlo model calculated by Hénon (1971 a). The initial state and the mass spectrum of Hénon's model is the same as in the N-body model P. An inspection of Fig. 3 reveals that Hénon's model evolves slightly slower than the N-body models. However, Hènon has used a large time step during the numerical integration for this model. A reasonable correction for this effect, derived from a study of the influence of the chosen step size on the evolution of other models (Hénon, 1972), would lead to a perfect agreement between the N-body models and

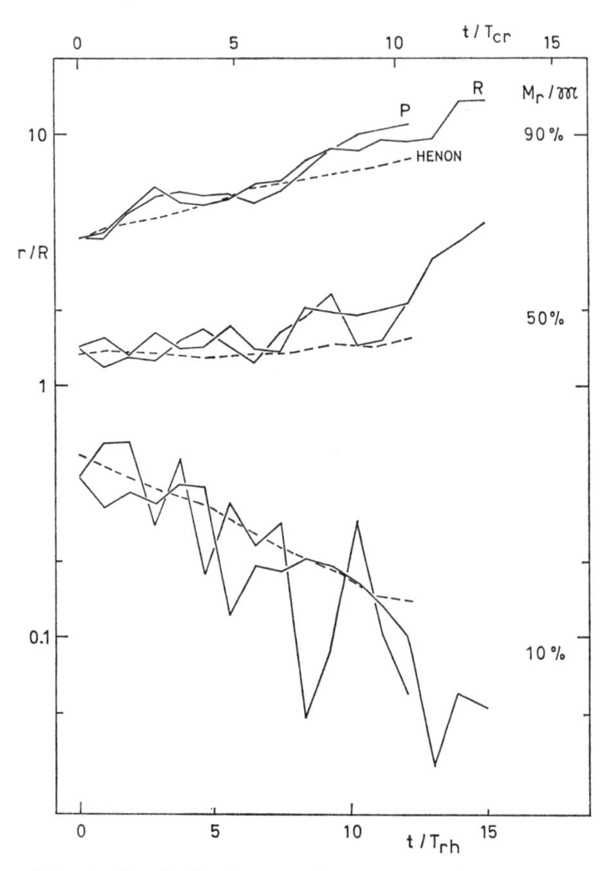

Fig. 3. Evolution of the density distribution. As Fig. 1, but now for clusters with a realistic mass spectrum

the Monte-Carlo result. The Monte-Carlo models of SPITZER and HART (1971b) cannot be used for a conclusive comparison. First, their initial state as well as their mass spectrum differ significantly from those of the N-body models. Second, the diffusion coefficients adopted by SPITZER and HART were independent of the velocity of a star. This assumption leads to a severe overestimate of the evolution rate, as pointed out by SPITZER and THUAN (1972).

Fig. 1 and 3 indicate that the Monte-Carlo description of the dynamical evolution of stellar systems, used by HÉNON and by SPITZER and his colleagues, is able to predict rather successfully the evolution of the density distribution observed in N-body experiments, even for clusters containing stars of unequal masses. Also for other overall properties of clusters, e.g. the degree of segregation of stars of different masses, no severe discrepancies have been found. Therefore, in this respect, the numerical N-body experiments seem to confirm the basic assumption of the considered statistical theories of stellar dynamics: namely, that the diffusion effect of independent distant two-body encounters is the main source of relaxation in stellar systems. However, in the following chapter, a discussion of escaping stars does reveal some discrepancies between the statistical theories and the numerical experiments. Those discrepancies cast a doubt whether the statistical theories in their present form are also able to predict correctly the long-term evolution and the final dissolution of star clusters.

b) Escape of Stars. Stars which escape from a cluster pose one of the main problems of the dynamical evolution of star clusters. The permanent loss of members, of total mass and of total energy does not permit that a star cluster reaches a final equilibrium. The escape of stars always drives the dynamical evolution of a star cluster and leads to a finite lifetime of the cluster. In this chapter, we shall consider isolated clusters. The effect of external fields on the escape of stars will be discussed in Chapter V. In isolated star clusters, escaping stars can be easily identified by their positive total energy,

$$E_i/m_i = \tfrac{1}{2}\dot{\boldsymbol{r}}_i^2 - \sum_{\substack{j=1\\ j\neq i}}^{N} \frac{G m_j}{|\boldsymbol{r}_j - \boldsymbol{r}_i|} > 0. \tag{6}$$

It very seldomly happens that a star which has once reached positive energy, is recaptured by the cluster during its last flight through the cluster.

What is the dynamical mechanism for producing escaping stars? The classical statistical theories of stellar dynamics (for references see Table 3) predict a slow diffusion of the energy of the stars due to distant encounters. In this picture, an escaper has slowly gained energy on the average, until its energy finally passes the critical limit $E_i = 0$. However, it has been pointed out by HÉNON (1960) on theoretical grounds that the diffusion process can hardly produce escapers in this way. The stars which have slowly reached nearly the limit $E_i = 0$, have very long orbital periods, they spend most of their time in the nearly encounterless halo of the cluster and hence their energy cannot anymore increase significantly within a reasonable interval of time.

HÉNON (1960, 1969) has advocated that in isolated clusters the escapers are mainly produced by single close two-body encounters. The numerical N-body experiments confirm essentially this prediction. We find that all the stars which escape from the star cluster models have gained their positive total energy rather suddenly during a very short event. This is illustrated in Figs. 4 and 5, where the total energy E_i of some escapers has been plotted as a function of time. The energy always jumps suddenly to positive values. The basic mechanism for this sudden change of the energy is in most cases a close encounter of the star which escapes afterwards, either with another single star (two-body encounter) or with a binary or a small group of stars in the cluster (multiple encounter). While two-body encounters are more frequent, their effectivness in producing escapers is limited, since two-body encounters are elastic and the escaping star can extract atmost the whole kinetic energy of the encountered star. Multiple encounters, however, can be highly superelastic. For instance, during an encounter with a binary, the binary may become more tightly bound. The positive difference between the former and later binding energy of the binary can be transfered to the escaping star, thus producing sometimes highly energetic escapers. Due to this effect, the mean energy of the escaping stars is about 3 or 4 times larger than the mean kinetic energy of bound

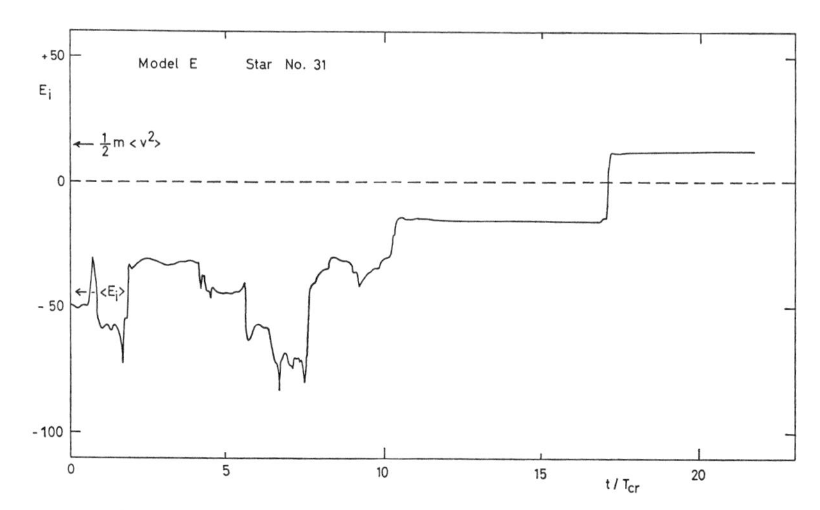

Fig. 4. Change of the total energy of an escaping star with time for a cluster with stars of equal mass

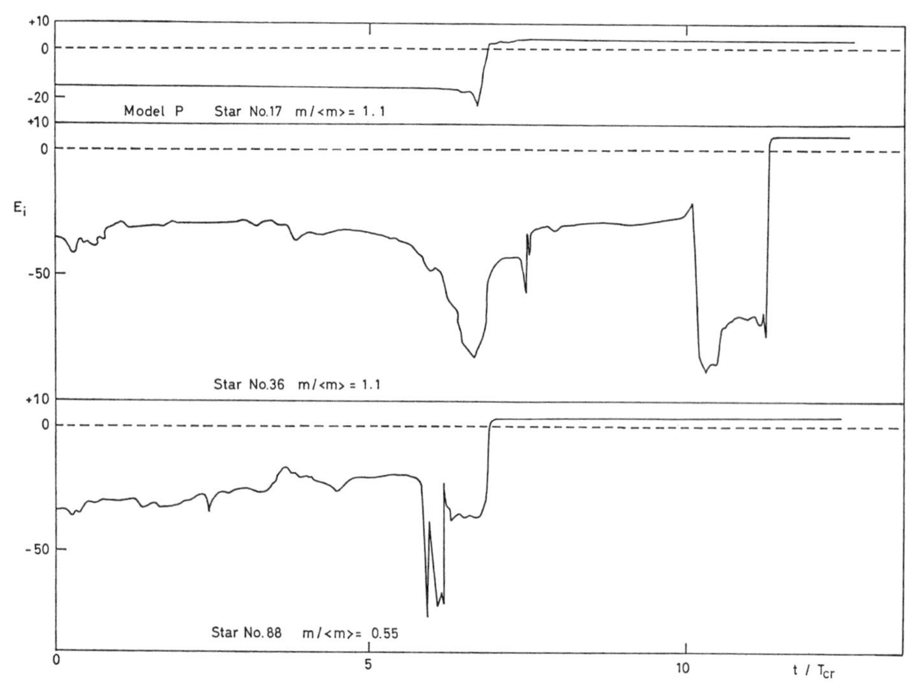

Fig. 5. Change of the total energy of some escaping stars with time for a cluster with a realistic mass spectrum

cluster members. Hence the escapers carry away an important amount of energy. Since the probability for a close encounter is much higher in the dense core of a cluster, most escapers are produced there. More complicated escape processes,

Table 3. Escape rate for isolated clusters (Stars of equal masses)

Number of escapers per crossing time $\dot{N}\,T_{cr}$ [stars]	Source		
0.70	SPITZER		1940
0.55	CHANDRASEKHAR		1942
1.18	CHANDRASEKHAR		1943
1.58	WHITE		1949
0.98	SPITZER and HÄRM		1958
0.31	KING		1958a
0.4	LARSON		1970b
0.055	SPITZER and THUAN		1972
0.027	HÉNON		1960
0.08	WIELEN,	Model HE	$N = 50$
0.06	WIELEN,	Model E	$N = 100$
0.13	WIELEN,	Model DE	$N = 250$
0.17	AARSETH,	Model III	$N = 250$

which include the exchange of components in binaries or the ejection of a member from a bound triple system, as described by VAN ALBADA (1968) for very small clusters ($N \sim 10$), are sometimes also observed in large systems ($N > 100$). The relative frequency of those complicated events is, however, very small for clusters containing a few hundred stars.

It may be argued that the diffusion process becomes more important for rich systems, because the theoretically predicted ratio between the frequency of distant and close encounters is proportional to log N. However, if less than 10% of the escapers are due to diffusion by distant encounters for $N = 500$, then the contribution of distant encounters remains also small for globular clusters ($N \sim 2 \cdot 10^5$), namely less than $(10\% \log(2 \cdot 10^5)/\log 500) \sim 20\%$.

c) Rate of Escape. We shall now discuss quantitatively the rate of escape observed in the N-body experiments. We measure the escape rate by the average number of stars which escape in one crossing time, i.e. $|dN/dt| \cdot T_{cr}$. Since we do not observed a significant variation of the escape rate with time, we form an average over the whole period of integration. HÉNON's theory (1960, 1969) predicts that the quantity $\dot{N}\,T_{cr}$ is independent of N. All the classical theories predict a weak dependence of $\dot{N}\,T_{cr}$ on N, namely $\dot{N} T_{cr} \propto \log_{10}(0.4\,N)$. The numbers quoted in Table 3 for the classical theories refer to $N = 100$.

In Table 3, we compare the various theoretical predictions with the experimental escape rates for clusters containing stars of equal masses. The experimental rates are highly uncertain because of the small number of observed escapers in the case of equal masses. Therefore, no conclusion can be reached whether or not the escape rate $\dot{N}\,T_{cr}$ depends on N in the range of N from 50 to 250. The agreement between the escape rates derived by AARSETH (1973) and WIELEN for $N = 250$ stars of equal masses is very satisfactory. Most of the classical theories of stellar dynamics predict escape rates which are much higher than those derived from the N-body experiments. The recent theoretical treatments based on the Fokker-Planck

Table 4. Escape rate for isolated clusters (Realistic spectrum of stellar masses)

Number of escapers per crossing time $\dot{N} T_{cr}$ [stars]	Source		
0.48	HÉNON		1969
1.0	WIELEN,	Model HP	$N = 50$
0.8	WIELEN,	Models P+P2+R	$N = 100$
1.1	WIELEN,	Models DP+DP2	$N = 250$
1.1	AARSETH,	Model IV	$N = 250$
0.6	WIELEN,	Models FP+FP2	$N = 500$
1.6	AARSETH,	Model V	$N = 500$

equation (LARSON, 1970b; SPITZER and THUAN, 1972; also HÉNON, 1971a) give highly discordant results for the escape rate, ranging from 0.4 to 0.055 to essentially zero. Even if these theories would predict the rate of escape correctly, the severe discrepancy between the observed and predicted mechanism of producing escapers should cast doubts on the significance of such an eventual coincidence. HÉNON's theory of escape (1960) predicts an escape rate for the Plummer model which is too small by a factor of about five, compared to the N-body experiments. What are possible reasons for this disagreement? HÉNON considers only two-body encounters between single stars, and neglects the presence of binaries. Furthermore, he does not take into account in his derivation the formation of the dense core of a cluster. This core may enhance the rate of escape significantly, via the increased density as well as by means of an increase in the relative frequency of multiple encounters.

In Table 4, we present the escape rates derived from star cluster models with a realistic spectrum of stellar masses. The experimental rates for these models are higher than for equal masses by a factor of about 10. Due to the higher number of observed escapers, these escape rates are relatively more accurate than for equal masses. Within the error limits (see Fig. 7), the escape rate, $\dot{N} T_{cr}$, does not depend significantly on N, in agreement with the general statement of Henon's theory of escape. The agreement between the experimental results derived by AARSETH (1973) and WIELEN is sufficiently good. The slight discordance for $N = 500$ may be caused by some differences between the actually used mass spectra (ratio of largest to smallest mass; continuous spectrum instead of mass groups).

In Table 4, the prediction of Henon's theory of escape has been evaluated for a Plummer model with strictly the same mass spectrum as used in our N-body experiments. HÉNON (1969) predicts that the rate of escape for the realistic mass spectrum should be larger than for equal masses by a factor of 18. This is not too far from the experimentally found increase (factor 10). On the contrary, most classical theories of stellar dynamics have, at least implicitly, assumed that the escape rate does not depend severely on the mass spectrum, since the authors applied their results for equal masses directly to real clusters. It must be regarded as a mere accident that the escape rates predicted by many classical theories for equal masses are in good agreement with the experimental rates for a realistic mass

Table 5. Relative rate of escape for stars of different masses $(\Delta N_m/N_m)/(\Delta N/N)$

$\frac{m}{\langle m \rangle}$	$\frac{N_m}{N}$	CHANDRASEKHAR (1942)	SPITZER and HÄRM (1958)	HÉNON (1969)	Isolated models	Model DG3
4.4	4%	$1 \cdot 10^{-8}$	~0	0.05	0.4:	0.3:
2.2	10%	$2 \cdot 10^{-4}$	~0	0.25	0.6:	0.4:
1.1	24%	0.27	0.23	0.71	1.0	1.1
0.55	62%	1.51	1.52	1.30	1.1	1.1

spectrum. Hénon's predicted escape rate is too low by a factor of two, judged from the N-body experiments. The agreement is here even better than for equal masses. The remaining discrepancy is astonishingly small, although HÉNON has ignored, for example, the mass segregation in the cluster.

How does the escape rate depend on the mass of the escaping stars? In Table 5 we give the relative rates of escaping stars as a function of the stellar mass. The relative rates are normalized with the total escape rate for the adopted mass spectrum. The experimental results for isolated clusters represent an average over many of our star cluster models, in order to improve the statistics. We find that the relative escape rate depends only weakly on the stellar mass, in contradiction to the theoretical predictions. The mass segregation certainly favours the escape of massive stars, since these stars spend more time in the dense core than the low-mass stars do. This may explain why Hénon's prediction, in which the segregation is neglected, fails to reproduce the experimental result for the relative escape rates of different masses.

Only stars with a mass significantly larger than the mean stellar mass escape slightly less frequently than the average stars. For small masses, the escape rate seems to be independent of the mass. We find essentially the same result when we consider more realistic star cluster models (e.g. model DG3 in Table 5) in which the galactic tidal field is taken into account. This indicates clearly that the mass function for low-mass stars in star clusters is not affected by the escape of stars. The relative scarcity of faint dwarfs observed in actual open star clusters, with respect to the luminosity function for field stars in the solar vicinity, must be caused by special circumstances during the formation of the clusters, and not by the later dynamical evolution of the clusters. A discussion of this problem and a comparison of N-body experiments with the Hyades have been carried by AARSETH and WOOLF (1972).

V. Simulation of Open Star Clusters

a) Physical Effects for Open Clusters. The dynamical evolution of open star clusters is mainly caused by the following effects: (1) internal relaxation by encounters among the cluster stars, as in isolated systems; (2) gravitational tidal field of the Galaxy; (3) fluctuating gravitational fields of passing interstellar HI-clouds; (4) possible mass loss of evolved massive stars in the late phase of their internal evolution. The relative importance of these effects varies from cluster to cluster, according to the specific properties of the cluster like total mass and median radius.

In the simulation of the dynamical evolution of open star clusters, we assume that the cluster moves in the galactic plane in a circular orbit around the galactic center. Small deviations from such a simple galactic orbit should not be important for the dynamical evolution of the cluster, since it is only the gradient of the gravitational field over the cluster which determines the tidal effect. Hence it is sufficient to use a linear approximation for the galactic field, corresponding to the following potential:

$$\begin{aligned} V_{\mathrm{gal}}(\boldsymbol{r}) = V_0 &+ \tilde{R}_0\,\omega_0^2(\tilde{R}-\tilde{R}_0) \\ &- 0.5\,\omega_0(3A+B)(\tilde{R}-\tilde{R}_0)^2 \\ &+ \left(2\pi\,G\rho_0+\omega_0(A+B)\right)z^2. \end{aligned} \tag{7}$$

Here, $\tilde{R}$ is distance from the galactic center and z that from the galactic plane, ω_0 is the angular velocity of galactic rotation at the distance $\tilde{R}_0$, A and B are Oort's constants, and $\rho_0 \sim 0.15\,\mathcal{M}_\odot/\mathrm{pc}^3$ is the local galactic mass density which mainly determines the galactic force in the z-direction.

Up to now, the effect of passing HI-clouds has been neglected in the N-body experiments carried out by AARSETH and by WIELEN. BOUVIER and JANIN (1970) have experimentally studied the cloud effect for very small systems only ($N=25$). Because of the small N and since they have neglected the tidal field of the Galaxy, their results cannot be directly applied to actual open clusters. At the moment, the effect of the HI-clouds can only be inferred from theoretical studies (SPITZER, 1958; BOUVIER, 1971; PRATA, 1971a, b; WIELEN, 1971a; SPITZER and CHEVALIER, 1973).

At present, our knowledge on the occurence of a significant mass loss of evolving stars is rather vague. The possibilities range from no mass loss (black holes) to slow processes (stellar winds) to explosions (supernovae). The effect of a sudden mass loss of massive stars, after a time derived from stellar evolution theory, on the dynamical evolution of N-body models has been studied by WIELEN (1968, 1969) and by AARSETH (1973). The adopted lifetimes of stars as a function of their mass are indicated in Fig. 9.

What are the essential differences between the dynamical evolution of the isolated N-body star cluster models and of the more realistic ones? The tidal effect of the galactic field (and very probably also that of HI-clouds) affects mainly the halo of the cluster. The halo becomes flattened in the z-direction and elongated in the direction from galactic center to anticenter. The evolution of the core of the cluster and the mass segregation are not severely altered by the external field. The galactic tidal field facilitates the escape of stars drastically, as discussed in detail in the following sections. The mass loss of evolving stars occurs usually in the core of the cluster, because of the segregation of the massive stars. Hence the mass loss leads to a larger expansion of the cluster than it was theoretically expected (VON HOERNER, 1958; OORT and VAN HERK, 1959) for clusters without a mass sedimentation. The escape rate is not significantly changed by a simulated mass loss, because of some compensating effects. Therefore, and because of the basic uncertainties of the mass loss mechanism, we shall concentrate in the following on star cluster models without a mass loss of evolving stars.

b) Relevant Astronomical Observations. What astronomical observations give relevant information about the dynamical evolution of open clusters? For the presently observed density distribution or the mass segregation etc., it is difficult to separate the consequences of a dynamical evolution from the conditions at the epoch of formation of the cluster. One might hope to set up a dynamical evolutionary sequence of star clusters by ordering observed clusters according to their age τ derived from stellar evolution theory. However, it has been shown (WIELEN, 1971b) that the absolute age τ is only very weakly correlated with the relative dynamical age, defined by the ratio of τ to the total lifetime of the cluster. This effect is caused by the dependence of the rate of dynamical evolution on the widely varying initial properties (total mass, median radius, etc.) of open clusters.

The most direct information on the time scale of the dynamical evolution is provided by the observed distribution of ages of open clusters (WIELEN, 1971b). There is no indication that the age distribution of open clusters within 1000 pc from the sun is severely affected by observational selection effects. If the rate of formation of clusters is constant in time, then the observed decrease of the frequency $\nu(\tau)$ of clusters with increasing age τ must be due to the dynamical dissolution of clusters. In this way, we deduce statistically that 50% of newly formed open clusters disintegrate within $2 \cdot 10^8$ years, 10% have a total lifetime longer than $5 \cdot 10^8$ years, and only 2% live longer than $1 \cdot 10^9$ years. Hence, the typical lifetime is short, but there exists a wide spread in the individual lifetimes. The deduced total lifetimes can serve as a powerful observational test of a theory of the dynamical evolution of star clusters. Therefore, we shall concentrate in the following sections on the total lifetimes of clusters predicted by the N-body experiments.

c) Rate of Escape. In most of the N-body experiments, we have not followed the dynamical evolution up to the total dissolution of the star cluster model, because the computing time would be too long. Only a few models, e.g. G and G3, have lost nearly all of their members at the end of the integration time. Hence, we shall usually first derive the rate escape, $\dot{N}\, T_{cr}$, and then estimate from this quantity the total lifetime. This seems possible, since the rate $\dot{N}$ of escaping stars is approximately constant in time, i.e., the number of bound stars in a cluster decreases almost linearly with time.

How can we identify an escaper in a non-isolated star cluster model? It turns out that, instead of the energy used in isolated models, we have now to consider the Jacobian constant C_i of a star:

$$\begin{aligned} C_i = \tfrac{1}{2} \boldsymbol{v}_{\text{corot}}^2 - \sum_{\substack{j=1 \\ j \neq i}}^{N} \frac{G m_j}{|\boldsymbol{r}_j - \boldsymbol{r}_i|} \\ - 2A\,\omega_0 (\tilde{R}_i - \tilde{R}_0)^2 \\ + \big(2\pi G \rho_0 + \omega_0 (A+B)\big)\, z_i^2 . \end{aligned} \tag{8}$$

Here, $\boldsymbol{v}_{\text{corot}}$ is the velocity of the star in the corotating $\xi\eta\zeta$-system explained below. The quantity C_i is constant during the motion of a star, if there are no encounters and if the density distribution of the cluster is stationary in the $\xi\eta\zeta$-system.

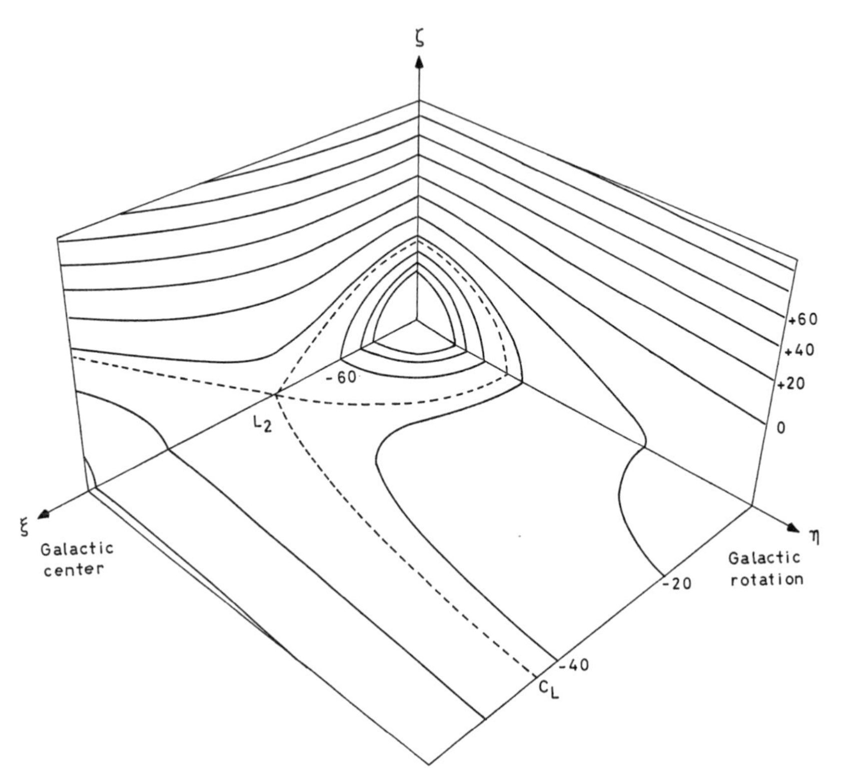

Fig. 6. Equipotential surfaces for a star cluster in the galactic tidal field. The surfaces are symmetric in ξ, η and ζ. The critical surface ($C = C_L$) is indicated by the dashed curves

In Fig. 6, we show the equipotential surfaces, sometimes called zero-velocity surfaces, around a cluster which is embedded in the galactic tidal field. These surfaces correspond to constant values of C_i for $\boldsymbol{v}_{\text{corot}} = 0$. For simplicity, the gravitational potential of the cluster has been approximated by that of a mass point at the center. The ξ-axis points towards the galactic center, the η-axis in the direction of galactic rotation, and the ζ-axis to the north galactic pole. The equipotential surfaces near the core of the cluster are nearly spheres, as in isolated systems. With increasing distance from the center, they become more and more distorted. The last closed equipotential surface passes through the two Lagrangian points L_1 and L_2. This critical surface corresponds to the Jacobian constant

$$C_L = -\tfrac{3}{2}(4A\,\omega_0\, G^2 \mathcal{M}^2)^{1/3}. \tag{9}$$

The distance of the Lagrangian points L_1 or L_2 from the cluster center is given (KING, 1962) by

$$\xi_L = (G\mathcal{M}/4A\,\omega_0)^{1/3}, \tag{10}$$

and is often called the "tidal radius" of the cluster. For $\mathcal{M} = 500\,\mathcal{M}_\odot$, we find $\xi_L = 11$ pc. Since the equipotential surfaces are open for $C > C_L$, we count stars with $C_i > C_L$ as escapers (instead of $E_i > 0$ for isolated systems). Of course, it can take

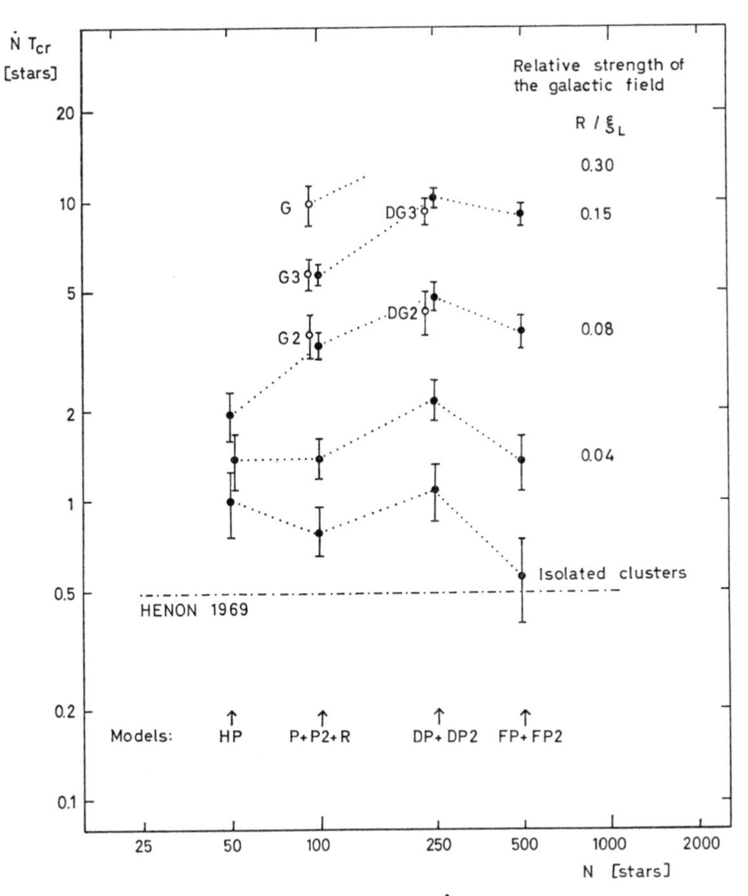

Fig. 7. The number of escaping stars per crossing time, $\dot{N}T_{cr}$, as a function of the number N of stars and of the relative strength of the galactic tidal field, R/ξ_L, according to N-body experiments

a long time until the escaper finds the "hole" in the surface. We have found experimentally that almost all stars with $C_i > C_L$ really leave the cluster after some time.

The mechanism of the production of stars which escape from clusters embedded in the tidal field, is basically the same as for isolated clusters: namely, close encounters with other cluster stars. The tidal field cannot "pull out" stars from the cluster, if their Jacobian constant is less than C_L. The tidal field lowers only the energy limit for escapers from zero to C_L. The "excitation" of the escapers must come from encounters (or perhaps also from passing HI-clouds). Since E_i/m_i is nearly identical with C_i in the core of a cluster, where the escapers are produced, we can rather successfully predict the escape rate of non-isolated clusters from an isolated star cluster model, if we count the stars with $E_i/m_i > C_L$ as escapers. This works very well as long as the main body of the cluster is not strongly perturbed by the tidal field.

In Fig. 7 we summarize our experimental results for the number of escaping stars per crossing time, $\dot{N}\,T_{cr}$. The open circles represent N-body models in which

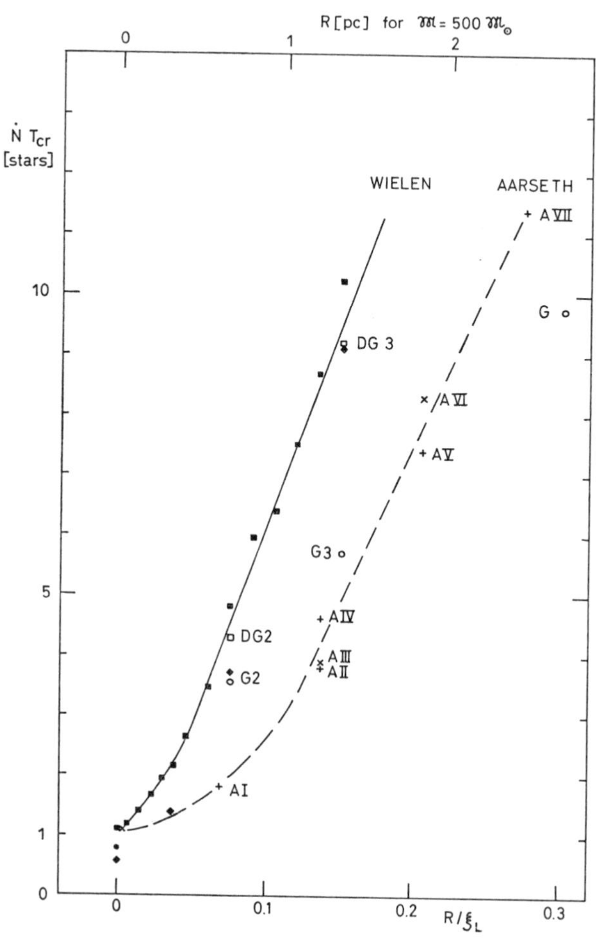

Fig. 8. Comparison of the escape rates for non-isolated star cluster models, as derived by AARSETH and WIELEN. Aarseth's models ($N=250$): + with mass loss, × without mass loss. Wielen's models (without mass loss): □ and ■ $N=250$, ○ and ● $N=100$, ♦ $N=500$

the tidal field was consistently included; the filled circles are based on isolated models with an assumed energy cut-off as described above. The error bars are estimated statistically from the number of escapers which has been used for deriving the escape rate. The relative strength of the galactic field compared to the internal field of the cluster can be measured by the ratio of the median radius R of the cluster to its tidal radius ξ_L. The results in Fig. 7 show clearly the drastic increase of the escape rate in the presence of the galactic tidal field. The escape rate $\dot{N}\, T_{cr}$ does not seem to vary significantly with N over the range from 100 to 500.

In Fig. 8 we compare our results with AARSETH's set (1973) of non-isolated models containing 250 stars. While there is good agreement for isolated clusters, the escape rates derived by AARSETH for star clusters in a tidal field are systematically smaller than our values by a factor of about 2. Since the shape of the curves in Fig. 8 is otherwise very similar, the discrepancy can be probably explained by the different criteria used for identifying escapers. AARSETH (1973) uses a more conservative criterion; for instance, he accepts only stars outside of $2\xi_L$ as escapers. While AARSETH tends to underestimate the number of escapers, our criterion $C_i > C_L$ gives an upper limit for the escape rate. However, passing interstellar HI-clouds should rapidly remove the outermost halo of a cluster, which is mainly formed by stars with $C_i > C_L$ having not found the hole in the equipotential surfaces. If this mechanism operates, then our escape rates may be rather representative (or even too small) for actual open clusters.

d) Experimental Total Lifetimes. From the experimental escape rate, we can derive the evaporation time of a cluster:

$$T_{\mathrm{ev}} = N/\dot{N}. \tag{11}$$

Since $\dot{N}$ is rather constant during the dynamical evolution, the evaporation time is nearly identical with the total lifetime of a star cluster, if we insert for N in Eq. (11) the initial number of stars in the cluster. The evaporation time of a cluster depends mainly on three overall properties of the cluster: (1) the total number of stars N, (2) the total mass $\mathcal{M}$, and (3) the median radius R of the cluster. For actual open clusters, $\mathcal{M}$ and N are strongly correlated. We assume here the relation $\mathcal{M} = 0.5\,\mathcal{M}_\odot\,N$, based on the mean mass of the stars in the Hyades (VAN ALTENA, 1966). Hence we have essentially only two free parameters, say $\mathcal{M}$ and R.

In Fig. 9 we have plotted the evaporation time of a cluster, T_{ev} in years, as a function of the median radius of the cluster in projection, R in parsec, for star clusters containing 100 stars ($\mathcal{M} = 50\,\mathcal{M}_\odot$), 250 stars ($\mathcal{M} = 125\,\mathcal{M}_\odot$) or 500 stars ($\mathcal{M} = 250\,\mathcal{M}_\odot$). These experimental evaporation times are based on the escape rates, $\dot{N}$, found in our N-body star cluster models identified in Fig. 9. From each isolated model, we derive a whole sequence of quasi-non-isolated models by using the appropriate energy cut-off for the escape limit as determined by the chosen radius R. Both Figs. 7 and 9 show that this crude procedure of the energy cut-off represents the results derived from the consistent non-isolated models (G, G2, G3, DG2, DG3) extremely well.

From Fig. 9 we see that for fixed $\mathcal{M}$ and N, the evaporation time T_{ev} increases with the radius R of the cluster, in spite of the increasing importance of the galactic tidal field. This is due to the fact that the crossing time T_{cr} increases always faster with R than the escape rate $\dot{N}\,T_{\mathrm{cr}}$. For fixed R, the evaporation time T_{ev} increases with increasing N. As a typical example, we find for model DG3 ($N = 250$ stars, $\mathcal{M} = 125\,\mathcal{M}_\odot$, $R = 1.0$ pc) a total lifetime of about $2.5 \cdot 10^8$ years. As mentioned in Section c, AARSETH (1973) derives for a similar cluster (his models II, III and IV) a longer total lifetime of about $5 \cdot 10^8$ years.

Because many actual open clusters contain more than 500 stars, we shall try to extrapolate our experimental results to somewhat higher values of N. There exist two possibilities for extrapolating the escape rates $\dot{N}\,T_{\mathrm{cr}}$, shown in Fig. 7 for $N \leqq 500$, to larger N: either we derive the N-dependence of the quantity $\dot{N}\,T_{\mathrm{cr}}$ from the experimental data for $N \leqq 500$ and extrapolate this empirical fit to

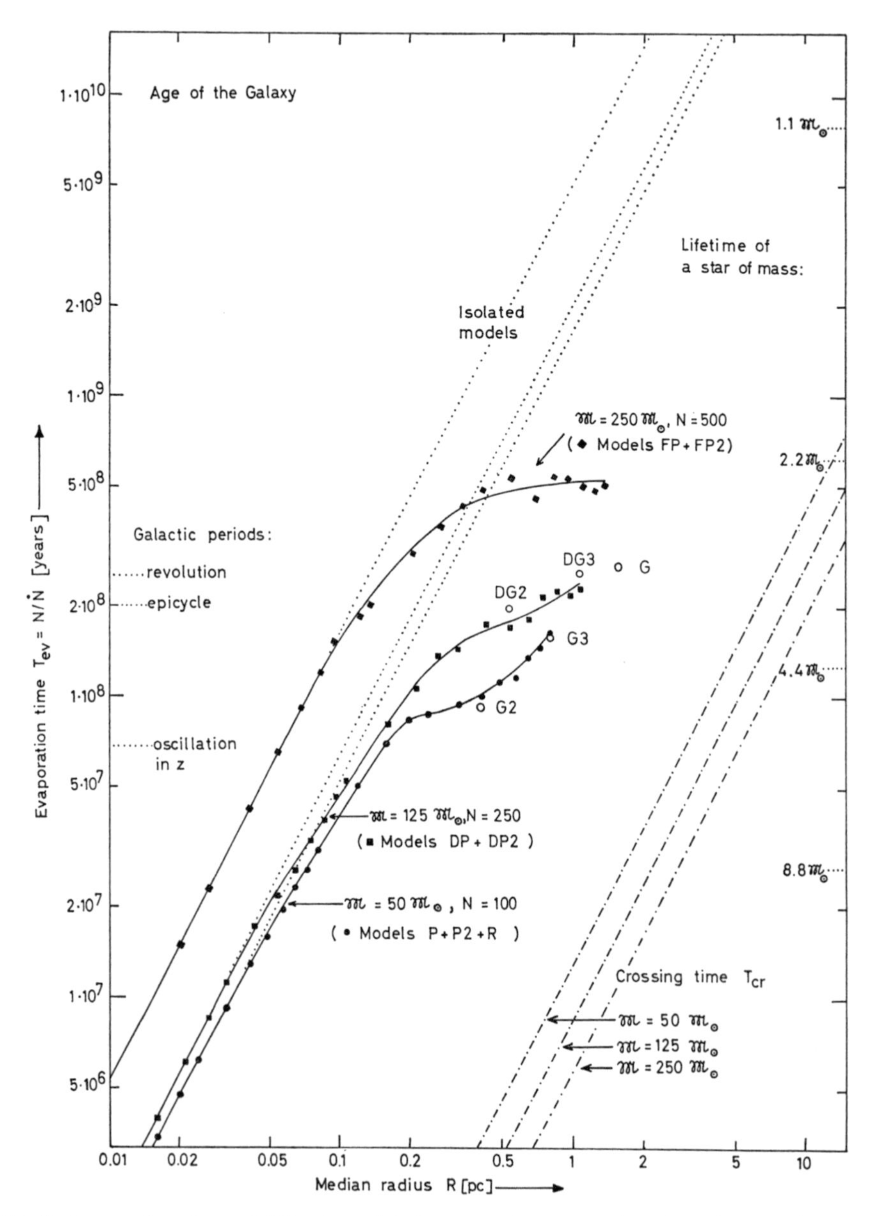

Fig. 9. Evaporation times of N-body models for open clusters as a function of the median radius R of the cluster for various total masses $\mathcal{M}$ (shown by the symbols and the fitted solid lines)

$N > 500$, or we adopt a theoretically derived dependence of $\dot{N}\,T_{cr}$ on N. Henon's theory of escape predicts that the quantity $\dot{N}\,T_{cr}$ should be independent of N, at least for isolated clusters. The experimental results shown in Fig. 7 support this hypothesis, even for non-isolated models with a fixed value of R/ξ_L. Therefore,

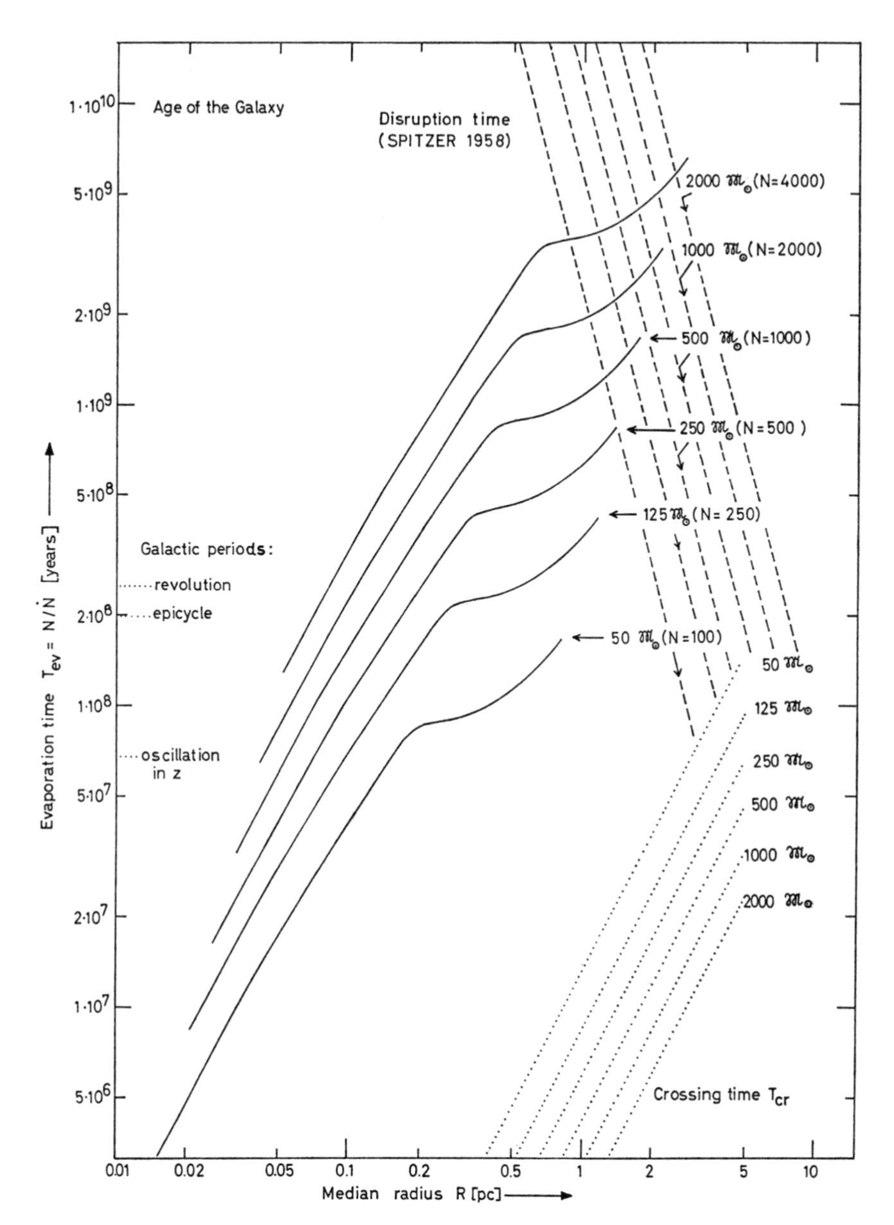

Fig. 10. Extrapolated evaporation times (solid lines) and disruption times due to interstellar clouds (dashed lines) for open clusters as a function of the median radius R of the cluster for various total masses $\mathscr{M}$

we assume for the extrapolation that $\dot{N}\,T_{cr}$ is independent of N for each value of R/ξ_L. In order to compute the evaporation time T_{ev} for $N>500$, we have to take into account that the total mass $\mathscr{M}$ increases proportional to N, and $\mathscr{M}$ enters into T_{cr} and ξ_L. The final result of the extrapolation from a value N_1 to another N_2

corresponds to a shift of the curve $T_{ev}(R)$ for N_1, shown in Fig. 9 for some values of N, to the right by a factor $(N_2/N_1)^{1/3}$ and then upwards by N_2/N_1. The results of such a procedure, starting from the experimental results for $N_1=100$, are presented in Fig. 10.

e) Comparison of Total Lifetimes. Can the numerical experiments explain the total lifetimes deduced from the observed age distribution of open clusters? Before we can answer this question we have to know the initial total masses $\mathcal{M}$ and the median radii R of open clusters. Most observed young open cluster have total masses between 100 $\mathcal{M}_\odot$ and a few 1000 $\mathcal{M}_\odot$ and median radii in projection between 0.5 pc and 3 pc. A typical open cluster may be characterized by $\mathcal{M}=500\,\mathcal{M}_\odot$, $N=1000$ stars and $R=1$ pc.

Figure 10 indicates that the wide spread of the observed total lifetimes can be explained as due to the variety of total masses and radii of open clusters. It seems to be more difficult to explain the observed typical lifetime of open clusters, about $2\cdot 10^8$ years. For the typical values of $\mathcal{M}$, N and R, namely 500 $\mathcal{M}_\odot$, 1000 stars and 1 pc, the slight extrapolation of our experimental results for $N=500$ by a factor of 2 in N leads to a total lifetime of about 10^9 years. What are the possible reasons for this discrepancy by a factor of about five? The extrapolation from 500 to 1000 stars cannot be severely in error. The evaporation time would overestimate the total lifetime of a cluster, if the dynamical evolution speeds up with time as predicted by KING (1958b) and VON HOERNER (1958). Our star cluster models, however, do not support this theoretical prediction. A possible error source are the adopted typical values of $\mathcal{M}$, N and R. Furthermore, the total lifetimes deduced from the age distribution of open clusters may be too small, because of undetected selection effects or incorrect age calibrations. The effect of passing interstellar HI-clouds on the dynamical evolution of open clusters is discussed below. Considering all the uncertainties in the experimental and observed values of the lifetimes of open clusters, we should be rather satisfied about the degree of agreement between the N-body experiments and the observations. It seems quite probable that the observed age distribution of open clusters can be explained by the experimental N-body simulation of the dynamical dissolution of star clusters.

SPITZER (1958) has pointed out that the gravitational shocks, which a cluster suffers from passing HI-clouds, may be efficient in disrupting open clusters. We have plotted Spitzer's results for the disruption time due to clouds as dashed lines in Fig. 10. More recent studies by SPITZER and CHEVALIER (1973) have only slightly changed the final result for the time scale of dissolution of clusters due to the passing clouds. Furthermore, the numerical experiments of BOUVIER and JANIN (1970) at least do not contradict the theoretical prediction. A comparison of the experimental evaporation times (solid lines in Fig. 10) with the disruption time due to clouds indicates that the gravitational shocks of the clouds may be important for disrupting open clusters with a median radius R larger than about 2 pc. Due to the tidal field, most of these large clusters are, however, gravitationally only weakly bound or already beyond the verge of dynamical stability. This may be seen from Fig. 6, where the critical surface has a minimal distance from the center of the cluster of about 0.5 ξ_L or 5 pc for a typical cluster with $\mathcal{M}=500\,\mathcal{M}_\odot$.

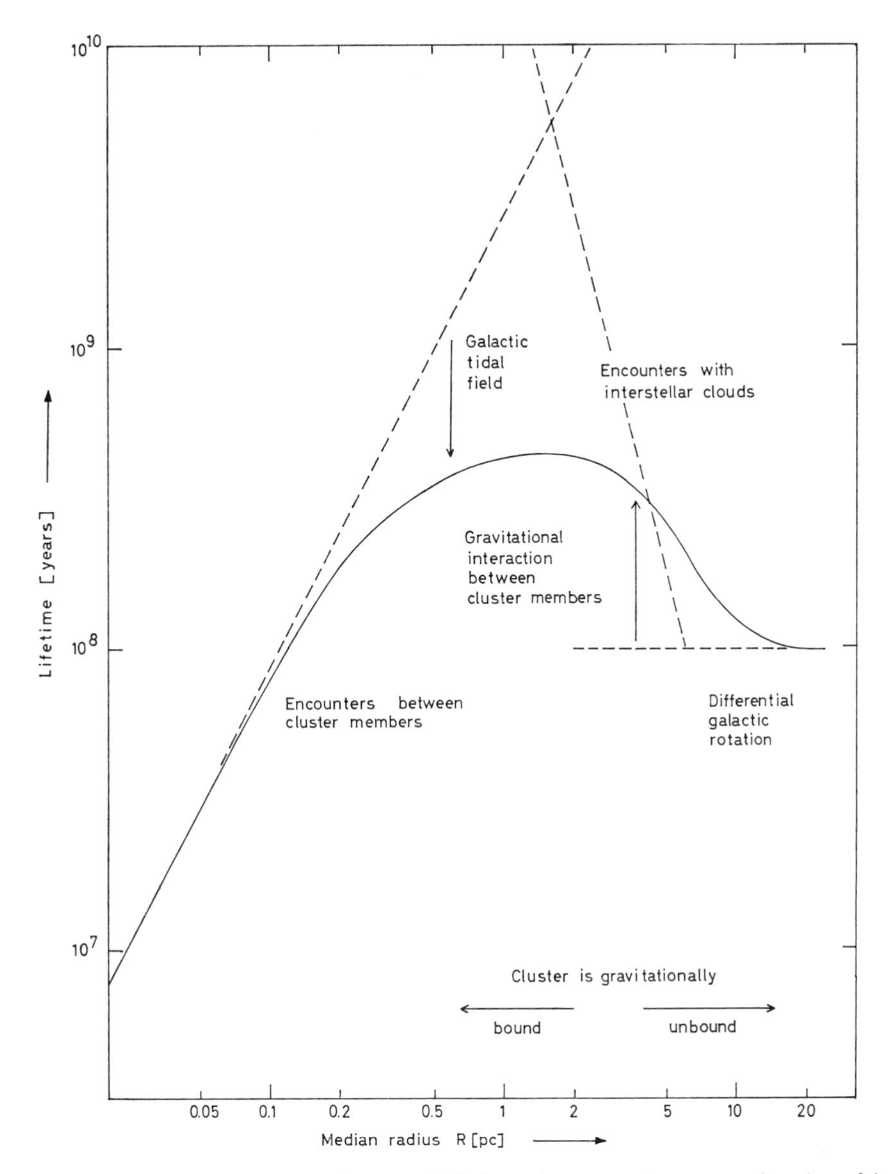

Fig. 11. Schematic representation of the total lifetime of an open cluster as a function of its median radius R

It is indicated by Fig. 10 that for a typical open cluster with $\mathscr{M} = 500\,\mathscr{M}_\odot$ and $R = 1$ pc, the disrupting effect of passing HI-clouds is small compared to the evaporation process due to stellar encounters and facilitated by the galactic tidal field. The gravitational interaction with the interstellar clouds may, however, very effectively disperse the halo of escapers or nearly escaping stars around a stable open cluster, and the clouds may strongly enhance the dissolution of large, weakly bound clusters. In any case, it would be highly desirable to include the effect of the passing clouds into realistic N-body experiments for star cluster models containing a few hundred stars. Unfortunately, the data on the dynamically relevant properties of interstellar clouds are not very precisely known.

In Fig. 11 we give a schematic representation of how the total lifetime depends on the median radius R of an open cluster for fixed initial values of $\mathcal{M}$ and N (The variation with $\mathcal{M}$ or N may be inferred from Fig. 10). Very small clusters ($R<0.1$ pc) are practically isolated and their dissolution is only caused by encounters between the cluster stars. For larger clusters, the energy cut-off provided by the galactic tidal field strongly enhances the evaporation. Clusters with $R>2$ pc may be severely affected by encounters with interstellar clouds. Clusters with $R>3$ pc are already gravitationally unbound in the galactic tidal field, and still larger clusters are rapidly dissolved by the shearing effect of differential galactic rotation. The longest lifetimes occur for median radii of about 1 to 3 pc. This favourable range of radii depends only slightly on the total mass (KING, 1958c). The median radii (VAN DEN BERGH and SHER, 1960) of old open clusters fall exactly into this range: M67 with $R=2.2$ pc and NGC 188 with 2.4 pc. The main reason that these old open clusters still do exist must be, however, their sufficiently large initial total mass, according to Fig. 10.

VI. Simulation of Clusters of Galaxies

The N-body experiments can also give valuable information on the dynamical evolution of many clusters of galaxies. The mass-point approximation should not seriously affect the results, although the galaxies are relatively extended in comparison to their mean distances and encounters may be inelastic because of tidal effects in the galaxies. The main problem in the dynamics of clusters of galaxies is posed by the "missing mass": the total mass of a cluster derived from the application of the virial theorem is usually much larger (up to two orders of magnitude) than the total mass estimated from the counted galaxies via a reasonable mass-to-luminosity ratio.

AARSETH and SASLAW (1972) have studied in detail the dynamics of clusters of galaxies by means of N-body experiments under various basic assumptions about the state of these systems. They find that the virial theorem should give rather correct total masses, if the clusters are really bound systems. Dynamical fluctuations, projection effects and incomplete data introduce a typical uncertainty of about 50% in the computed total mass. An observational selection of heavy, luminous members produces an underestimate of the total mass by a factor of about two. Therefore, the missing mass may be even larger than usually supposed. If one does not postulate a lot of unseen matter in clusters of galaxies, in the form of gas or stars or dwarf galaxies, etc., then the clusters should be expanding systems, violating the virial theorem drastically. AARSETH and SASLAW investigate the dynamical evolution of expanding clusters which have already initially positive total energy, and they simulate systems in which the expansion is caused by a rapid mass loss of some active galaxies. They propose several methods for distinguishing between the different hypotheses. Of course, as long as the problem of the missing mass has not really been clarified by observations, the dynamical history and the ultimate fate of actual clusters of galaxies cannot be implied by N-body experiments. A more or less continuous distribution of hidden mass in the cluster may even destroy the principal applicability of the present N-body experiments on clusters of galaxies, because most of the total mass would be then either essentially encounterless or follow hydrodynamical laws rather than the simple N-body motions.

References

[1.] IAU Symposium No. 25: The Theory of Orbits in the Solar System and in Stellar Systems. Ed. G. Contopoulos. London-New York: Academic Press 1966.
[2.] Colloque sur les méthodes nouvelles de la Dynamique stellaire. Bull. Astron. (3) **2**, 1 (1967).
[3.] Symposium on Computer Simulation of Plasma and Many-Body Problems. NASA Special Report SP-153, 1967.
[4.] IAU Colloquium on the Gravitational Problem of N Bodies. Bull. Astron. (3) **3**, 1 (1968).
[5.] IAU Colloquium No. 10: Gravitational N-body Problem. Ed. M. Lecar. Dordrecht-Holland: D. Reidel 1972. Also partly reprinted in: Astrophys. Space Sci. **13**, 279 (1971) and **14**, 3 (1971).

Aarseth, S. J.: Monthly Notices Roy. Astron. Soc. **126**, 223 (1963).
Aarseth, S. J.: Monthly Notices Roy. Astron. Soc. **132**, 35 (1966).
Aarseth, S. J.: Bull. Astron. (3) **3**, 105 (1968).
Aarseth, S. J.: Monthly Notices Roy. Astron. Soc. **144**, 537 (1969).
Aarseth, S. J.: Astrophys. Space Sci. **13**, 324 (1971a)=[5], p. 88.
Aarseth, S. J.: Astrophys. Space Sci. **14**, 20 (1971b)=[5], p. 29.
Aarseth, S. J.: Astrophys. Space Sci. **14**, 118 (1971c)=[5], p. 373.
Aarseth, S. J., Saslaw, W. C.: Astrophys. J. **172**, 17 (1972).
Aarseth, S. J., Woolf, N. J.: Astrophys. Letters **12**, 159 (1972).
Aarseth, S. J.: In A. Beer (ed.), Vistas in Astronomy, vol. 15, p. 13, Oxford: Pergamon Press 1973.
Albada, T. S. van: Bull. Astron. Inst. Neth. **19**, 479 (1968).
Altena, W. F. van: Astron. J. **71**, 482 (1966).
Bergh, S. van den, Sher, D.: Publ. David Dunlap Obs. **2**, 203 (1960).
Bettis, D. G., Szebehely, V.: Astrophys. Space Sci. **14**, 133 (1971)=[5], p. 388.
Bouvier, P., Janin, G.: Astron. Astrophys. **9**, 461 (1970).
Bouvier, P.: Astron. Astrophys. **14**, 341 (1971).
Chandrasekhar, S.: Principles of Stellar Dynamics. Chicago: Chicago University Press 1942.
Chandrasekhar, S.: Astrophys. J. **98**, 54 (1943).
Hayli, A.: Astron. Astrophys. **7**, 17 (1970).
Hénon, M.: Ann. Astrophys. **23**, 668 (1960).
Hénon, M.: Ann. Astrophys. **27**, 83 (1964).
Hénon, M.: Astron. Astrophys. **2**, 151 (1969).
Hénon, M.: Astrophys. Space Sci. **13**, 284 (1971a)=[5], p. 44.
Hénon, M.: Astrophys. Space Sci. **14**, 151 (1971b)=[5], p. 406.
Hénon, M.: Private communication (1972).
Hénon, M.: In Proceedings of the Third Advanced Course in Astronomy and Astrophysics: Dynamical Structure and Evolution of Stellar Systems. Saas-Fee, Switzerland 1973 (to be published).
Hoerner, S. von: Z. Astrophys. **44**, 221 (1958).
Hoerner, S. von: Z. Astrophys. **50**, 184 (1960).
King, I.: Astron. J. **63**, 109 (1958a).
King, I.: Astron. J. **63**, 114 (1958b).
King, I.: Astron. J. **63**, 465 (1958c).
King, I.: Astron. J. **67**, 471 (1962).
Larson, R. B.: Monthly Notices Roy. Astron. Soc. **147**, 323 (1970a).

LARSON, R. B.: Monthly Notices Roy. Astron. Soc. **150**, 93 (1970b).
MICHIE, R. W.: Ann. Rev. Astron. Astrophys. **2**, 49 (1964).
MILLER, R. H.: Astrophys. J. **140**, 250 (1964).
MILLER, R. H.: J. Comput. Phys. **8**, 449 (1971).
OORT, J. H., HERK, G. VAN: Bull. Astron. Inst. Neth. **14**, 299 (1959).
PRATA, S. W.: Astron. J. **76**, 1017 (1971a).
PRATA, S. W.: Astron. J. **76**, 1029 (1971b).
SPITZER, L. JR.: Monthly Notices Roy. Astron. Soc. **100**, 396 (1940).
SPITZER, L. JR.: Astrophys. J. **127**, 17 (1958).
SPITZER, L. JR.: In D. J. K. O'CONNELL (ed.), Study Week on Nuclei of Galaxies, p. 443. Amsterdam-London: North-Holland 1971.
SPITZER, L. JR., CHEVALIER, R. A.: Astrophys. J. **183**, 563 (1973).
SPITZER, L. JR., HÄRM, R.: Astrophys. J. **127**, 544 (1958).
SPITZER, L. JR., HART, M. H.: Astrophys. J. **164**, 399 (1971a).
SPITZER, L. JR., HART, M. H.: Astrophys. J. **166**, 483 (1971b).
SPITZER, L. JR., SHAPIRO, S. L.: Astrophys. J. **173**, 529 (1972).
SPITZER, L. JR., THUAN, T. X.: Astrophys. J. **175**, 31 (1972).
WHITE, M. L.: Astrophys. J. **109**, 159 (1949).
WIELEN, R.: Veröffentl. Astron. Rechen-Inst. Heidelberg No. 19 (1967a).
WIELEN, R.: Bull. Astron. (3) **2**, 117 (1967b).
WIELEN, R.: Bull. Astron. (3) **3**, 127 (1968).
WIELEN, R.: Habilitationsschrift, University of Heidelberg (1969).
WIELEN, R.: Astrophys. Space Sci. **13**, 300 (1971a) = [5], p. 62.
WIELEN, R.: Astron. Astrophys. **13**, 309 (1971b).

Discussion

BLAAUW, A.:
How were the double stars handled in you calculations?

WIELEN, R.:
I have not introduced a special treatment of binaries into my integration scheme. The individual time steps save automatically a significant amount of computing time if binaries are present. However, the most economic way of handling close binaries is the regularization of the two-body motion (AARSETH, 1971c; BETTIS and SZEBEHELY, 1971).

SEITTER, W.:
My question concerns the concentration of massive stars towards the center of clusters. Dr. PENDL at Bonn has looked at the distribution of giants in a number of open clusters and in *no* case finds a concentration, but instead a smooth distribution over the cluster or even a ring-like structure. Can you explain this effect?

WIELEN, R.:
The observational evidence for or against the segregation of stars of different mass in actual star clusters is somewhat controversial. While the effect is observed in many clusters, other clusters do not show it. The reason may be the different dynamical ages of clusters which are only loosely correlated with the absolute age (see Section V b). A ring-like structure has never been found in N-body experiments.

A Comparative Study of Periodic Orbits ($m < 4$) in Various Three-Dimensional Models of Our Galaxy

By L. MARTINET and F. MAYER

Observatoire de Genève, Sauverny, Switzerland

Abstract

The paper presents some general conclusions deduced from a systematic exploration of m-periodic orbits ($m<4$) calculated in the SCHMIDT models (1956 and 1965) of the Galaxy.

During the last years, several three-dimensional mass models of the Galaxy were constructed from the observational data. One of the main aims may be to study the motions of old stellar populations, by computing the orbits in the stationary and axisymmetric potential corresponding to the mass model considered. However, the galactic potential remains badly known and it is not evident that the results which are obtained are singificant and anyway useful for the applications to the old stellar populations. Nevertheless, theoretically, for a potential of galactic type we can reduce the problem to the typical one of dynamical system with two degrees of freedom. Theory and numerical experiments have already given some very general results on orbital properties in these systems. Therefore, it is instructive to see for example if slightly different potentials give the same results as regards the position of the periodic and tube orbits in our Galaxy. Then it should be possible to precise for example for what initial conditions we obtain tube orbits in connection with different resonances in the system. Work in progress at the Geneva Observatory is a preliminary approach to these questions. The details of our exploration of periodic and non-periodic orbits in given galactic potentials will be published elsewhere (MARTINET and MAYER, 1973). Here we only give some conclusions deduced from a systematic exploration of m-periodic orbits ($m<4$). We use the SCHMIDT (1956) potential (noted I henceforth) represented by an interpolation formula due to OLLONGREN and the SCHMIDT (1965) potential (II) with a central nucleus of radius $=0.4$ kpc instead of the central point introduced by the author. Our discussion is based on (a) the so-called diagram of characteristics and (b) on the rotation number r_{opc} of the central periodic orbit. In (a) each symmetric periodic orbit is represented by a point $(E, \tilde{\omega})$ where E is the energy and $\tilde{\omega}$, the value of the distance to the axis of symmetry of the system corresponding to the intersection of the given orbit in the galactic plane, with $\dot{\tilde{\omega}}=0$. Each family of periodic orbits may be identified by the number of rotation $r=2-\dfrac{n}{m}$ for orbits which close after n oscillations along the $\tilde{\omega}$-axis and m oscillations along the z-axis in the "reduced" potential $V(\tilde{\omega}, z)=\Phi(\tilde{\omega}, z)+A^2/2\tilde{\omega}^2$ (Φ is the given Schmidt potential and A, the angular momentum). The "central" periodic orbit is the symmetric orbit which crosses the $\tilde{\omega}$-axis always at the same point. Its rotation number r_{opc}, mentioned above, is defined as the limit of the rotation number of nearby non-periodic orbits in the $(\tilde{\omega}, \dot{\tilde{\omega}})$ plane (CONTOPOULOS, 1971). The study of the variation of r_{opc} as a function of E for any given A reveals the existence of the various families of periodic orbits which we could find in the $(E, \tilde{\omega})$ diagram on condition that the characteristics of these families intersect the characteristic of the central family.

From a comparison in (I) and (II), it appears that: 1) Generally the same families of resonant periodic orbits exist with the same form of the characteristics, for slightly different angular momentum. We followed the evolution of the characteristics $\frac{1}{1}$, $\frac{2}{3}$, $\frac{1}{2}$, in both potentials when A is varying. It is to be pointed out that a closed characteristic of a resonant family which intersects two times the characteristic of the central family for a given value of A can change into a characteristic without any connection with the characteristic of the central family when A increases. 2) The central family is always stable for energies less than -100 (unit $100\ \text{km}^2 \cdot \text{s}^{-2}$) except on small ranges of values of E where the stability is lost as a consequence of an intersection of the central characteristic with that of another 1-periodic family. As $E > -100$ the central family becomes unstable only in (I) and seems to remain stable until $E = 0$ in (II).

We can also compare the results obtained from (II) with two different laws of density for the outer shell of the model: a) $\rho \div \tilde{\omega}^{-4}$ and b) $\rho \div \tilde{\omega}^{-6}$. We observe that: 1. For small values of $E(E < -600)$, r_{opc} decreases as E increases but its variation is the same in both of these cases and so we have the same critical points on the central characteristic in the $(E, \tilde{\omega})$ diagram. 2. For any A, the minimum of r_{opc} is larger in the case (a), what slightly changes the value of A for which we have disparition of the various families of resonant periodic orbits. But this does not concern the important low order resonances. 3. For the highest values of E, the orbits penetrate more and more in the outer regions and consequently, r_{opc} increases more rapidly and the central periodic orbit is no longer stable for E approaching to 0 in the case (b), at least for values of A larger than 140.

Further we note that a change in the radius of the nucleus from 0.1 to 0.5 kpc in (II) does modify practically nothing in the chart of the characteristics.

Finally, for the minimum value of the energy $E = E_c$, corresponding to the circular motion for a given A, the rotation number $(r_{\text{opc}}) = 2 - \sqrt{A_c/B_c}$ where

$$A_c = \frac{3A^2}{\tilde{\omega}_c^4} + \left(\frac{\partial^2 \Phi}{\partial \tilde{\omega}^2}\right)_{\tilde{\omega} = \tilde{\omega}_c,\, z = 0}, \qquad B_c = \left(\frac{\partial^2 \Phi}{\partial z^2}\right)_{\tilde{\omega} = \tilde{\omega}_c,\, z = 0}.$$

We point out that the first exploration of periodic orbits in a highly perturbed potential by CONTOPOULOS was based on the particular case $\sqrt{A_c}/\sqrt{B_c} = 4/3$. It is very far from the values which we generally have here, since we find that $\sqrt{A_c}/\sqrt{B_c} = 2 - (r_{\text{opc}})_c$ is practically constant (≈ 0.4) for $A > 40$. So the Contopoulos potential $V = \frac{1}{2}[A_c(\tilde{\omega} - \tilde{\omega}_c)^2 + B_c z^2] - \varepsilon(\tilde{\omega} - \tilde{\omega}_c) z^2$ with $\sqrt{A_c}/\sqrt{B_c} = 4/3$ could fit ours in the vicinity of a circular orbit located near the center of the system and so can be considered as a particular case of the present exploration.

References

CONTOPOULOS, G.: Astron. J. **76**, 147 (1971).
MARTINET, L., MAYER, F.: To be published (1973).
SCHMIDT, M.: Bull. Astron. Inst. Neth. **13**, 15 (1956).
SCHMIDT, M.: In A. BLAAUW and M. SCHMIDT (eds.), Galactic Structure, p. 513. Chicago-London: Chicago University Press 1965.

Discussion

CONTOPOULOS, G.:
I am happy to see this systematic study by MARTINET and MAYER. When one calculates orbits in a particular model of the Galaxy he cannot be sure that all the types of orbits he finds exist in the real Galaxy. I may remind you of the separable model of HORI (Publ. Astron. Soc. Japan **14**, 353, 1962); where no tube orbits appear at all. We have reasons to believe that separable models are quite exceptional. However, only a comparison of various models can give us some confidence about the existence and importance of various types of tube orbits in our Galaxy.

Author Index

Numbers in *italics* refer to the page on which the reference is listed

Aarseth, S. J. 329, 330, 339, 340, 341, 342, 346, 347, 352, *353*, 354
Ables, H. D. 299, 303, *325*
Ackermann, G. 192, *205*
Agekyan, T. A. 276, 277, *277*
Ahnert, P. 49, *52*
Aitken, D. 248
Albada, T. S., van 10, 12, 13, *15*, 329, 339, *353*
Alexander, J. B. 14, *15*
Alfvén, H. 289, *289*
Allen, C. W. 94, *112*, 135, 136, *139*, 167, *172*
Allen, D. A. *207*, 216, *230*
Allen, L. R. *187*
Aller, L. H. *181*, *206*
Altena, W. F., van 347, *353*
Altenhoff, W. J. 141, 142, 146, 147, 155, *155*, *156*, 213, *230*
Andrews, P. J. *15*
Andriesse, C. D. 247
Anosova, Zh. P. 276, 277, *277*
Aoki, S. 84, 85, *87*
Arenstorf 275
Armstrong, K. R. *240*
Arp, H. 165, *172*, 292, 305, *322*
Arrhenius, G. 289, *289*
Asteriadis, G. 17, 19, *29*
Auer, L. H. 184, *187*
Augason, G. C. *230*
Aumann, H. H. 192, 199, 204, *205*, *206*, 210, 225, *230*, *241*, 246, *247*

Baade, W. 292, 297, *322*
Bacik, H. *263*
Baglin, A. *15*
Bahcall, J. N. *231*
Bahner, K. 18, 19, *29*
Baker, D. J. *241*
Baker, N. 10, 11, 12, 13, *15*, 15
Baker, P. L. 121, *123*
Bakos, G. A. 52, 60, 64, *64*, 278
Baldwin, J. E. *156*
Ball, J. A. 142, 145, 146, *155*
Bappu, M. K. V. 185, *187*
Barmore, F. E. 139
Barrett, A. H. *113*
Bartlett, T. J. *206*
Basinska, E. 165
Baum, W. A. *263*
Becker, W. *127*
Becklin, E. E. 8, 9, 100, 109, *112*, 157, 160, *163*, 197, 198, 200, 201, 202, 203, 205, *205*, *206*, *207*, 210, 214, *230*, 248, *324*
Beckman, W. A. 238, *240*
Beer, A. *240*, *324*, *353*
Beer, R. 237, *240*
Bell, R. L. 235, *241*
Bensammar 215
Bergh, S. van den *206*, 292, 293, 299, 318, *322*, 352, *353*
Bernstein, S. 302, *324*
Bertola, F. 305, *322*
Bettis, D. G. *353*, 354
Biermann, P. 13, *15*
Binnendijk, L. 53, *58*
Biraud, Y. 215
Birkhoff, G. D. 275, 277, *277*
Blaauw, A. 66, 67, 70, 71, *74*, 75, *127*, *172*, *324*, 354, *357*
Blackwell, D. E. 270, *272*
Bless, R. C. 186, *187*
Boer, K. S. de 187, *187*
Böhm-Vitense, E. 38, *38*
Bolton, J. G. *163*
Bonneau, M. 67, 70, *74*
Bonsack, W. H. 131, 132, *139*
Booth, R. S. *206*
Borgman, J. 188, 195, *206*, 207, 208, 231, 233, 238, *240*, 241, 247, 248, 254
Boulesteix, J. 325
Bouvier, P. 342, 350, *353*
Bozis, G. 279, 280, 282, *282*
Brandt, J. C. 299, *322*
Brault, J. W. 258, *263*
Breger, M. 3, 4, 7, 8, 9, *15*, 195, *206*, *207*

Bregman, J. *15*, *207*
Breuer, H. D. 107, *113*
Brezgunov, V. N. 148, *155*
Brillouin, L. 36, *38*
Brooke, A. L. *15*, *207*
Brosche, P. 320, *322*
Buhl, D. 160, *163*
Burbidge, E. M. 153, *155*, 181, *181*, *207*
Burbidge, G. R. 153, *155*, *181*, 229, *230*
Burke, B. F. *130*, *156*
Butler, H. E. *139*

Cameron, A. G. W. 168, *172*, 318, *325*
Cameron, M. J. 153, *155*
Cannon, A. J. 65, 68, *74*
Capps, R. W. 238, *240*
Carbieri, C. 254
Carbon, D. F. *177*
Carlborg, N. 289
Carlton, N. P. 238, *241*
Caroff, L. J. *230*
Carranza, G. *323*
Carrasco, L. *207*
Carruthers, G. R. 91, *113*
Carswell, R. *207*
Cashman, R. J. *240*
Catchrole, R. M. *15*
Chabbal, R. *139*
Chandrasekhar, S. 37, 38, *38*, 339, 341, *353*
Chapman, S. 165, *172*
Chasmar, R. P. *241*
Chazy, J. 275, 277, *277*
Chevalier, C. 7, *15*
Chevalier, R. A. 342, 350, *354*
Cheung, A. C. 89, *113*, 160, *163*
Christy, R. F. 10, *15*, 15, 33, *35*
Churchwell, E. B. 140, 150, 155, *155*
Clark, B. G. 96, *113*
Clegg, P. E. *230*, 236, *240*
Clube, S. V. M. 72, *74*
Coblentz, W. W. 233, *240*
Code, A. D. 137, *139*, 186, *187*
Cohen, M. 190, 195, 196, 198, *206*
Cohen, R. J. 115
Coleman, I. *240*
Connes, J. 188, 205, 233, 237, 238, *240*
Connes, P. 205, 233, 237, *240*
Contopoulos, G. *127*, 279, 313, 320, *322*, *353*, 356, *357*, 357
Core, G. D. *206*
Cork, G. M. W. *241*
Courtès, G. 247, 293, 310, *323*, 325
Cowling, T. G. 165, *172*
Crampin, D. J. 321, *323*
Cruickshank, D. P. 233, *240*

Daghesamansky, R. D. 155
Daintree, E. J. 119
Damme, K. J. van *163*
Danielsson, L. 283, *289*
Davies, R. D. 68, 72, *74*, 115, 116, 117, *118*, 124, 126, *126*, 127, 197, *206*
Davis, J. *187*
Deharveng, J. M. 293, *323*
Denis, J. 36, 37, *38*
Dickel, H. R. 105, *113*
Dieckvoss, W. 84, 85, *87*
Dieter, N. H. 125, *126*
Diethelm, R. 49, *52*
Dixon, M. E. *74*, 313, *323*
Donn, B. *113*
Doucet, C. 60, *64*
Downes, D. 142, 143, 146, 147, 148, 155, *155*, 204, *206*, *230*
Duchesne, M. *163*
Dugan, R. S. 47, 48, 49, *52*
Duinen, R. J. van 243
Dumitrescu, A. 53, *59*
Dürbeck, H. 41
Duursma, T. 247
Dyck, H. M. 4, *15*, 195, *206*, 238, *240*

Eddington, A. S. 312, 316, *323*
Eddy, J. A. 233, *240*
Edison 233
Eelsalu, H. 310, 312, 313, *323*
Efremov, Yu. N. *29*, 31, 32, 33, 34, *35*
Eggen, O. J. 17, *29*, 63, 64, *64*, 68, *74*, 314, 320, *323*
Einasto, J. 291, 293, 294, 295, 296, 297, 298, 299, 301, 304, 310, 314, 316, 317, 318, 320, 321, *323*, 325
Einasto, L. 322, *323*
Ekers, R. D. 147, *155*, 204, *206*
Elsässer, H. 193, 195, *207*
Emery, R. J. *113*, 147, *155*, *206*, 209, *230*, *240*, *247*
Epchtein, N. 215
Erickson, E. F. 215, 216, 224, *230*
Evans, D. S. *323*, *324*
Evans, W. D. 139

Fatchikin, N. V. 83, 84, 85, *87*
Fay, T. D. 265, 267, 271, *272*
Feast, M. W. *15*
Fedorovich, V. P. *29*, *35*
Fehrenbach, Ch. *81*
Feiter, L. de 187
Fellgett, P. B. 233, 238, *240*, 242
Fernie, J. D. 17, 29, *29*, 31, 32, 34, *35*, 68, *74*
Field, G. B. 97, *113*, 123, *123*
Fitch, W. S. 33, *35*
FitzGerald, M. P. 68, *74*
Flin, P. 49, *52*
Fomalont, E. B. 106, *113*, 154, *155*

Forbes, F. F. *230*, *241*
Ford, W. K., Jr. 308, *324*
Forrest, W. J. 196, 198, 199, *206*
Fortini, P. 164
Fowler, W. A. *181*
Fox, D. R. 314, *323*
Frederick, C. L. *113*, 147, *155*, 192, *206*, 209, *230*, *240*, *247*
Freeman, K. C. *324*
Fricke, K. 9, *15*
Fricke, W. 83, 84, 85, *87*
Frogel, J. A. *206*, *207*
Frolov, M. S. *29*, *35*
Furniss, I. 211, *230*

Gahm, G. F. 4, *15*
Gammon, R. H. 237, *240*
Ganesh, K. L. 185, *187*
Ganson, A. *263*
Gaposchkin, S. *35*
Gardner, F. F. 72, *74*, *156*, *163*
Garmire, G. 214, *230*
Garz, T. 180, *181*
Gaustadt, J. E. *15*, 192, *206*, *240*
Gay, J. 215
Geballe, T. R. *207*, 237, *240*, *241*
Gehrz, R. D. 196, *206*
Geisel, S. L. *15*
Gentien, E. P. *113*
Georgelin, Y. *323*
Germann, R. *52*
Gezari, D. Y. *206*, *230*
Giacaglia, G. E. O. *277*, *282*
Gillespie, C. M., Jr. *247*
Gillett, F. C. *15*, 193, 196, 198, 199, 204, 205, *206*, *207*, 225, 237, *240*
Giuli, R. T. *187*
Giver, L. P. *230*
Glicker, S. *113*
Goad, L. *155*, *230*
Goldberg, L. 180, *181*, 199, *206*
Goldreich, P. 9, *15*
Goldsmith, D. W. *123*
Gottlieb, C. A. *155*
Gould, B. A. 65, *74*
Grasdalen, G. L. 192, *206*, *207*
Gratton, L. 16
Greenberg, J. M. *113*, *206*, *207*, *241*
Greenstein, J. L. 131, 132, *139*, *324*
Griffin, R. F. 175, *177*, 302, *323*
Groot, G. de 247, *247*
Groot, T. de 189
Grossman, K. 313, *323*
Groth, H. G. 180, *181*
Gualdi, C. 164
Gunn, J. E. *324*
Gurevič, L. E. 319, *323*

Habing, H. J. 95, 96, *113*, *123*, 125, *126*
Hachenberg, O. 120
Hackwell, J. A. 196, *206*
Hadjidemetriou, J. 278
Hammerschlag, A. 187, *187*
Hanbury Brown, R. 186, *187*
Härm, R. 339, 341, *354*
Harper, D. A., Jr. 151, 152, *155*, 198, 199, *206*, 210, 211, 213, *230*, 239, *240*
Harrington, R. S. 277, *277*
Harris, D. H. 195, *206*
Harris, J. *241*
Hart, M. H. 330, 333, 336, *354*
Harten, R. H. 70, 72, *74*
Hartwick, F. D. A. 317, *323*
Harwit, M. O. *206*, *230*, *241*
Hawkins, M. R. S. 255, 264
Hayli, A. 329, *353*
Heeschen, D. S. 68, *74*
Heiles, C. 100, *113*, 117, 118, *118*
Helmerhorst, Th. J. 243
Hénon, M. 280, *282*, 330, 333, 334, 335, 337, 339, 340, 341, *353*
Henry, J. C. *113*
Herbig, G. H. 130, 131, 136, *139*, 187, *187*, 195, 197, 198, 203, *206*
Herbison-Evans, D. *187*
Herk, G. van 342, *354*
Herschel, J. F. W. 65, *74*
Herschel, Sir W. 232, *240*
Hertzsprung, E. 23, *29*
Hill, E. R. 310, *323*
Hilliard 267, 268
Hoag, A. A. 269
Hobbs, L. M. 73, *74*, 131, 135, 137, 139, *139*
Hobbs, R. W. 147, *155*
Hodge, P. W. 299, 304, *323*
Hoekstra, R. 182, 187, *187*
Hoerner, S. von 314, 319, 320, *323*, 328, 342, 350, *353*
Hoffmann, W. F. 112, *113*, 141, 142, 145, 147, 151, 152, *155*, 192, 198, 199, 200, 201, 204, *206*, 209, 213, 214, 215, 216, 224, 225, 226, *230*, 236, 239, *240*, 243, 246, 247, *247*
Hofmeister, E. 6
Högbom, J. 157, 162, *163*
Höglund, B. 142, *155*
Högner, W. 304, *324*
Hollenbach, D. 99, 107, *113*
Hollinger, J. P. 142, *155*
Holmberg, E. 311, 318, *323*
Holtz, J. Z. *207*, *241*
Holweger, H. *181*
Hoof, A. van 38, *38*
Hori, G.-I. 357
Houck, J. R. 201, *206*, 225, 226, *230*, *241*, *247*
Houghton, J. 235, *240*

Houziaux, L. *139*
Hoyle, F. *181*, 321, *323*
Hubble, E. 68, *74*
Hube, D. P. 314, *323*
Hube, J. O. 17, *29*
Hucht, K. A. van der 182, 187, *187*
Hudson, R. D. 235, *240*
Huizinga, J. S. 236, *240*
Hulst, H. C. van de *113*, 195, 201, *206*, *207*, 218, *230*, *241*
Humphrey, J. N. 235, *240*
Hunter, C. 126, *127*
Hurley, M. 37, *38*
Hyland, A. R. 197, *206*, *207*

Iben, I. 2, *15*, 57, 58, *58*
Iguchi, T. *155*
Innanen, K. A. 314, 321, *323*
Ir, W.-H. 283, *289*
Iriarte, B. *29*
Isles, J. *52*
Iwanowska, W. 165, *172*, 172, 173

Jager, C. de 16, *74*, 113
Janin, G. 342, 350, *353*
Jeans, J. H. 312, *323*
Jeffers, W. Q. 188, *207*
Jefferts, K. B. *113*, *163*, *207*, *231*
Jenkins, E. B. 94, *113*, 187, *187*
Jennings, R. E. *230*
Jôeveer, M. 310, 314, *323*
Johnson, H. L. *15*, 19, *29*, 189, 193, 195, 196, *206*, 233, 237, 238, 239, *240*, *241*
Johnson, H. M. 153, *155*, 304, *323*
Johnson, M. W. 95, 96, *113*
Johnston, K. J. 129, *130*
Joyce, R. R. 199, *206*, *207*, 216, 226, 228, 229, *230*
Jones, B. 248
Jones, F. E. *241*

Kaasik, A. 322
Kahan, E. *263*
Kaifu, N. 142, 150, *155*
Kalamees, P. 322
Kamperman, T. M. 187, *187*
Katchalov, V. 184, *187*
Kato, T. *155*
Kellermann, K. I. 229, 230, *230*
Kepner, M. 125, *127*
Kerr, F. J. 106, 145, 147, 155, *155*, 162, *324*
Khogali, A. *263*
Kholopov, P. N. *29*, *35*
King, I. R. 292, 298, *323*, *324*, 339, 344, 350, 352, *353*
Kinglesmith, D. A. 184, *187*
Kinman, T. D. 292, *323*
Kippenhahn, R. 1, 6, 9, 13, *15*, 15, 16, 57, *58*, 155, 172
Kizilirmak, A. 49, *52*
Kleinmann, D. E. *15*, 198, 199, 204, 205, *206*, 210, 214, 216, 226, 228, *230*, *241*
Klemola, A. R. 83, 84, 85, 86, *87*
Klemperer, W. 107, *113*
Knacke, R. F. *15*, *207*
Knapp, G. R. 145, 147, *155*
Knight, C. A. *130*
Knowles, S. H. *130*
Kock, M. *181*
Kondo, Y. 184, *187*
Koning, A. 247
Koornneef, J. 195, *206*,
Kopal, Z. 44, 48, *52*, 53, 54, 55, 56, 57, *58*
Kordylewski, K. 49, *52*
Kreiner, J. M. 50, 52, *52*
Krishna Swamy, K. S. 151, *155*, 199, *206*
Kristian, J. *206*
Kronberg, P. P. 205, *206*
Kruit, P. C. van der 205, *206*
Kuhi, L. V. 3, 4, *15*, *207*
Kuiper, G. P. 233, 239, *240*
Kukarkin, B. V. 19, 23, *29*, 32, 34, *35*, 297, 298, 319, *323*
Kukarkina, N. P. *29*, *35*
Kumsischwili, Ja. J. 49, *52*
Kunz, L. W. *230*
Kurochkin, N. E. *29*, 32, *35*
Kuroczkin, D. 172, *172*
Kutner, M. *113*, 117, *118*, *207*, *231*
Kutuzov, S. A. 294, 310, *323*
Kuzmin, G. G. 297, 310, 312, 313, 314, 319, 322, *323*

Lallemand, A. 162, *163*
Lamers, H. J. 182, 187, *187*
Langley 233
Larson, R. B. 5, 6, *15*, 333, 334, 339, 340, *353*, *354*
Latham, A. S. *15*
Laubscher, R. E. 84, *87*
Law, S. K. *207*
Lebovitz, N. R. 37, 38, *38*
Lecar, M. *353*
Lee, O. J. 192, *206*
Lee, T. A. 195, 201, *206*
Leighton, R, B. 192, *207*
Léna, P. 235, 236, *240*
Lequeux, J. 151, *155*, 229, *230*
Lesh, J. R. 67, 71, *74*
Liebert, J. *207*
Liller, M. H. 302, 305, *324*
Lilley, A. E. 68, *74*, *155*
Lillie, C. F. 94, 95, *114*, 136, *139*, 225, *231*
Lin, C. C. 73, *74*

Lindblad, B. 82
Lindblad, P.O. 65, 68, 70, 72, *74*, 75
Lloyd Evans, T. *15*
Lo, K.Y. *130*
Locher, K. *52*
Lockhardt, P. *123*
Lockman, F.J. 142, 145, 146, *156*
Lockwood, G.W. 198, *206*
Lodén, L.O. 80, *81*
Lomb, N.R. 38, *38*
Low, F.J. 3, *15*, 151, 152, *155*, 189, 192, 198, 199, 200, 201, 203, 204, 205, *205*, *206*, *207*, 210, 212, 213, 214, 216, 225, 226, 228, *230*, 233, 234, 235, 236, 238, 239, *240*, *241*, *247*
Lowinger, Th. 109, 110, 111, 140, 150, 152, 154, *156*
Luyten, W.J. 249
Lynden-Bell, D. 147, *155*, 204, *206*, 298, 320, 322, *323*, *324*
Lynds, B.T. 68, 72, *74*, 115, 117, *118*, *206*
Lyutyi, V.M. *52*

Mack, J.E. 132, *139*
Maillard, J.-P. *240*
Magalaschwili, N.L. 49, *52*
Malik, G.M. 32, *35*
Mallia, E.A. *272*
Manning, W.H. 39
Manno, V. 239, *240*, *241*, *247*
Maran, S.P. *155*
Markelhs, V.V. 277
Marks, A. 49, *52*
Martin, A.H.M. 147, *155*, 204, *206*, 212, *230*
Martin, T.Z. 235, *241*
Martinet, L. 355, *357*, 357
Martynov, D.Ya. *52*
Masheder, M.R.W. *206*
Mathewson, D.S. 72, 73, *74*
Matsuda, T. 317, *324*
Matthews, H.E. 115, 116, 117, *118*
Mavridis, L.N. 17, 18, 19, *29*, 30, *113*, *156*, *230*, 325
Maxwell, A. 146, 148, *155*, *230*
Mayall, N.U. 165, *172*, 292, *324*
Mayer, F. 355, *357*, 357
Mayer, P. 49, *52*
McCammon, D. 237, *241*
McCarthy, M.F. 41, 64, 82
McClure, R.D. 165, *172*, 292, *324*
McCuskey, S.W. 68, 72, *74*
McGee, J.D. 256, *265*
McGee, R.X. 106, 110, *113*, 142, *156*
McMullan, D. *263*
McNutt, D.P. 132, *139*
McVittie, G.G. *324*
Mebold, U. 120, 123, *123*
Medvedeva, G.I. *29*, *35*
Meeks, M.L. *113*, *155*
Mehra, R. 283
Melloni 232
Méndez, M.E. 237, *240*
Mentall, J.E. *113*
Menzies, J.W. *15*
Merrill, K.M. *206*, *207*
Mertz, L. 233, 237, *241*
Mezger, P.G. 88, 99, 100, 101, 102, 103, 104, 107, 110, *113*, 114, 130, 140, 142, 145, 155, *155*, *156*, 199, *207*, 226, *230*, 231, 242
Michie, R.W. 330, *354*
Middlehurst, B.M. *206*
Mihalas, D.M. 136, *139*, 184, *187*, 268
Miles, B.M. *187*
Miller, J.S. 198, *206*
Miller, R.H. 327, *354*
Mills, B.Y. 142, *156*
Milne, D.K. *156*
Milton, A.F. 236, *241*
Minkowski, R. 299, *324*
Mitchell, R.I. *29*, *240*
Modali, S.B. *155*
Modisette, J.L. *187*
Moesta, H. 107, *113*
Monnet, G. *323*
Moolenaar, W.H. *247*
Moorwood, A.F.M. *230*
Mord, A.J. *230*
Morgan, B.L. *263*
Morgan, W.W. 82, 153, *156*, 165, *172*, 292, *324*
Morris, M. *114*
Morton, D.M. 136, *139*
Moser, J.K. 277, *277*
Müller, E.A. *181*
Munch 127
Münch, G. 127, *241*
Murray, B.C. 233, 235, *241*
Murray, C.A. 310, *324*
Murray, J.D. *123*

Neugebauer, G. 100, 109, *112*, 157, 160, *163*, 188, 192, 200, 201, 202, 203, 204, *205*, *206*, *207*, 214, 226, 228, *230*, *241*, *324*
Ney, E.P. *206*, 216, *230*
Nicholls, D.C. 73, *74*
Nicholson, S.B. 233, *241*
Nikolov, N.S. 31, 32, *35*
Nobili, L. 232
Nolt, I.G. 235, *241*
Nordström, B. 76, *81*
Norton, R.H. *240*

O'Brien, P.A. *156*
Oburka, O. 49, *52*
O'Connell, D.J.K., S.J. 49, *52*, *354*

O'Dell, C.R. 199, *206*
Oke, J.B. 60, *64*
Okuda, H. 151, *156*
Oleak, H. 302, 305, 322
Oliver, M. *263*
Ollongren, A. 312, 313, *324*, 355
Olthof, H. 205, 222, *230*, 243, 247
Oort, J.H. 109, *113*, 125, *127*, 127, 140, 155, *156*, 162, 167, 172, *172*, 173, 207, 297, 298, 299, 308, 310, 314, 317, *324*, 342, *354*
Osterbrock, D.E. 229, *230*
Ostriker, J.D. 322, *324*
Ovenden, M.W. 290, *290*
Ozernoj, L.M. 298, 322, *324*

Paczynski, B. 13, *15*, 58, *59*
Palmer, P. *114*, 158, *163*
Papa, D.C. *130*
Papadopoulos, G.D. 129, *130*, 130
Parenago, P.P. 319, *324*
Park, W.M. 215, 216, *230*
Parker, R.A.R. 229, *230*
Pastoriza, M.G. 153, *156*
Pauliny-Toth, I.I.K. 229, 230, *230*
Pauls, T.A. 111, 112, 140
Payne-Gaposchkin, C. 34, *35*
Peat, D.W. *177*
Pecker, J.-C. 188, 195, *207*, *323*
Peebles, P.J.E. 322, *324*
Pellet, A. 293, *323*
Pendl, E.S. 354
Penston, M.V. 198, *207*
Penzias, A.A. *113*, *156*, 159, *163*, *207*, *231*
Perek, L. 167, 168, *172*, *206*, 291, *324*
Perova, N.B. *29*, *35*
Peter, H. *52*
Peters, C.F. 276, 277, *277*
Petford, A.D. *272*
Petit, M. 33, *35*
Petrosian, V. 222, *231*
Pettit, E. 233, *241*
Pfeiffer 271
Piazzi Smyth, C. 232, 239, *241*
Pipher, J.L. *206*, *230*, *247*
Pishmish, P. 111, 112, 155
Pismis, P. 74
Plaskett, J.S. 67, *74*
Plavec, M. 49, 51, *52*
Pohl, E. 49, *52*
Popovici, C. 49, *52*, 53, *59*, 59
Popper, D.M. 53, *59*
Porter, F.C. *207*
Pottasch, S.R. 151, 152, *156*, *187*, 198, 199, 209, 222, *231*, 231, 247
Pourcelot, A. *323*
Prata, S.W. 342, *354*
Pratt, N.M. 253
Price, S.D. 192, *207*
Pritchet, C.J. *206*

Quist, T.M. 235, *241*

Racine, R. 292, *324*
Radford, H.E. *155*
Radhakrishnan, V. 123, *123*
Rank, D.M. 89, *113*, *163*, 199, 200, *207*, 237, *240*, *241*
Reddish, V.C. 249, 254, 318, *324*
Redman, R.D. 174, 177, *177*
Rego, M.E. 175, *177*
Reifenstein, E.C.III. 142, 146, 147, *156*
Richter, J. *181*
Richter, N. 304, *324*
Rieke, G.H. 200, 201, 203, 204, *207*, *230*, 234, 235, 236, 238, 239, *240*, *241*
Rijsbergen, R. van 187, 254
Rinehart, R. *155*, *230*
Ring, J. 239, *240*, *241*, *247*
Roberts, M.S. 142, 145, 146, *156*, 307, 308, 309, 310, 311, *324*
Roberts, P.H. *38*
Roberts, W.W. 73, *74*
Robinson, B.J. 106, 110, *113*, 142, *156*, 158, *163*
Robinson, L.T. 49, *52*
Rodgers, A.W. *324*
Roesler, F.L. 132, 139, *139*
Roosen, R.G. 299, *322*
Rootsmäe, T. 319, *324*
Rougoor, G.W. 109, 110, *113*, 140, *156*
Roy, A.E. 290, *290*
Roxburgh, H. 9, *15*
Rubin, V.C. 308, *324*
Rümmel, U. 297, 299, 322, *323*
Rydgren, A.E. 68, *187*
Rysbergen 59

Saar, E. 318, 322, *324*
Sabbata, V. de 164
Sakhibulin, N. 185, *187*
Salpeter, E.E. 98, 107, 108, *113*, 116, 117, *118*, *231*, 318, *324*
Sancini 119
Sandage, A.R. 201, *207*, 298, 299, 318, 320, *323*, 324
Sanders, R.H. 109, 110, 111, *113*, 140, 150, 152, 154, *156*
Sandqvist, A. 72, 106, *113*, 118, 119, 157, 159, 162, *163*
Sanduleak, N. 310, 317, *324*
Saslaw, W.C. 154, *156*, 352, *353*
Savage, B.D. 94, *113*, 137, *139*
Schild, R. 195, *207*
Schmidt, M. 124, *127*, *172*, 313, 317, *324*, 355, *357*

Schraml, J. 145, *156*, 199, *207*
Schubert, G. 9, *15*
Schwartz, P. R. *130*
Schwarzschild, M. 9, *15*, 302, 319, *324*
Schweizer, F. *207*
Scoville, N. Z. 101, 105, 106, 110, 111, 112, *113*, 142, 148, 149, 150, 154, *156*, 162, *163*
Seaman, C. H. *240*
Seaton, M. J. 93, *113*
Seitter, W. 39, 41
Sengbusch, K. von 11, 12, 13, *15*, 15
Serkowski, K. 4, *15*, *207*
Sersic, J. L. 153, *156*
Sèvre, F. 215
Shapiro, I. I. *130*
Shapiro, S. L. 311, *324*, 333, *354*
Shapley, H. 65, 68, *74*
Shapley, M. B. 48, *52*, 54, 57, *58*
Sharov, A. S. 292, 299, *324*
Shaw, J. A. 238, *240*
Sher, D. 352, *353*
Shivanandan, K. 233, 239, *241*
Shobbrook, R. R. 38, *38*
Siegel, C. L. 277, *277*
Simon, M. *206*, *230*
Simon, R. 37, *38*
Slingerland, J. H. *206*
Smeyers, P. 36, 37, *38*
Smith 271
Smith, F. G. 142, *156*
Smith, H. E. *207*
Smith, L. F. 155
Smith, M.W. *187*
Smith, R. A. 235, *241*
Smith, S. D. *240*
Smyth, M. J. 207, 232, 237, *241*, 241, 242, 264
Snijders, M. A. J. *187*
Snyder, L. E. 89, 90, 91, *113*, *156*, 160, *163*
Soifer, B.T. *206*, *230*, 246, *247*
Solomon, P. M. *113*, 142, *156*, *163*
Spicer, W. E. 235, *241*
Spinrad, H. 165, *172*, 174, 177, *177*, 188, 201, 202, *207*, 292, 298, 299, 309, *324*
Spitzer, L., Jr. 131, *139*, 154, *156*, 221, *231*, 319, *324*, 330, 333, 334, 336, 339, 340, 341, 342, 350, *354*
Stair, A.T. *241*
Standish, E. M., Jr. 275, 277, *277*
Stein, W. A. 13, *15*, 193, 199, 205, *206*, *207*, 229, *230*, 238, *240*, *241*
Steinmetz, D. L. *240*
Stepien, K. 15, 114
Stief, L. J. 106, 107, *113*
Stobie, R. S. 10, *15*
Stokes, N. R. *324*
Stratton, F. J. M. 39
Strecker, D.W. *206*
Strittmatter, P. A. 155, 205, *207*
Strom, K. M. *15*, *177*, 195, *207*
Strom, S. E. 3, *15*, 176, *177*, 195, *207*
Strömgren, B. 186, *187*, 313, *322*
Strong, J. 233, *241*
Sullivan, W.T. III. *130*
Sundman, A. 76, 80, *81*, 82
Svetchnikoff, M. A. 54, 55, 56, 57, *59*
Swift, C. D. *230*
Szebehely, V. 273, 275, 276, 277, *277*, 278, *353*, 354

Tafferts, K. B. *156*
Talbot, R. J. 2, *15*
Tammann, G. A. 292, 299, *324*
Taylor, B. J. 165, *172*, 174, 177, *177*, 292, *324*
Teich, M. C. 236, *241*
Thaddeus, P. 105, *113*, 117, *118*, *156*, *163*, 198, *207*, 216, *231*
Thornton, D. C. *113*, *163*
Thuan, T. X. 333, 334, 336, 339, 340, *354*
Tinsley, B. M. 318, *324*
Todoran, I. 30, 42, 48, 49, *52*
Toombs, R. I. 198, *207*
Toomre, A. 126, *127*
Torres-Peimbert, S. 177, *177*
Townes, C. H. *113*, *163*, *207*, *241*
Traat, P. 322
Traub, Wes. 139
Treder, H. J. 322
Treffers, R. R. *240*
Trimble, V. 14, *15*
Truran, J.W. 168, *172*, 318, *325*
Tsioumis, A. 17, *29*
Tsvetkov, T. 31
Turlo, Z. 114, 253, 272
Turner, B. E. *114*

Udal'tsov, V. A. *155*
Unsöld, A. 93, 94
Unsov, V.V. 298, *324*
Utiger, H. E. 131, 272

Vallace, T. *241*
Vallak, R. 145, *155*
Vanasse, G. A. 237, *241*
Vasilevskis, S. 83, 84, 85, 86, *87*
Vaucouleurs, G. de 259, 299, 302, 303, 304, 305, *325*
Vaughan, A. H., Jr. 266, *272*
Vennik, J. 322
Vetesnik, M. 292, *325*
Vickers, D. G. *230*
Voelcker, K. 193, 195, *207*
Vogel, U. 123, *123*
Vorontson-Velyaminon, B. A. 298, *325*
Vries, M. de 195, 205, *207*

Wal, P. B. van der 247
Walker, M. F. 2, 4, *15*, *163*
Wall, J.V. 305, *325*
Walmsley, C. M. 150, *155*
Walter, K. 51, *52*
Wampler, E. J. 229, *231*
Wannier, P. 125, *127*
Warner, B. 13, *15*, 186, *187*
Watson, W. D. 107, 108, *113*, 116, 117, *118*
Wayman, P. A. 314, *325*
Weaver, H. F. 125, *127*
Webbink, R. F. 188, *207*
Webster, W. J. 147, *156*
Weedman, D.W. 153, *156*
Weigert, A. 57, *58*
Weinreb, S. 89, *113*
Wehinger, A. P. 264
Welch, W. J. *113*, *130*, *163*
Weliachew, L. N. 106, *113*, 154, *155*
Werlinger, A. P. 254
Werner, M.W. 98, *113*
Werner, W. 187, *187*
Westerhout, G. 142, *156*
Westerlund, B. E. *81*, 305, *325*
Westphal, J. A. *207*, 207, *241*
Weymann, R. J. 238, *241*
White, M. L. 339, *354*
White, O. R. 258, *263*
Whittle, R. P. J. 123
Wickramasinghe, N. C. 107, *113*, 151, *156*, 204, *207*
Wielen, R. 326, 329, 331, 339, 340, 342, 343, 346, *354*, 354
Wierstra, T. 247
Wiese, W. L. 186, *187*
Wijnbergen, J. J. 243, *247*
Wildey, R. L. 233, 235, *241*
Williams, B. F. 235, *241*
Williams, P. M. 81, 173, 174, 175, *177*
Wilson, A. J. 115
Wilson, R.W. 104, *113*, *127*, *156*, *163*, *207*, *231*
Wilson, T. L. 142, 146, *156*
Wilson, W. *240*
Wink, J. E. 147, *156*
Wing, R. F. 188, *207*
Winnewisser, G. 89, 92, 93, 107, *113*
Wisniewski, W. Z. *29*
Wisse, M. *15*
Wisse, P. N. J. *15*
Witt, A. N. 94, 95, 96, *113*, *114*, 225, *231*
Witteborn, F. C. *230*
Woerden, H. van 293, *324*, *325*
Wolf, B. 178, 180, *181*
Wolfe, W. L. 235, *241*
Wollman, E. R. *240*
Woolf, N. J. 188, 190, 193, 195, 196, 198, 199, 200, 201, 202, 203, 204, *206*, *207*, 225, 236, 237, 238, *241*, 248, 341, *353*
Wright, K. *38*
Wrixon, G.T. *127*
Wyller, A. A. 265, 267, *272*, 272
Wynn-Williams, C. G. 147, *156*, 210, *230*, *231*

Yekovleva, A. 184, *187*
Yost, J. *15*, *207*
Young, D. S. de 311, *325*
Young, E.T. 204, *207*
Young, J.W. *324*
Yuan, C. 313, *323*
Yun, H. S. 267, *272*

Zaitseva, G.V. 49, *52*
Zappala, R. R. 197, 198, *206*
Zikides, M. 279
Zirin, H. 127
Zuckerman, B. 96, 106, *114*, 154, *156*, 158, 162, *163*
Zuidema, N. 247

Subject Index

Apsidal motion in close binaries 42ff.
Asteroids, capture by the Earth 283ff.
Atmospheric windows in the infrared 189

Bartol Coudé Observatory, URSIES 265ff.
Becklin's object 3ff., 198

β Canis Majoris stars 36ff.
Cepheids, pulsation amplitude 31ff.
—, stability of light curves 17ff.
Close binaries, apsidal motion 42ff.
Clusters of galaxies, simulation 352f.
Commensurable mean motions and solar system dynamics 290
Compact HII regions, broadband spectral measurements 214ff.
—, interpretation of infrared measurements 218ff.
—, relationship between infrared flux and radio emission 212ff.
—, size of the emitting regions 210ff.

Diffusion of elements in the Galaxy 165ff.
— in the stars 7ff.

Eclipsing binaries, radius-luminosity relation 53ff.
Electronographic images, analysis 255ff.
Evolution of galaxies 316ff.
—, pre-main sequence 7
Extragalactic sources, infrared emission 204f., 226ff.

Galactic Center, comparison with the nuclei of external galaxies 153
—, gas 108ff.
—, infrared emission 151ff., 200ff., 225f.
—, ionized gas 140ff.
—, lunar occultations 157ff.
—, model 164
—, 8–13 μm spectrum 248
—, thermal radio emission 151ff.
Galactic kinematics from stellar proper motions with respect to galaxies 83ff.
Galactic models and stellar orbits 291ff.
—, three-dimensional 355ff.
Galactic populations, kinematical properties 312ff.
Galactic structure, large-scale 108ff.
Galaxies, evolution 316ff.
Galaxy, chemical evolution 178ff.
GALAXY machine 249ff.
Gould belt, expansion 67ff.
—, origin 73
—, structure 65ff.

H II regions, broadband spectral measurements 214ff.
—, infrared emission 198f.
—, interpretation of infrared measurements 218ff.
—, relationship between infrared flux and radio emission 212ff.
—, size of the emitting regions 210ff.
High-velocity clouds, origin 126
—, positions 125f.
—, relationship to spiral arms 124ff.
—, velocities 125f.

Infrared astronomy, detectors 234ff.
—, instruments 237ff.
—, observatories 239
—, techniques 188ff., 232ff.
—, telescopes 238
Infrared emission, extragalactic sources 204f.
—, galactic center 151ff., 200ff.
—, H II regions 198f., 243ff.
—, planetary nebulae 199f.
—, starlike objects 193ff.
—, stars 193ff.
Infrared observations, airborne 239f.
—, interpretation 209ff.
—, space 239f.
Infrared sources, distribution over the sky 192f.
Intercloud HI-gas, brightness temperature 120ff.
—, homogeneity 120ff.
—, kinetic temperature 120ff.

—, velocity dispersion 120ff.
Interferometer measurements, very-long baseline 129
Intermediate-velocity clouds, relationship to spiral arms 128
Interstellar clouds, densities 115ff.
—, kinetic temperature 117
—, molecules in 115ff.
—, neutral hydrogen in 115ff.
—, physical state 102ff.
—, star formation and 96ff.
Interstellar extinction law 193ff.
Interstellar ionized gas, dynamics 148ff.
—, physical state 150f.
Interstellar lithium abundance 131ff.
Interstellar matter, UV radiation field and 93ff.
—, X-ray heating 123
Interstellar molecules, catalogue 90ff.
—, excitation temperature 115
—, formation 115f.
—, lifetime 117
Interstellar radiation field 106ff.
Interstellar sodium 137f.

Large Magellanic Cloud, gravitational effect on the Galaxy 126f.
Lithium abundance, chondritic meteorites 131, 135
—, cosmic 135
—, interstellar 131ff.
—, T Tauri stars 131, 135
Lunar occultations of the galactic center 157ff.

Magellanic Clouds, abundance determination of A-supergiants 178ff.
Magnitude systems, Johnson 189
Masers, densities 115
—, diameters 115
—, H_2O 115, 129f.
—, OH 115, 129f.
—, pumping mechanisms 115
Meteorites, chondritic 131, 135
Molecular clouds, dynamics 148ff.

N-body problem for star cluster 326ff.
Newcomb lunisolar precission 84
Newcomb planetary precission 84
Nova Cephei 1971 41
— Delphini 1967 39ff.
— Herculis 1934 39
— Herculis 1963 41
— Vulpeculae 1968 I 41
Nuclear disk of the Galaxy, kinematics 140ff.
—, physical condition 140ff.

OAO 137
Oort's constants 85f.

PEPSIOS 131ff.
Periodic orbits in three-dimensional models of the Galaxy 355ff.
— of the restricted three-body problem 279ff.
Photographic plates, high-speed evaluation 249ff.
Planetary nebulae, infrared emission 199f.
Pre-main sequence contraction 7
Proper motions with respect to galaxies 83ff.
Protostellar clouds 104f.

Radio recombination line observations of discrete sources in the galactic center 142f.
— of extended thermal sources 148
Radio sources, brightness temperature 115
—, variability 129f.
Radio spectroscopy, importance 88f.
—, instruments 89f.
R Coronae Borealis stars models 13f.
Restricted three-body problem, periodic orbits in the 279ff.
RR Lyrae stars, pulsation 9ff.

δ Scuti stars 7
Shells around compact HII regions 105ff.
Solar system, dynamic studies 290
Space distribution of stars in the Carina-Centaurus region 76ff.
Spiral Structure of the Galaxy, inner 125
—, outer 124ff.
—, relationship to the high-velocity clouds 124ff.
—, tilt 124ff.
Star formation 96ff.
Stellar metallicity and stability 7ff.
Stellar orbits and galactic models 291ff.
— and kinematic properties of galactic populations 312ff.
Stellar rotation and non-radial oscillations 36ff.
— and stability 7ff.
Stellar stability 7ff.
Super-metal-rich stars, narrow-band observations 174ff.
—, status 177

Thermal radio emission from the galactic center 151ff.
Three-body problem 273ff.
Tidal action and non-radial oscillations of stars 36ff.
Toro, resonance capture by the Earth 283ff.

T Tauri stars, energy distribution 2f.
—, inflow of matter 4
—, infrared excess 3
—, lithium lines 131, 135
—, mass loss 4
—, UV excess 3

Ultraviolet radiation field and interstellar matter 93ff.
Ultraviolet stellar spectra, high resolution 182ff.
URSIES and the Bartol Coudé Observatory 265ff.

Variable stars and stellar evolution 1ff.
Visual binaries, kinematic properties 60ff.

X-ray heating of interstellar matter 123